Environmental Technologies to Treat Nitrogen Pollution

Integrated Environmental Technology Series

The *Integrated Environmental Technology Series* addresses key themes and issues in the field of environmental technology from a multidisciplinary and integrated perspective.

An integrated approach is potentially the most viable solution to the major pollution issues that face the globe in the 21st century.

World experts are brought together to contribute to each volume, presenting a comprehensive blend of fundamental principles and applied technologies for each topic. Current practices and the state-of-the-art are reviewed, new developments in analytics, science and biotechnology are presented and, crucially, the theme of each volume is presented in relation to adjacent scientific, social and economic fields to provide solutions from a truly integrated perspective.

The *Integrated Environmental Technology Series* will form an invaluable and definitive resource in this rapidly evolving discipline.

Series Editors

Dr Ir Piet Lens, Sub-department of Environmental Technology, IHE-UNESCO, P.O. Box 3015, 2601 DA Delft, The Netherlands (p.lens@unesco-ihe.org).

Published titles

Biofilms in Medicine, Industry and Environmental Biotechnology: *Characteristics, analysis and control*
Decentralised Sanitation and Reuse: *Concepts, systems and implementation*
Environmental Technologies to Treat Sulfur Pollution: *Principles and engineering*
Phosphorus in Environmental Technology: *Principles and applications*
Soil and Sediment Remediation: *Mechanisms, techniques and applications*
Water Recycling and Resource Recovery in Industries: *Analysis, technologies and implementation*

Environmental Technologies to Treat Nitrogen Pollution

Francisco J. Cervantes

Publishing

London · New York

Published by   **IWA Publishing**
                      Alliance House
                      12 Caxton Street
                      London SW1H 0QS, UK
                      Telephone: +44 (0)20 7654 5500
                      Fax: +44 (0)20 7654 5555
                      Email: publications@iwap.co.uk
                      Web: www.iwapublishing.com

First published 2009
© 2009 IWA Publishing

Printed by Lightning Source

British Library Cataloguing in Publication Data
A CIP catalogue record for this book is available from the British Library

Library of Congress Cataloging-in-Publication Data
A catalog record for this book is available from the Library of Congress

ISBN: 1843392224
ISBN 13: 9781843392224

Contents

List of Contributors

Dr. Madan Tandukar
Research Associate
Environmental Engineering Program, School of Civil and Environmental
Engineering, Georgia Institute of Technology
311 Ferst Drive, Atlanta, GA 30332-0512, USA

Dr. Spyros G. Pavlostathis
Professor and Director
Environmental Engineering Program, School of Civil and Environmental
Engineering, Georgia Institute of Technology
311 Ferst Drive, Atlanta, GA 30332-0512, USA

Peter van der Maas PhD MSc
WLN
PO Box 26, 9470 AA Zuidlaren, The Netherlands
Tel.: +31 50 4022116
m.: +31 6 20614637
Fax.: +31 50 4094274
Email: p.vandermaas@wln.nl

Dr. Mahmut Altınbaş
Istanbul Technical University (ITU), Department of Environmental
Engineering, Insaat Fakultesi, Maslak 34469, Istanbul/Turkey
E-mail: maltinbas@ins.itu.edu.tr
Tel.: +90 212 2856542
Fax.: +90 212 2853781

F. Cuervo-López
Universidad Autónoma Metropolitana-Iztapalapa. Depto.
Biotecnología-CBS. Av. San Rafael Atlixco No 186-Col.
Vicentina D.F., México
Tel.: (00 52 55) 5804 6408
Email: fmcl@xanum.uam.mx

S. Martínez Hernández
Universidad Autónoma Metropolitana-Iztapalapa. Depto.
Biotecnología-CBS. Av. San Rafael Atlixco No 186-Col.
Vicentina D.F., México
Tel.: (00 52 55) 5804 6408
Email: serb@xanum.uam.mx

A.-C. Texier
Universidad Autónoma Metropolitana-Iztapalapa. Depto.
Biotecnología-CBS. Av. San Rafael Atlixco No 186-Col.
Vicentina D.F., México
Tel.: (00 52 55) 5804 6408
Email: actx@xanum.uam.mx

J. Gómez
Universidad Autónoma Metropolitana-Iztapalapa. Depto.
Biotecnología-CBS. Av. San Rafael Atlixco No 186-Col.
Vicentina D.F., México
Tel.: (00 52 55) 5804 6408
Email: dani@xanum.uam.mx

Nicolas BERNET
INRA, UR50, Laboratoire de Biotechnologie de l'Environnement,
Avenue des Etangs, Narbonne, F-11100, France
Email: bernet@supagro.inra.fr

Mathieu SPÉRANDIO
Professor
INSA Toulouse Laboratoire d'Ingénierie des Systèmes Biologiques
et Procédés UMR5504, UMR792, CNRS, INRA, INSA
135 avenue de Rangueil, 31077 Toulouse Cedex 04, France
Tel.: +33 (0)5 61 55 97 55
Fax.: +33 (0)5 61 55 97 60
Email: mathieu.sperandio@insa-toulouse.fr

Mónica Figueroa Leiro
Ph student
Department of Chemical Engineering, School of Engineering. University of Santiago de Compostela
Rua. Lope Gómez de Marzoa s/n. E-15782 Santiago de Compostela, Spain
Tel.: +34 981 563100 Ext 16739
Fax.: +34 981 528050
E-mail: Monica.figueroa@usc.es

José Ramón Vázquez Padín
Ph student
Department of Chemical Engineering, School of Engineering. University of Santiago de Compostela
Rua. Lope Gómez de Marzoa s/n. E-15782 Santiago de Compostela, Spain
Tel.: +34 981 563100 Ext 16739
Fax.: +34 981 528050
E-mail: Jose.vazquez.padin@usc.es

Ángeles Val del Río
Ph student
Department of Chemical Engineering, School of Engineering. University of Santiago de Compostela.
Rua. Lope Gómez de Marzoa s/n. E-15782 Santiago de Compostela, Spain
Tel.: +34 981 563100 Ext 16739
Fax.: +34 981 528050
E-mail: Mariaangel.val@rai.usc.es

Nicolás Morales Pereira
Ph student
Department of Chemical Engineering, School of Engineering. University of Santiago de Compostela
Rua. Lope Gómez de Marzoa s/n. E-15782 Santiago de Compostela, Spain
Tel.: +34 981 563100 Ext 16739
Fax.: +34 981 528050
E-mail: Nicolas.morales@rai.usc.es

Isaac Fernández Rodríguez
Ph student
Department of Chemical Engineering, School of Engineering. University of Santiago de Compostela
Rua. Lope Gómez de Marzoa s/n. E-15782 Santiago de Compostela, Spain
Tel.: +34 981 563100 Ext 16739
Fax.: +34 981 528050
E-mail: Isaac.fernandez@usc.es

José Luis Campos Gómez
Assistant professor
Department of Chemical Engineering, School of Engineering. University of Santiago de Compostela
Rua. Lope Gómez de Marzoa s/n. E-15782 Santiago de Compostela, Spain
Tel.: +34 981 563100 Ext 16777
Fax.: +34 981 528050
E-mail: Joseluis.campos@usc.es

Anuska Mosquera-Corral
Assistant professor
Department of Chemical Engineering, School of Engineering. University of Santiago de Compostela
Rua. Lope Gómez de Marzoa s/n. E-15782 Santiago de Compostela, Spain
Tel.: +34 981 563100 Ext 16779
Fax.: +34 981 528050
E-mail: Anuska.mosquera@usc.es

Ramón Méndez Pampín
Professor
Department of Chemical Engineering, School of Engineering. University of Santiago de Compostela
Rua. Lope Gómez de Marzoa s/n. E-15782 Santiago de Compostela, Spain
Tel.: +34 981 563100 Ext 16793
Fax.: +34 981 528050
E-mail: Ramon.mendez.pampin@usc.es

Andrew Thornton
Senior Process Engineer
Terriers House 201 Amersham Road High Wycombe Bucks HP13 5AJ
Tel.: +44 (0) 7747 773616
Email: Andrew.thornton@mwhglobal.com

Prof Simon A. Parsons
Head of Centre for Water Science
Professor of Water Sciences Centre for Water Science, School of Applied
Sciences Cranfield University, Cranfield, Bedfordshire MK43 0AL, UK
Email: s.a.parsons@cranfield.ac.uk
Tel.: +44 (0)1234 758311
754108 (Jackie Major – PA/Administrator)

Dr. Bernhard Wett
ARAconsult
Unterbergerstr.1, 6020 Innsbruck Austria
E-mail: wett@araconsult.at

Prof. Dr.-Ing. Norbert Jardin
Ruhrverband Kronprinzenstraße 37, 45128 Essen Germany
E-mail: nja@ruhrverband.de

Dimitrios Katehis, PhD, PE
Principal Technologist
CH2M Hill USA
Email: Dimitrios.Katehis@CH2M.com

Preface

Nitrogen production continues to increase every year and it is dominated by agricultural activities, but fossil fuel energy plays a major role as well. Besides the nitrogen pollution linked to intensive agriculture and combustion of fossil fuels, an important source of this contamination comes from industries demanding different nitrogenous compounds. For instance, ammonium is used as a raw material to create multiple products, such as nylon, plastics, resins, glues, animal/fish/shrimp feed supplements, and explosives. The acceleration of the nitrogen cycle caused by anthropogenic activities has certainly fulfilled essential requirements to sustain an increasing global population, such as providing enough food and fuel. Nevertheless, the nitrogen cycle has shifted from how to promote food and fuel production to a realisation that intensification of these activities damages environmental systems. Furthermore, the global nitrogen demand for nutritional purposes does not always match the nitrogen production in many regions. Indeed, in some parts of the world, nitrogen has been used to create an excess of food, while also contributing to a host of environmental problems. In contrast, other world regions lack sufficient nitrogen to meet even the most basic caloric demands of hundreds of millions of people. Therefore, two paradigms are currently considered when nitrogen management is tackled. Firstly, efficient nitrogen removing technologies should be implemented in those sectors generating high concentrations of nitrogenous compounds. Otherwise, pollution prevention approaches should also be executed. Contamination linked to recalcitrant nitrogenous compounds, such as azo dyes, nitroaromatics and aromatic amines, requires special attention given the outsized impacts of these contaminants on public health and on the environment. Secondly, as an essential nutrient, nitrogen should be recovered

© 2009 IWA Publishing. *Environmental Technologies to Treat Nitrogen Pollution: Principles and Engineering*, Edited by Francisco J. Cervantes. ISBN: 9781843392224. Published by IWA Publishing, London, UK.

from concentrated streams generated in several industrial sectors and connected to those activities demanding large amounts of this element, such as intensive agriculture.

This book provides a comprehensive compilation describing the main anthropogenic activities, which contribute to the pollution of different environments by both inorganic and organic nitrogenous compounds. The main ecological, toxicological and economical impacts of nitrogen pollution will also be described, as well as its effects on public health (chapter 1). The text also describes the fundamental aspects of biological treatment processes involved in nitrogen removal from wastewaters, such as nitrification (chapter 2), denitrification (chapter 3) and anaerobic ammonium oxidation (anammox, chapter 4). Furthermore, design criteria of several distinct biological wastewater treatment systems to remove nitrogen from municipal wastewater (chapter 5) and high-strength effluents, such as sludge liquors (chapter 6) and leachates (chapter 8) are also explained. The book also includes the fundamentals and design criteria of wastewater treatment processes applied for the removal of recalcitrant nitrogenous compounds from industrial wastewaters, such as those generated from the production of textiles, explosives and pharmaceuticals (chapter 7). The principles, process control parameters and application of struvite precipitation, as a strategy to recover nitrogen from concentrated effluents, are illustrated on chapter 9. Chapter 10 shows the main factors to consider during the application of ion exchange processes for the removal of ammonium from contaminated waters. The application of different treatment systems for the removal of nitrogen from industrial flue gases is described in chapter 11. Meanwhile, the simultaneous removal of nitrogenous and sulphurous contaminants from industrial wastewaters by different technologies is discussed in chapter 12. Finally, chapter 13 deals with the application of aerobic granular treatment systems for the removal of nitrogen from wastewaters.

All contributions have been prepared by recognised specialists, who have collected their experience through several years of professional practice. The fundamentals, operational parameters and design criteria are illustrated in several case studies documenting the application of several distinct treatment processes for the removal or recovery of nitrogenous compounds from domestic and industrial effluents.

I would like to express my appreciation for all contributors, who compiled their manuscripts in a very enthusiastic and punctual way. I am deeply impressed for the high quality of every contribution to this book. Thank all for your great effort.

Finally, I would like to dedicate this book to my wife, Liz, and to our sons: Eliot, Alfonso and Román. Thank you for inspiring me every day and for gratifying my life with your love.

San Luis Potosí, Mexico; January 2009
Francisco J. Cervantes

1

Anthropogenic sources of N-pollutants and their impact on the environment and on public health

F.J. Cervantes

1.1 INTRODUCTION

Humans are altering the global cycle of nitrogen (Figure 1.1) via combustion of fossil fuels, production of nitrogen fertilisers, cultivation of nitrogen-fixing legumes, and other unrestrained actions (Vitousek *et al.* 1997). The large acceleration of the nitrogen cycle caused by these anthropogenic activities has certainly fulfilled essential requirements to sustain an increasing global population, such as providing enough food. Nevertheless, the nitrogen cycle

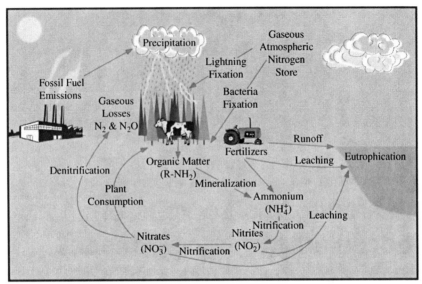

Figure 1.1. Anthropogenic and natural processes involved in the nitrogen cycle (Pidwirny 2006)

has shifted from how to promote food production to a realization that agricultural intensification damages environmental systems (Galloway *et al.* 2008). Furthermore, the global nitrogen demand for nutritional purposes does not match the nitrogen production in many regions. Indeed, in some parts of the world, nitrogen has been used to create an excess of food and a growing prevalence of unhealthy diets, while also contributing to a host of environmental problems (Vitousek *et al.* 1997; Galloway *et al.* 2004; Jin *et al.* 2005). In contrast, other world regions lack sufficient nitrogen to meet even the most basic caloric demands of hundreds of millions of people (Sanchez and Swaminathan 1995).

Nitrogen creation continues to increase every year. It is dominated by agricultural activities, but fossil fuel energy plays a major role as well. From 1860 to 1995, energy and food production increased steadily on both an absolute and a per capita basis; nitrogen creation also increased from ~15 Tg N in 1860 to 156 Tg N in 1995. The change was enormous, and it increased further from 156 Tg N yr^{-1} in 1995 to 187 Tg N yr^{-1} in 2005, in large part because cereal and meat production increased by 20% and 26%, respectively, during this period (Galloway *et al.* 2008).

Besides the nitrogen pollution linked to intensive agriculture and combustion of fossil fuels, an important source of this contamination comes from industries demanding different nitrogenous compounds. For instance, NH_3 is used as a raw material to create multiple products, such as nylon, plastics, resins, glues, melamine, animal/fish/shrimp feed supplements, and explosives (Galloway *et al.* 2008). In 2005, ~23 Tg N was used for the production of different chemicals (Prud'homme 2007), but little is known about the fate of nitrogenous compounds derived from these industrial activities.

This chapter underlines the main anthropogenic activities, which contribute to the pollution of aquatic environments by both inorganic and organic nitrogenous compounds. The main ecological, toxicological and economical impacts of nitrogen pollution will also be described, as well as its effects on public health. Lastly, the chapter will describe different strategies to prevent nitrogen pollution.

1.2 MAJOR ANTHROPOGENIC SOURCES OF NITROGENOUS POLLUTANTS

1.2.1 Inorganic nitrogen pollution

Ammonium (NH_4^+), nitrite (NO_2^-) and nitrate (NO_3^-) constitute the most common nitrogenous inorganic compounds found in aquatic ecosystems (Wetzel 2001; Rabalais 2002). These ions can naturally be present as a result of atmospheric deposition, surface and groundwater runoff, dissolution of nitrogen-rich geological deposits, N_2 fixation by certain prokaryotes and biological degradation of organic matter (Camargo and Alonso 2006). However, during the past two centuries, and especially over the last five decades, humans have substantially altered the global nitrogen cycle, increasing both the availability and the mobility of nitrogen over large regions of Earth (Camargo and Alonso 2006). Consequently, in addition to natural sources, inorganic nitrogen can enter aquatic ecosystems through point and nonpoint sources originated from human activities (Table 1.1). Nonpoint sources are generally considered of greater relevance than point sources as they are larger and more difficult to control. Moreover, anthropogenic inputs of particulate nitrogen and organic nitrogen to the environment can also result in inorganic nitrogen pollution (Smil 2001). Chapters 5 and 6 describe the current technologies applied for the removal of inorganic nitrogen from municipal and high-strength wastewaters, respectively.

1.2.2 Organic nitrogen pollution

Several industrial sectors require different N-substituted aromatic compounds, such as nitroaromatics, azo dyes and aromatic amines, for the intensive

Table 1.1. Major anthropogenic sources of inorganic nitrogen in aquatic ecosystems (Camargo and Alonso 2006)

Point sources
− Wastewaters from livestock (cattle, pigs, chickens) farming
− N releases from aquaculture (fish, shrimps, prawns) operations
− Municipal sewage effluents (including effluents from sewage treatment plants without tertiary treatments in their facilities)
− Industrial wastewater discharges
− Runoff and infiltration from waste disposal sites
− Runoff from operational mines, oil fields, and industrial sites lacking sewage installations
− Overflows of combined storms and sanitary sewers

Nonpoint sources
− Widespread cultivation of N_2-fixing crops species, and subsequent N mobilisation
− Use of animal manures and inorganic N fertilisers, and the subsequent runoff from agriculture
− Runoff from burned forests and grasslands
− Urban runoff from sewered and unsewered areas
− Septic leachate and runoff from failed septic systems
− Runoff from construction sites and abandoned mines
− N loadings to groundwater, and subsequently, to receiving surface water bodies (rivers, lakes, estuaries, coastal zones)
− Emissions to the atmosphere of reduced (from volatilisation of manure and fertilisers) and oxidised (from combustion of fossil fuels) N compounds, and subsequent deposition over surface waters
− Other activities contributing to N mobilisation (from long-term storage pools) such as biomass burning, land clearing and conversion, and wetland drainage

production of dyes, explosives, pesticides and pharmaceutical products. These chemicals are intentionally designed to remain unaffected under conventional product service conditions and it is this property, linked with their toxicity to microorganisms, which makes their biodegradation difficult. The release of these recalcitrant N-pollutants into the environment may create serious health and environmental problems. Certainly, many N-substituted aromatic pollutants have been shown to be toxic for different aquatic species and to have mutagenic or carcinogenic activity. Moreover, due to their hydrophobic nature, many recalcitrant N-pollutants have the risk to bio-accumulate in the food chain (Razo-Flores *et al.* 1997; Pinheiro *et al.* 2004; Dos Santos *et al.* 2007). Chapter 7 describes the main environmental problems associated with organic N-pollutants, as well as the technologies for the treatment of the main industrial effluents containing these contaminants (originated from the production of textiles, explosives and pharmaceuticals).

1.3 IMPACT OF NITROGEN POLLUTION

A recent global assessment has pointed out three major environmental problems caused by nitrogen pollution in aquatic ecosystems: (1) it can drastically decrease the pH of freshwater ecosystems without much acid-neutralising capacity, leading to acidification of these water bodies; (2) it can stimulate or enhance the development, maintenance and proliferation of primary producers, resulting in eutrophication of aquatic ecosystems; and (3) it can also impair the ability of aquatic animals to survive, grow and reproduce as a consequence of direct toxicity of inorganic nitrogenous compounds. In addition, inorganic nitrogen pollution of ground and surface waters can induce adverse effects on human health and economy (Camargo and Alonso 2006). Furthermore, several organic nitrogenous contaminants generated by different industrial sectors are associated with toxicity, carcinogenesis, mutagenesis and allergies in living organisms (see further details in chapter 7).

1.3.1 Ecological effects

1.3.1.1 Acidification of freshwater ecosystems

Nitrogen dioxide (NO_2) and nitrogen oxide (NO), commonly known as NOx pollutants, have been identified as major acidifying pollutants in lakes and streams (Schindler 1988; Baker et al. 1991). The sources and atmospheric effects of NOx pollutants are described in chapter 11. Once emitted into the atmosphere, these gaseous contaminants can undergo chemical reactions leading to the production of the strong acids HNO_2 and HNO_3 (Hales 1982; Stepanov and Korpela 1997; Princiotta 1982), which in turn increase the concentration of H^+ in freshwater environments without enough acid-neutralising capacity (Camargo and Alonso 2006). In fact, HNO_3 is now playing an increasing role in the acidification of freshwater environments considering that SO_2 emissions have significantly been decreased during the last two decades, while emissions of NOx pollutants have gone unchecked (Vitousek et al. 1997; Stoddard et al. 1999; Skjelkvale et al. 2001). NH_4^+ can also contribute to the acidification process through nitrification (Vitousek et al. 1997). Currently, eastern North America and northern and central Europe are major acidified regions on Earth with several lakes and streams presenting pH values ranging from 4.5 to 5.8 (Camargo and Alonso 2006).

Table 1.2. Adverse effects of anthropogenic acidification on freshwaters (Camargo and Alonso 2006)

- Depression of net photosynthesis in planktonic and attached algae
- Reduction of net productivity inplanktonic and attached algae
- Increased bioaccumulation of aluminium and other trace metals in aquatic (especially submerged) macrophytes
- Increased abundance of filamentous green algae no longer attached to the substratum (metaphyton)
- Declined species diversity in phytoplankton and periphyton communities, with the loss of sensitive species
- Disruption of ionic regulation, especially loss of body sodium and failure to obtain sufficient calcium, in molluscs, insects, crustaceans, fish, and amphibians
- Respiratory and metabolic perturbations in molluscs, insects, crustaceans, fish, and amphibians
- Increased bioaccumulation and toxicity of aluminium and other trace metals in insects, crustaceans, fish, and amphibians
- Arrested development of fish and amphibian embryos, presenting in some cases skeletal deformities
- Hatching delay of fish and amphibian eggs
- Disruption of molting and emergency in insects and crustaceans
- Reduced growth rates in cladocerans, fish and amphibians
- Reduced efficiency and activity of grazing zooplankton (cladocerans), producing ramifying effects on the phytoplankton community
- Reduced efficiency or activity of prey capture by copepods, planarians, and fish, producing ramifying effects on pray populations and on populations of other predators
- Increased migration of aquatic insects (caddisfly larvae) from their retreat and capture nets
- Increased drift behaviour of benthic invertebrates to be transported downstream
- Avoidance of acid spawning sites by insects, fish, and amphibians
- Declined species diversity in zooplankton, macrobenthic, fish, and amphibian communities

Anthropogenic acidification of water ecosystems is associated with a number of adverse effects on primary and secondary producers (Table 1.2). The most evident effects of anthropogenic acidification are biotic impoverishments, particularly concerning invertebrates and fish, which have been detected in many atmospherically acidified freshwater ecosystems in North America and Europe (Krause-Dellin and Steinberg 1986; Huckabee *et al.* 1989; Morris *et al.* 1989; Schindler *et al.* 1989; Fjellheim and Raddum 1990; Meriläinen and Hynynen 1990; Ormerod *et al.* 1990; Cummins 1994; Allan 1995). Key microbial processes essential for nutrients cycling and ecosystems functioning, such as nitrification and leaf-litter decomposition, can also be inhibited or altered as a consequence of decreased pH values (Camargo and Alonso 2006).

1.3.1.2 Eutrophication of aquatic ecosystems

High levels of NH_4^+, NO_2^- and NO_3^-, derived from human activities, can stimulate or enhance the development, maintenance and proliferation of primary producers (phytoplankton, benthic algae, macrophytes), contributing to the widespread phenomenon of anthropogenic eutrophication of aquatic ecosystems (Camargo and Alonso 2006). In fact, over the last four decades, human activities are estimated to have increased N fluxes (mainly as NO_3^-) into the coastal waters of north-eastern USA by 6- to 8-fold, into the coastal waters of the Gulf of Mexico by 4- to 5-fold, and into the European rivers draining to the North Sea by 6- to 20-fold, linking these increased N fluxes to increased numbers of algal blooms (Vitousek *et al.* 1997; Smith 2003; Camargo and Alonso 2006).

Recent nutrient criteria to prevent and manage eutrophication of aquatic environments include both phosphorus and nitrogen, owing to the increasing link between nitrogen pollution and cultural eutrophication (Camargo and Alonso 2006).

Eutrophication of aquatic ecosystems can cause ecological and toxicological effects that are either directly or indirectly related to the proliferation of primary producers (Table 1.3), which in turn lead to low oxygen levels in bottom waters and sediments of eutrophic and hypereutrophic aquatic ecosystems with low water turnover rates (Camargo and Alonso 2006).

Anoxic conditions promoted by eutrophication of aquatic ecosystems cause extensive kills of both invertebrates and sensitive fish species, with significant reduction in area of suitable habitat for food, growth and reproduction of these aquatic organisms (Diaz and Rosenberg 1995; Wetzel 2001; Anderson *et al.* 2002; Breitburg 2002). Furthermore, the decline in dissolved oxygen concentrations can also promote the accumulation of toxic reduced compounds, such as hydrogen sulphide (H_2S), resulting in higher adverse effects on aquatic animals (Diaz and Rosenberg 1995; Wetzel 2001; Breitburg 2002). H_2S is a very toxic compound that can cause acute mortalities in aquatic animals at relatively low concentrations by affecting the nervous system (Ortiz *et al.* 1993). Moreover, mass occurrences of toxic algae, currently a global phenomenon that appears to be favoured by nutrient pollution, can significantly contribute to the extensive kills of aquatic animals (Camargo and Alonso 2006).

Table 1.3. Ecological and toxicological effects of eutrophication (Camargo and Alonso 2006)

- Reductions in water column transparency and light availability
- Increased sedimentation of organic matter
- Decreased concentration of dissolved oxygen in bottom waters and sediments
- Accumulation of reduced (toxic) compounds (*e.g.* H_2S) in bottom waters and sediments
- P releases from sediments that can reinforce eutrophication
- Increased biomass and productivity of phytoplankton
- Shifts in phytoplankton composition to bloom-forming species, which may be toxic
- Increased biomass, and changes in productivity and species composition, of freshwater macrophytes, often with proliferation of duckweeds
- Increased biomass and productivity, and shifts in species composition, of marine macroalgae communities, being usually favoured fast-growing seaweeds at the expense of sensitive seagrasses
- Losses of species diversity in phytoplankton, periphyton, macrophyte and macroalgae communities
- Increased biomass, and changes in productivity and species composition, of zooplankton, being usually favoured invertebrate grazers (*e.g., Daphnia*) at the expense of other trophic groups
- Changes in biomass, productivity and species composition of benthic invertebrates and fish, often with major mortality events in sensitive populations and reduction in the area of suitable habitat for reproduction. Tolerant grazers (*e.g., Lymnaea*) may proliferate at the expense of other trophic groups
- Losses of species diversity of zooplankton, benthic invertebrates and fish communities
- Reduction in the health and size of marine coral populations, often with large increases in cover and biomass of fleshy macroalgae (*e.g., Lobophorä*)
- Losses of species diversity in marine coral communities
- Alterations in the food web structure of freshwater, estuarine, and coastal marine ecosystems, with ramifying effects on every trophic level

1.3.2 Toxicological effects

High concentrations of NH_4^+, NO_2^- and NO_3^-, originated from anthropogenic activities, can impair the capacity of aquatic organisms to survive, grow and reproduce, resulting in direct toxicity of these inorganic nitrogenous compounds. Unionised ammonia (NH_3) is very toxic to aquatic animals, particularly to fish, whereas ionised ammonia (NH_4^+) is nontoxic or apparently less toxic (Russo 1985; Adams and Bealing 1994; Constable *et al.* 2003). NH_3 can also be inhibitory for *Nitrosomonas* and *Nitrobacter* bacteria (Anthonisen *et al.* 1976; Russo 1985) causing inhibition of the nitrification process. Consequently, increased accumulation of NH_4^+ (plus NH_3) due to inhibition of nitrification may intensify the toxicity effects to bacteria and aquatic animals (Russo 1985).

A number of mechanisms have been identified as responsible for the toxicological effects of ammonia: (1) asphyxiation caused by damage to the gill epithelium; (2) progressive acidosis and reduction in blood oxygen-carrying capacity caused by suppression of Krebs cycle and stimulation of glycolisis; (3) Inhibition of ATP production and depletion of ATP in the basilar region of the brain caused by uncoupling oxidative phosphorylation; (4) disruption of blood vessels and osmoregulatory activity upsetting the liver and kidneys, (5) repression of immune system increasing the susceptibility to bacterial and parasitic diseases (Camargo and Alonso 2006).

Moreover, NH_4^+ can contribute to NH_3 toxicity by reducing the internal Na^+ to fatally low concentrations (Russo 1985; Augspurger *et al.* 2003). These negative physiological effects can result in decreased feeding activity, fecundity and survivorship, decreasing populations sizes of aquatic animals (Constable *et al.* 2003; Alonso and Camargo 2004; Alonso 2005). Additive toxicity or even synergistic effects have been observed through the chemical interaction of ammonium with other chemical pollutants, such as copper, cyanide, phenol, zinc and chlorine (Alabaster and Lloyd 1982; Russo 1985; Adams and Bealing 1994).

On the basis of acute and chronic toxicity data, water quality criteria, ranging 0.05–0.35 mg NH_3-N/L for short-term exposures and 0.01–0.02 for long-term exposures, have been estimated and recommended to protect sensitive aquatic animals (Camargo and Alonso 2006).

Regarding the toxicological effects imposed by nitrite, another important inorganic nitrogenous contaminant commonly found in aquatic environments, it has been reported that, as in the case of NH_3, it can also cause toxicity to *Nitrosomonas* and *Nitrobacter* bacteria, thus inhibiting the nitrification process (Anthonisen *et al.* 1976; Russo 1985). Unionised nitrite (HNO_2) has been identified as the most inhibiting species for nitrifying bacteria (Anthonisen *et al.* 1976; Russo 1985). Nevertheless, since the nitrite concentration is usually much higher than HNO_2 in most aquatic ecosystems, nitrite ions are considered to be the major responsible for nitrite toxicity to aquatic microorganisms and animals (Anthonisen *et al.* 1976; Russo 1985; Lewis and Morris 1986; Eddy and Williams 1987; Tahon *et al.* 1988; Chen and Chen 1992; Jensen 2003).

A number of mechanisms of nitrite toxicity on aquatic animals, particularly on fish and crayfish, have been identified. Nitrite can transform oxygen-carrying pigments to forms that are incapable of transporting oxygen, causing hypoxia and ultimately death (Camargo and Alonso 2006). In fish, incorporation of nitrite into the red blood cells is associated with the oxidation of iron atoms ($Fe^{+2} \rightarrow Fe^{+3}$), which promotes the conversion of haemoglobin into methemoglobin that is unable to deliver oxygen to body tissues because of its high dissociation constant (Russo 1985; Lewis and Morris 1986; Eddy and Williams 1987; Jensen 2003). Similarly, in crayfish, entry of nitrite into the blood plasma is linked to the oxidation of copper atoms ($Cu^{+1} \rightarrow Cu^{+2}$), whereby functional hemocyanin is converted into methemocyanin that cannot bind reversibly to molecular oxygen (Tahon *et al.* 1988; Chen and Chen 1992; Jensen 2003).

The following toxic effects of nitrite on fish and crayfish have also been found (Camargo and Alonso 2006): (1) depletion of extracellular and intracellular Cl^- levels causing severe electrolyte imbalance; (2) depletion of intracellular K^+ and elevation of extracellular K^+ levels affecting membrane potentials, neurotransmission, skeletal muscle contractions, and heart function; (3) formation of *N*-nitroso compounds that are mutagenic and carcinogenic; (4) damage to mitochondria in liver cells causing tissue oxygen shortage; and (5) repression of immune system decreasing the tolerance to bacterial and parasitic diseases.

Anthropogenic discharges containing large concentrations of nitrite have been associated with fish kills in aquatic ecosystems (Lewis and Morris 1986; Philips *et al.* 2002), although direct toxicity of nitrite has only been evidenced by laboratory studies (Russo 1985; Lewis and Morris 1986; Eddy and Williams 1987; Jensen 2003, Alonso 2005). Among the different taxonomic groups of freshwater invertebrates and fish that have been exposed to nitrite toxicity, certain crustaceans (decapods, amphipods), insects (ephemeropterans), and fish (salmonids, cyprinids) seem to be the most sensitive, exhibiting acute toxicities (96-hour LC_{50}) lower than 3 mg NO_2^--N/L and short-term safe levels (96-hour $LC_{0.01}$) lower than 0.25 NO_2^--N/L (Camargo and Alonso 2006). On the basis of acute toxicity data, Alonso (2005) has recently estimated water quality criteria, ranging 0.08–0.35 mg NO_2^--N/L, which may be adequate to protect sensitive aquatic animals, at least during short-term exposures.

Nitrate, the second most abundant nitrogenous compound in aquatic environments (after ammonium), has a similar toxicological mechanism to that of nitrite. Certainly, nitrate can modify oxygen-carrying pigments (hemoglobin, hemocyanin) to forms that are incapable of carrying oxygen (methemoglobin, methemocyanin) (Jensen 1996; Scott and Crunkilton 2000; Cheng *et al.* 2002). In fact, nitrate must be firstly converted into nitrite under internal body conditions before it becomes toxic (Cheng and Chen 2002). However, due to the low branchial permeability to nitrate ions, the nitrate uptake in aquatic animals is more limited than nitrite uptake, which contributes to the relatively low toxicity of nitrate (Camargo and Alonso 2006).

Toxicity of nitrate in aquatic environments has traditionally been considered to be irrelevant (Russo 1985; Camargo *et al.* 2005); however, several laboratory studies have shown that a nitrate concentration of 10 mg NO_3^--N/L (US federal maximum level for drinking water) can adversely affect, at least during long-term exposure, sensitive aquatic animals (Camargo and Alonso 2006). Among the different taxonomic groups of freshwater invertebrates and fish that have been exposed to nitrate toxicity, certain caddisflies, amphipods and salmonids seem to be the most sensitive, exhibiting short-term safe levels (120-hour $LC_{0.01}$) and no-observed effect concentrations (after 30-day) lower than 5 mg NO_3^--N/L (Camargo and Alonso 2006).

The sensitivity of certain amphibians to nitrate toxicity may be high and similar of sensitive freshwater invertebrates and fish species. Current field data suggest that nitrogen-based fertilisers may be contributing to the decline of amphibian populations in agricultural areas because of impaired swimming ability, decreased body size, and reduced fecundity and survivorship (Hecnar 1995; Birge *et al.* 2000). On the basis of toxicity data, Camargo *et al.* (2005) has recently proposed a maximum level of 2 mg NO_3^--N/L for the protection of sensitive aquatic animals.

1.3.3 Effects on human health

Ingestion of drinking water resources contaminated with nitrite can induce methemoglobinemia in humans. Moreover, nitrate-contaminated waters can also promote the same health problem by the reduction of nitrate to nitrite in the digestion tract. Methemoglobinemia is characterised by an obstruction of the oxygen-carrying capacity of hemoglobin, resulting in methemoglobin (Craun *et al.* 1981; Nash 1993; Ayebo *et al.* 1997; Knobeloch *et al.* 2000; Fwetrell 2004; Greer and Shannon 2005). Decreased oxygen transport is clinically noted when methemoglobin levels reach 10% or more (Knobeloch *et al.* 2000; Fwetrell 2004). Typical symptoms of methemoglobinemia are cyanosis, headache, stupor, fatigue, tachycardia, coma, convulsions, asphyxia and ultimately death (Camargo and Alonso 2006). Infants less than 4 months of age seem more vulnerable to this toxicity because of their greater fluid intake relative to body weight, their higher proportion of readily oxidisable fetal hemoglobin, their lower methemoglobin reductase activity (a red blood cell enzyme that converts methemoglobin back to hemoglobin), their higher gastric pH (which allows greater invasion of bacteria and enhances conversion of ingested nitrate into nitrite), and their increased susceptibility to gastroenteritis (Camargo and Alonso 2006). More than 3000 cases of methemoglobinemia have been reported worldwide since 1945, mainly associated with consumption of water from private wells containing high nitrate concentrations (>10 NO_3^--N/L) (Nash 1993; Ayebo *et al.* 1997; Knobeloch *et al.* 2000; Wolfe and Patz 2002).

Contradictory conclusions have been raised regarding the maximum nitrate and nitrite concentrations that should be established for drinking waters in order to prevent methemoglobinemia in humans. For instance, on the basis of case data, the US Environmental Protection Agency has recommended maximum contaminant levels of 10 mg NO_3^--N/L and 1 mg NO_2^--N/L to prevent this health problem (US Environmental Protection Agency, 1986; Wolfe and Patz 2002). However, Fwetrell (2004) has recently indicated that nitrate (and nitrite) may be one of a number of co-factors that play a complex role in causing methemoglobinemia and, consequently, it would be inappropriate to attempt to link illness rates with drinking water nitrate levels. The last observation is further emphasised by a recent report indicating that breastfeeding infants are not at risk of methemoglobinemia even when mothers ingest water with nitrate concentrations as high as 100 mg NO_3^--N/L (Greer and Shannon 2005).

Ingested nitrate and nitrite can also promote the formation of *N*-nitroso compounds (*e.g.* nitrosamines), which are among the most potent of the recognised carcinogens in mammals (Nash 1993; Knobeloch *et al.* 2000; Wolfe and Patz 2002; Fwetrell 2004). Carcinogenicity by di-methyl-nitrosamines may result from short-term exposure to a single large dose or from chronic exposure to relatively small doses, with DNA alkylation as the main carcinogenic mechanism (Manahan 1992; Wolfe and Patz 2002). Endogenous formation of nitrosamines has also been documented after long-term consumption of drinking water with nitrate concentrations even below the maximum contaminant level of 10 mg NO_3^--N/L (Van Maanen *et al.* 1996; Townsend *et al.* 2003).

A number of major health problems have also been associated with the ingestion of nitrate and nitrite by humans. Indeed, further scientific evidences suggest that ingested nitrate and nitrite might result in mutagenicity, teratogenicity and birth defects (Dorsch *et al.* 1984; Luca *et al.* 1987), contribute to the risks of non-Hodgkin's lymphoma (Ward *et al.* 1996), coronary heart disease (Cerhan *et al.* 2001), and bladder and ovarian cancers (Meyer *et al.* 2001), play a role in the etiology of insulin-dependent diabetes mellitus (Virtanen *et al.* 1994; Parslow *et al.* 1997; Van Maanen *et al.* 2000) and in the development of thyroid hypertrophy (Van Maanen *et al.* 1994), or cause spontaneous abortions (Centers for Disease Control and Prevention 1996) and respiratory tract infections (Gupta *et al.* 2000). Long-term consumption of drinking water with nitrate concentrations even below the maximum contaminant level of 10 mg NO_3^--N/L has been linked to higher risks for non-Hodgkin's lymphoma (Ward *et al.* 1996) and for bladder and ovarian cancers (Weyer *et al.* 2001).

Indirect health hazards can also occur as a consequence of toxic algal bloom promoted by eutrophication in aquatic ecosystems. An extensive description of these toxicological effects has been reported by Camargo and Alonso (2006).

1.3.4 Effects on human economy

Human sickness and death, resulting directly (*e.g.* ingested nitrate and nitrite from polluted drinking water) or indirectly (*e.g.* aerosol exposure to algal toxins, consumption of contaminated seafood causing poisoning syndromes) from inorganic nitrogen pollution, can have high economical expenses because of lost wages and work days, and because of medical treatment and investigation (Camargo and Alonso 2006). The annual average estimate of public health costs of illnesses due to shellfish poisoning in USA could range from US$400 thousand to US$2 million (Todd 1989; Hoagland *et al.* 2002). Furthermore, individuals who become sick and experience strong pain and suffering could develop further psychological disorders with long-term medical treatments.

Extensive kills of aquatic animals associated with inorganic nitrogen pollution may result in much higher economic costs because of reduced yields of desirable fish and shellfish species, and because of subsequent contraction of consumer demand usually associated with closing of commercially important fisheries (Van Dolah *et al.* 2001; Hoagland *et al.* 2002; Philips *et al.* 2002; Smith 2003). For instance, commercial harvest losses of clams, oysters, scallops and finfish, representing about US$8 million, were documented during a bloom of the toxic algae *Karenia brevis* in North Caroline (Tester *et al.* 1991). Moreover, Hoagland *et al.* (2002) estimated total economic effects of $18 million arising from the deaths of farmed Atlantic salmon killed by phytoplankton blooms during 1987, 1989 and 1990 in Washington (USA).

Camargo and Alonso (2006) have underlined other collateral effects on the economy caused by nitrogen pollution: (1) taste and odour problems in drinking water supplies; (2) additional improvements in water treatment procedures (*e.g.* activated carbon adsorption, chemical coagulation, liming, ozonation, ultrafiltration); (3) increased expenditures in monitoring and management of water pollution, including cleanup and restoration procedures; (4) decreases in the perceived aesthetic value of water bodies; (5) reduced use of degraded aquatic ecosystems for recreation and tourism.

1.4 APPROACHES TO PREVENT NITROGEN POLLUTION

Industrialisation has commonly been considered incompatible with the conservation and improvement of the environment for many years (Graedel and Allenby 1995). However, the focus on this issue has progressively changed. In fact, the higher demand to provide a suitable life quality for the Earth's citizens will involve even more industrial activities, which may contribute to serious environmental problems. Thus, providing a sustainable world will require closer industry-environment interactions (Cervantes and Pavlostathis 2006).

Since regulation on pollution was implemented, both industry and government regulatory agencies have focused their efforts on the reduction of toxic wastes by controlling discharges at the point where they enter the environment. Thus, end-of-pipe treatments were initially adopted for most industrial processes (Khan *et al.* 2001). However, the advent of strict environmental legislation in recent years, combined with the ineffectiveness and relatively high cost of several end-of-pipe treatment technologies, have, in many cases, resulted in making this concept inadequate to deal with the magnitude and complexity of environmental deterioration. Therefore, during the last few years, the concept of 'industrial ecology', which mission is to design zero-emission industrial processes by focusing on cleaner production and waste minimization, was implemented in several industrial sectors (Ayres and Simonis 1994; Graedel and Allenby 1995; Ayres and Ayres 1996; Allenby 1999). These pollution prevention concepts have significantly contributed not only to reducing pollution, but also to improving environmental performance, raising profitability and enhancing competitiveness in several industries (My Dieu 2003). For further details, Cervantes and Pavlostathis (2006) have recently described the main pollution prevention concepts, analysing their strengths and weaknesses. In this section, the main strategies to prevent nitrogen pollution of aquatic ecosystems will be discussed.

1.4.1 End-of-pipe treatments

The intention of this chapter is not to provide detailed description of all the technologies available to treat wastewaters containing nitrogenous contaminants. In the following chapters of this book, the different technologies available to treat low-strength (chapter 5) and high-strength (chapter 6) wastewaters containing inorganic nitrogenous pollutants will be described. Moreover, chapter 7 describes the technological concepts to treat recalcitrant nitrogenous pollutants.

However, it is important to emphasise that, even though end-of-pipe pollution control strategies have certainly contributed to reduce negative environmental consequences of industrial processes, they focus on the symptoms and not the origin of environmental problems (Khan *et al.* 2001). Other important disadvantages of end-of-pipe pollution control methods are: (1) they are not adequate to allow an efficient use of limited resources; (2) they cause greater consumption of materials and energy, more capital expenditure and more work hours compared to measures taken at source; (3) their use generally creates new environmental problems, such as the need for disposing wastes from treatment facilities. Moreover, despite the fact that chemical and biological treatment technologies are effective in limiting or reversing nitrogen pollution dispersal, it is important to realize that many of these technologies also have undesirable side-effects. For example, stripping of ammonia from concentrated wastewaters may derive in air contamination if preventive measures are not taken during the treatment process (Zhang and Lau 2007).

Coagulation processes are among the most efficient physical-chemical processes to remove recalcitrant nitrogenous pollutants (*e.g.* azo dyes) from industrial wastewaters; however, large volumes of solid wastes should be handled during the application of these treatment processes, which represents a serious drawback (Dos Santos *et al.* 2007). For example, a textile factory located in north-western Mexico, which mainly applies reactive azo compounds during dying processes, discharges an average of 400 m^3/d containing 1000 mg chemical oxygen demand (COD)/L, 350 mg biochemical oxygen demand (BOD)/L and 300 mg suspended solids (SS)/L. This textile effluent is treated by a physical-chemical process demanding large amounts of $Al(SO_4)_3$, as well as anionic and cationic polymers, generating ~20 Tons of solid waste residues monthly (Cervantes, unpublished data).

To some extent, even biological treatment processes could originate collateral environmental problems during the removal of nitrogenous pollutants from contaminated waters. For instance, incomplete denitrification may promote the discharge of treated effluents with important levels of nitrite or the release of nitrous oxide (N_2O) to the atmosphere due to a number of operational aspects. In fact, nitrogen-removing wastewater treatment plants (WWTP) have been identified as major N_2O sources, which significantly contribute to the global warming. For instance, a comprehensive study documented the release of N_2O from a WWTP located in Durham (New Hampshire) quantifying emissions of 3.5×10^4 g N_2O yr^{-1}, equivalent to a generation of 3.2 g N_2O person^{-1} yr^{-1} (Czepiel *et al.* 1995).

Furthermore, during the reductive decolourisation of azo dyes in anaerobic WWTP, carcinogenic aromatic amines are produced, which demand the application of an aerobic post-treatment (Field *et al.* 1995). In fact, the anaerobic decolourisation of azo dyes has been identified as the rate-limiting process of the overall conventional treatment train (anaerobic followed by aerobic treatment processes), due to electron transfer limitations (Van der Zee *et al.* 2001), which may cause poor rector performance or even collapse (Van der Zee *et al.* 2001a). Chapter 7 describes several strategies to improve the anaerobic decolourisation of azo dyes.

1.4.2 Cleaner production

Cleaner production, also referred to as 'waste minimisation', differs from end-of-pipe treatments in that it minimises wastes and emissions by reducing them at their sources. Cleaner production can generally be defined as the continuous application of an integrated preventive environmental strategy to production processes in order to avoid wastes and emissions at the source, to preserve energy and raw materials, to eliminate the use of toxic materials and to improve working conditions (Cervantes and Pavlostathis 2006). Cleaner production contributes to optimisation of resources, therefore reflecting environmental improvement on financial and economic benefits, as well as on technological progress.

Regarding contamination of waters by nitrogenous compounds, a number of measures have been identified within the cleaner production concept. For example, in the textile sector, which is one of the industries that discharges large quantities of recalcitrant nitrogenous compounds (*e.g.* aromatic amines, azo dyes) through their wastewaters, the use of non-toxic chemicals during dying processes has been proposed as a cost-effective, preventive strategy (Cervantes and Pavlostathis 2006). In the same industrial sector, correct scheduling of the process in view of equipment cleaning can also reduce waste generation. For example, preparing light paints before dark ones, or arranging fabric requiring similar dyeing and finishing process in sequential order, will make cleaning of vats unnecessary before starting a new batch (My-Dieu 2003).

Recycling, recovery and reuse of materials is the next most preferable strategy of cleaner production as many waste materials generated can be reused either onsite, in the original process with or without treatment to remove impurities, or offsite, in other plants. For instance, in the finishing stage of clothing manufacture, the amount of dyestuffs and printing pastes discharged with wastewater can be reduced by internal recycling. It is possible to apply the recycled printing pastes in processes where a lower quality is acceptable. Also, less current colours can be mixed to darker or black colours (My-Dieu 2003). These strategies decrease the emission of recalcitrant nitrogenous pollutants from the textile industry.

Recovery of nutrients, such as nitrogen and phosphorus, from concentrated effluents like piggery wastewaters, in the form of struvite ($MgNH_4PO_4 \cdot 6H_2O$), is also a common practice in some industrialised countries in which the mineral is being applied as a fertilizer (Doyle and Parsons 2002). Thus, contributing to a sustainable management of these nitrogen-rich wastewaters (see further details in chapter 9).

1.4.3 Industrial ecology

The concept of industrial ecology includes the transformation of the traditional model of industrial activity into a more integrated model – an industrial ecosystem, in which wastes from one process can serve as raw material for the others. Industrial ecology is an innovative strategy for sustainable industry involving design of industrial systems to minimise waste and maximise the cycling of materials and energy (Karamanos 1995). Moreover, industrial ecology strives to optimise resource flows rather than just preventing pollution, and to promote sustainability rather than only reduce risks (Olderburg and Geiser 1997).

As nitrogen is a primary nutrient demanded in several productive activities (*e.g.* agriculture) and, on the other hand, is a major contaminant discharged in concentrated wastewaters, such as those generated from farms and aquaculture, strategies to connect these dynamic sectors to achieve a sustainable nitrogen management is required.

A very promising approach to recover ammonium from concentrated wastewaters is by the production of struvite ($MgNH_4PO_4 \cdot 6H_2O$), a mineral that has been suggested to display excellent fertiliser qualities under specific conditions when compared with standard fertilisers (Ghosh *et al.* 1996). In fact, parallel recovery of both nitrogen and phosphorus occur through struvite precipitation, which explains the superior fertilising capacity of this mineral compared to conventional fertilisers. Recent studies have suggested that recovery of nutrients via struvite production from concentrated wastewaters (*e.g.* piggery effluents) could fulfil major nutritional requirements in regions with intensive farming and agricultural activities (Cervantes *et al.* 2007). Chapter 9 describes further details on the recovery of nitrogen via struvite precipitation.

REFERENCES

Adams, N. and Bealing, D. (1994) Organic pollution: biochemical oxygen demand and ammonia. In *Handbook of ecotoxicology*. Oxford. p. 264–285.

Alabaster, J.S. and Lloyd, R. (1982) Water quality criteria for freshwater fish. 2nd. Edition. London. Butterworths.

Allan, J.D. (1995) Stream ecology: structure and function of running waters. London. Chapman and Hall.

Allenby, B.R. (1999) Industrial ecology: Policy framework and implementation. Prentice Hall, New Jersey.

Alonso, A. (2005) Valoración de la degradación ambiental y efectos ecotoxicológicos sobre la comunidad de macroinvertebrados bentónicos en la cabecera del río Henares. Doctoral dissertation. Universidad de Alcalá, Alcalá de Henares (Madrid), Spain.

Alonso, A. and Camargo J.A. (2004) Toxic effects of unionized ammonia on survival and feeding activity of the freshwater amphipod *Eulimnogammarus toletanus* (Gammaridae, Crustacea). *Bull. Environ. Contam. Toxicol.* **72**, 1052–1058.

Anderson, D.M., Gilbert, P.M. and Burkholder, J.M. (2002) Harmful algal blooms and eutrophication: nutrient sources, composition and consequences. *Estuaries* **25**, 704–726.

Anthonisen, A.C., Loehr, R.C., Prakasam, T.B.S. and Srinath, E.G. (1976) Inhibition of nitrification by ammonia and nitrous acid. *J. Water Pollut. Control Fed.* **48**, 835–852.

Augspurger, T., Keller, A.E., Black, M.C., Cope, W.G. and Dwyer, F.J. (2003) Water quality guidance for protection of freshwater mussels (Unionidae) from ammonia exposure. *Environ. Toxicol. Chem.* **22**, 2569–2575.

Ayebo, A., Kross, B., Vlad, M. and Sinca, A. (1997) Infant methemoglobinemia in the Transylvania region in Romania. *Int. J. Occup. Environ. Health* **3**, 20–29.

Ayres, R.U. and Ayres L.W. (1996) IE: Towards closing the materials cycle. Edward Edgar.

Ayres, R.U. and Simonis, U.E. (1994) Industrial metabolism: Restructuring for sustainable development. University Press, New York.

Baker, L.A., Herlihy, A.T., Kaufmann, P.R. and Eilers, J.M. (1991) Acidic lakes and streams in the United States: the role of acidic deposition. *Science* **252**, 1151–1154.

Birge, W.J., Westerman, A.G. and Spromberg, J.A. (2000) Comparative toxicology and risk assessment of amphibians. In *Ecotoxicology of amphibians and reptiles.* Pensacola, SETAC. pp. 727–792.

Breitburg, D. (2002) Effects of hypoxia, and the balance between hypoxia and enrichment, on coastal fishes and fisheries. *Estuaries* **25**, 767–781.

Camargo, J.A. and Alonso A. (2006) Ecological and toxicological effects of inorganic nitrogen pollution in aquatic ecosystems: a global assessment. *Environ. International* **32**, 831–849.

Camargo, J.A., Alonso, A. and Salamanca, A. (2005) Nitrate toxicity to aquatic animals: a review with new data for freshwater invertebrates. *Chemosphere* **58**, 1255–1267.

Centers for Disease Control and Prevention (1996) Spontaneous abortions possibly related to ingestion of nitrate-contaminated well water in La Grande County, Indiana, 1991–1994. *Morb. Mortal Wkly. Rep.* **45**, 569–572.

Cerhan, J.R., Weyer, P.J., Janney, C.A., Lynch, C.F. and Folsom, A.R. (2001) Association of nitrate levels in municipal drinking water and diet with risk of coronary heart disease mortality: the Iowa Women's Health Study. *Epidemiology* **12**, S84–457.

Cervantes F.J. and Pavlostathis, S.G. (2006) Strategies for industrial water pollution control. In *Advanced Biological Treatment Processes for Industrial Wastewaters: Principles & Applications.* Cervantes F.J. and Pavlostathis, S.G. and Van Haandel, A.C. (Eds.). London. IWA Publishing. pp. 1–15.

Cervantes, F.J., Saldivar-Cabrales, J. and Yescas, J.F. (2007) Strategies for the utilization of piggery wastes on agriculture. *Rev. Latinoam. Rec. Nat.* **3**, 3–12.

Chen, J.C. and Chen S.F. (1992) Accumulation of nitrite in the heamolymph of *Panaeus monodon* exposed to ambient nitrite. *Comp. Biochem. Physiol.* **103C**, 477–481.

Cheng, S.Y. and Chen, J.C. (2002) Study of the oxyhemocyanin, deoxyhemocyanin, oxygen affinity and acid-base balance of *Marsupenaeus japonicas* following exposure to combined elevated nitrite and nitrate. *Aquatic Toxicol.* **61**, 181–193.

Cheng, S.Y., Tsai, S.J. and Chen, J.C. (2002) Accumulation of nitrate in the tissues of *Panaeus monodon* following elevated ambient nitrate exposure after different time periods. *Aquatic Toxicol.* **56**, 133–146.

Constable, M., Charlton, M., Jensen, F., McDonald, K., Craig, G. and Taylor, K.W. (2003) An ecological risk assessment of ammonia in the aquatic environment. *Hum. Ecol. Risk Assess* **9**, 527–548.

Craun, G.F., Greathouse, D.G. and Gunderson, D.H. (1981) Methemoglobin levels in young children consuming high nitrate well water in the United States. *Int. J. Epidemiol.* **10**, 309–317.

Cummins, C.P. (1994) Acid solutions. In *Handbook of ecotoxicology*. Oxford. Blackwell Scientific Publications. pp. 21–44.

Czepiel, P., Crill, P. and Harriss, R. (1995) Nitrous oxide emissions from municipal wastewater treatment. *Environ. Sci. Technol.* **29**, 2352–2356.

Diaz, R.J. and Rosenberg, R. (1995) Marine benthic hypoxia: a review of its ecological effects and the behavioral responses of benthic macrofauna. *Oceanogr. Mar. Biol. Ann. Rev.* **33**, 245–303.

Dorsch, M.M., Scragg, R.K.R., McMichel, A.J., Baghurst, P.A. and Dyer, K.F. (1984) Congenital malformations and maternal drinking water supply in rural south Australia: a case-control study. *Am. J. Epidemiol.* **119**, 473–486.

Dos Santos, A.B., Cervantes, F.J. and van Lier, J.B. (2007) Review paper on current technologies for decolourisation of textile wastewaters: perspectives for anaerobic biotechnology. *Biores. Technol.* **98**, 2369–2385.

Doyle, J.D. and Parsons, S.A. (2002) Struvite formation, control and recovery. *Wat. Res.* **36**, 3925–3940.

Eddy, F.B. and Williams, E.M. (1987) Nitrite and freshwater fish. *Chem. Ecol.* **3**, 1–38.

Field, J.A., Stams, A.J.M., Kato, M. and Schraa, G. (1995) Enhanced biodegradation of aromatic pollutants in cocultures of anaerobic and aerobic bacterial consortia. *Antonie van Leeuwenhoek* **67**, 47–77.

Fjellheim, A. and Raddum G.G. (1990) Acid precipitation: biological monitoring of streams and lakes. *Sci. Total Environ.* **96**, 57–66.

Fwetrell, L. (2004) Drinking-water nitrate, methemoglobinemia, and global burden of disease: a discussion. *Environ. Health Perspect.* **112**, 1371–1374.

Galloway, J.N., Dentener, F.J., Capone, D.G., Boyer, E.W., Howarth, R.W. Seitzinger, S.P., Asner, G.P., Cleveland, C.C., Green, P.A. Holland, E.A., Karl, D.M., Michaels, A.F., Porter, J.H., Townsend, A.R. and Vöosmarty, C.J. (2004) Nitrogen Cycles: Past, Present, and Future. *Biogeochemistry* **70**, 153–226.

Galloway, J.N., Townsend, A.R., Erisman, J.W., Bekunda, M., Cai, Z., Freney, J.R., Martinelli, L.A., Seitzinger, S.P. and Sutton, M.A. (2008) Transformation of the nitrogen cycle: recent trends, questions, and potential solutions. *Science* **320**, 889–892.

Ghosh, G.K., Mohan, K.S. and Sarkar, A.K. (1996) Characterization of soil fertilizer P reaction products and their evaluation as sources of P for gram (*Cicer arietinum* L.). *Nutr. Cycling Agroecosyst.* **46**, 71–79.

Graedel, T.E. and Allenby, B.R. (1995) Industrial Ecology. Prentice Hall, New Jersey.

Greer, F.R. and Shannon, M. (2005) Infant methemoglobinemia: the role of dietary nitrate in food and water. *Pediatrics* **116**, 784–786.

Gupta, S.K., Gupta, R.C., Gupta, A.B., Seth, A.K., Bassin J.K. and Gupta, A. (2000) A recurrent acute respiratory infections in areas with high nitrate concentrations in drinking waters. *Environ. Health Perspect.* **108**, 363–366.

Hales, J.M. (1982). The role of NOx as a precursor of acidic deposition. Proceedings of the US-Dutch International Symposium on Nitrogen Oxides, Maastricht, The Netherlands. Elsevier.

Hecnar, S.J. (1995) Acute And chronic toxicity of ammonium nitrate fertilizer to amphibians from southern Ontario. *Environ. Toxicol. Chem.* **14**, 2131–2137.

Hoagland, P., Anderson, D.M., Kaoru, Y. and White, A.W. (2002) The economic effects of harmful algal blooms in the United States: estimates, assessment issues, and information needs. *Estuaries* **25**, 819–837.

Huckabee, J.W., Mattice, J.S., Pitelka, L.F., Porcella, D.B. and Goldstein, R.A. (1989) An assessment of the ecological effects of acidic deposition. *Arch. Environ. Contam. Toxicol.* **18**, 3–27.

Jensen, F.B. (1996) Uptake, elimination and effects of nitrite and nitrate in freshwater crayfish (*Astacus astacus*). *Aquatic Toxicol.* **34**, 95–104.

Jensen, F.B. (2003) Nitrite disrupts multiple physiological functions in aquatic animals. *Comp. Biochem. Physiol.* **135A**, 9–24.

Jin, X., Xu, Q. and Huang C. (2005) Current status and future tendency of lake eutrophication in China. *Sci. China C Life Sci.* **48**, 948–954.

Karamanos, P. (1995) Industrial ecology: new opportunities for the private sector. *J. Industry and Environment* **18**(4), 38–44.

Khan, F.I., Natrajan, B.R. and Revathi, P. (2001) GreenPro: a new methodology for cleaner and greener process design. *Journal of Loss Prevention in the Process Industries* **14**, 307–328.

Knobeloch, L., Salna, B., Hogan, A., Postle, J. and Anderson, H. (2000) Blue babies and nitrate-contaminates well water. *Environ. Health Perspect.* **108**, 675–678.

Krause-Dellin, D. and Steinberg, C. (1986) Cladoceran remains as indicators of lake acidification. *Hydrobiologia* **143**, 129–134.

Lewis, W.M. and Morris, D.P. (1986) Toxicity of nitrite to fish: a review. *Trans. Am. Fish Soc.* **115**, 183–195.

Luca, D., Luca, V., Cotor, F. and Rileanu, L. (1987) In vivo and in vitro cytogenetic damage induced by sodium nitrite. *Mutat. Res.* **189**, 333–340.

Manahan, S.E. (1992) Toxicological chemistry. 2nd Edition. Boca Raton (FL). Lewis Publishers.

Meriläinen, J.J. and Hynynen, J. (1990) Benthic invertebrates in relation to acidity in Finnish forest lakes. In *Acidification in Finland*. Berlin. Springer-Verlag. pp. 1029–1049.

Morris, R., Taylor, E.W., Brown, D.J.A. and Brown, J.A. (1989) Acid toxicity and aquatic animals. Cambridge. Cambridge University Press.

My-Dieu, T.T. (2003) Greening food processing industry in Vietnam: Putting industrial ecology to work. Ph.D. thesis, Wageningen University, The Netherlands.

Nash, I. (1993) Water quality and health. In *Water in crisis: a guide to the world's freshwater resources*. New York. Oxford University Press. pp. 25–39.

Oldenburg, K.U. and Geiser, K. (1997) Pollution prevention and . . . or industrial ecology? *J. Cleaner Prod.* **5**(1–2), 103–108.

Ormerod, S.J., Weatherley, N.S. Merrett, W.J. (1990) Restoring acidified streams in upland Wales: a modeling comparison of the chemical and biological effects of liming and reduced sulphate deposition. *Environ. Pollut.* **64**, 67–85.

Ortiz, J.A., Rueda, A., Carbonell, G., Camargo, J.A., Nieto, F. and Tarazona, J.V. (1993) Acute toxicity of sulfide and lower pH in cultured rainbow trout and coho salmon. *Bull. Environ. Contam. Toxicol.* **50**, 164–170.

Parslow, R.C., McKinney, P.A., Law, G.R., Staines, A., Williams, R. and Bodanski, H.J. (1997) Incidence of childhood diabetes mellitus in Yorkshire, northern England, is associated with nitrate in drinking water: an ecological analysis. *Diabetologia* **40**, 550–556.

Philips, S., Laanbroek, H.J. and Werstraete, W. (2002) Origin, causes and effects of increased nitrite concentrations in aquatic environments. *Rev. Environ. Sci. Biotechnol.* **1**, 115–141.

Pinheiro, H.M., Touraud, E. and Thomas, O. (2004) Aromatic amines from azo dye reduction: status review with emphasis on direct UV spectrophotometric detection in textile industry wastewaters. *Dyes Pigments* **61**, 121–139.

Princiotta FT (1982). Stationary source NOx control technology overview. Proceedings of the US-Dutch International Symposium on Nitrogen Oxides, Maastricht, The Netherlands. Elsevier.

Rabalais, N.N. (2002) Nitrogen in aquatic ecosystems. *Ambio.* **31**, 102–112.

Razo-Flores, E., Donlon, B., Lettinga, G. and Field, J.A. (1997) Biotransformation and biodegradation of *N*-substituted aromatics in methanogenic granular sludge. *FEMS Microbiol. Rev.* **20**, 525–538.

Russo, R.C. (1985) Ammonia, nitrite and nitrate. In *Fundamentals of aquatic toxicology.* Washington, D.C. Hemisphere Publishing Corporation. pp. 455–471.

Sanchez, P.A. and Swaminathan, M.S. (2005) Hunger in Africa: the link between unhealthy people and unhealthy soils. *Lancet* **365**, 442–444.

Schindler, D.W. (1988) Effects of acid rain on freshwater ecosystems. *Science* **239**, 149–157.

Schindler, D.W., Kasian, S.E.M. and Hesslein, R.H. (1989) Losses of biota from America aquatic communities due to acid rain. *Environ. Monit. Assess* **12**, 269–285.

Scott, G. and Crunkilton, R.L. (2000) Acute and chronic toxicity of nitrate to fathead minnows (*Pimephales promelas*), *Ceriodaphnia dubia* and *Daphnia magna*. *Environ. Toxicol. Chem.* **19**, 2918–2922.

Skjelkvale, B.L., Stoddard, J.L. and Andersen, T. (2001) Trends in surface water acidification in Europe and North America (1989–1998). *Water Air Soil Pollut.* **130**, 787–792.

Smil, V. (2001) Enriching the Earth. Cambridge. The MIT Press.

Smith, V.H. (2003) Eutrophication of freshwater and coastal marine ecosystems: a global problem. *Environ. Sci. Pollut. R.* **10**, 126–139.

Stepanov AL and Korpela TK (1997). Microbial basis for the biotechnological removal of nitrogen oxides from flue gases. *Biotechnol. Appl. Biochem.* **25**, 97–104.

Stoddard, J.L., Jeffries, D.S., Lükewille, A., Clair, T.A., Dillon, P.J., Driscoll, C.T. et al. (1999) Regional trends in aquatic recovery from acidification in North America and Europe. *Nature* **401**, 575–578.

Tahon, J.P., Van Hoof, D., Vinckier, C., Witters, R., De Ley, M. and Lontie, R. (1988) The reaction of nitrite with the haemocyanin of *Astacus leptodactylus*. *Biochem. J.* **249**, 891–896.

Tester, P.A., Stumpf, R.P., Vurovich, F.M., Fowler, P.K. and Turner, J.T. (1991) An expatriate red tide bloom: transport, distribution, and persistence. *Limnol. Oceanogr.* **36**, 1053–1061.

Todd, E.C.D. (1989) Preliminary estimates of costs of food-borne disease in the United States. *J. Food Prot.* **52**, 595–601.

Townsend, A.R., Howarth, R.W., Bazzaz, F.A. and Booth, M.S., Cleveland, C.C., Collinge, S.K. et al. (2003) Human health effects of a global changing nitrogen cycle. *Front. Ecol. Environ.* **1**, 240–246.

Van der Zee, F.P., Bouwman, R.H.M., Strik, D.P.B.T.B., Lettinga, G. and Field, J.A. (2001a) Application of redox mediators to accelerate the transformation of reactive azo dyes in anaerobic bioreactors. *Biotechnol. Bioeng.* **75**, 691–701.

Van der Zee, F.P., Lettinga, G. and Field, J.A. (2001) Azo dye decolourisation by anaerobic granular sludge. *Chemosphere* **44**, 1169–1176.

Van Dolah, F.M., Roelke, D. and Greene, R.M. (2001) Health and ecological impacts of harmful algal blooms: risk assessment needs. *Hum. Ecol. Risk Assess.* **7**, 1329–1345.

Van Maanen, J.M.S., Albering, H.J., De Kok, T.M.C.M., Van Breda, S.G.I., Curfs, D.M.J., Vermeer, I.T.M. et al. (2000) Does the risk of childhood diabetes mellitus require revision of the guideline values for nitrate in drinking water? *Environ. Health Perspect.* **108**, 457–461.

Van Maanen, J.M.S., Van Dijk, A., Mulder, K., De Baets, M.H., Menheere, P.C.A., Van der Heide, D. et al. (1994) Consumption of drinking waters with high nitrate levels causes hypertrophy of the Thyroid. *Toxicol. Lett.* **72**, 365–374.

Van Maanen, J.M.S., Welle, I.J., Hageman, G., Dallinga, J.W., Mertens, P.L.J.M. and Kleinjans, J.C.S. (1996) Nitrate contamination of drinking wáter: relationship with HPRT variant frequency in lymphocyte DNA and urinary excretion of *N*-nitrosamines. *Environ. Health Perpect.* **104**, 522–528.

Virtasen, S.M., Jaakkola, L., Rasanen, L., Ylonen, K., Aro, A., Lounamaa, R. et al. (1994) Nitrate and nitrite intake and the risk for Type 1 diabetes in Finnish children. *Diabet. Med.* **11**, 656–662.

Vitousek, P.M., Aber, J.D., Howarth, R.W., Likens, G.E., Matson, P.A., Schindler, D.W. et al. (1997) Human alteration of the global nitrogen cycle: sources and consequences. *Ecol. Appl.* **7**, 737–750.

Ward, M.H., Mark, S.D., Cantor, K.P., Weisenburger, D.D., Correa-Villasenor, A. and Zahm, S.H. (1996) drinking water nitrate and the risk of non-Hodgkin's lymphoma. *Epidemiology* **7**, 465–471.

Wetzel, R.G. (2001) Limnology. 3rd. Edition. New York. Academic Press.

Weyer, P.J., Cerhan, J.R., Kross, B.C., Hallberg, G.R., Kantamneni, J., Breuer, G., Jones, M.P., Zheng, W. and Lynch, C.F. (2001) Municipal Drinking Water Nitrate Level and Cancer Risk in Older Women: The Iowa Women's Health Study. *Epidemiology* **12**, 327–338.

Wolfe, A.H. and Patz, J.A. (2002) Reactive nitrogen and human health: acute and long-term implications. *Ambio.* **31**, 120–125.

Zhang, W. and Lau, A. (2007) Reducing ammonia emission from poultry manure composting via struvite formation. *J. Chem. Technol. Biotechnol.* **82**, 598–602.

2
Principles of nitrifying processes

N. Bernet and M. Spérandio

2.1 INTRODUCTION

Nitrification is the biological oxidation of ammonia to nitrite and then to nitrate. This process was discovered at the end of the nineteenth century (Winogradsky 1890). It is carried out by two types of chemolithoautotrophic bacteria: ammonia oxidation is catalysed by ammonia-oxidizing bacteria (AOB) whereas nitrite oxidation is catalysed by nitrite-oxidizing bacteria (NOB). Compared with heterotrophic organisms, growth of nitrifying bacteria is slow and scarce, even in optimal conditions.

Nitrification is a process which occurs in natural environments like soils, continental and marine waters in which it plays a fundamental role in the nitrogen cycle. This process is the first step of nitrogen removal in biological wastewater treatment.

© 2009 IWA Publishing. *Environmental Technologies to Treat Nitrogen Pollution: Principles and Engineering*, Edited by Francisco J. Cervantes. ISBN: 9781843392224. Published by IWA Publishing, London, UK.

2.2 BIOCHEMISTRY OF NITRIFICATION

2.2.1 Ammonia oxidation

It is generally accepted that ammonia (NH_3) rather than ammonium (NH_4^+) is the substrate for AOB. Ammonia is oxidised according to the following reactions (Kowalchuk and Stephen 2001):

$$NH_3 + 2H^+ + 2e^- + O_2 \rightarrow NH_2OH + H_2O \qquad (2.1)$$

$$NH_2OH + H_2O \rightarrow HNO_2 + 4H^+ + 4e^- \qquad (2.2)$$

$$2H^+ + 0.5O_2 + 2e^- \rightarrow H_2O \qquad (2.3)$$

The first reaction is catalysed by an ammonia monooxygenase (AMO) and the second one by a hydroxylamine oxidoreductase (HAO).

Ammonia is used as an electron donor by the AOB and the final electron acceptor is oxygen. Two of the electrons produced in the second reaction are used to compensate for the electron input of the first reaction, whereas the other two are passed via an electron transport chain to the terminal oxidase, thereby generating a proton motive force (Kowalchuk and Stephen 2001). This proton motive force is used as the energy source for ATP production.

The following reaction gives the sum reaction of ammonia oxidation to nitrite:

$$NH_3 + 1.5O_2 \rightarrow HNO_2 + H_2O \qquad (2.4)$$

or
$$NH_3 + 1.5O_2 \rightarrow NO_2^- + H_3O^+ \qquad (2.5)$$

The standard free energy yield ($\Delta G^{\circ\prime}$) from the oxidation of ammonia is -275 kJ $\cdot$ mole^{-1}.

It can be seen from reaction (2.5) that ammonia oxidation produces acidity and that this reaction is highly oxygen consuming: 1.5 mole oxygen per mole ammonia, which is 3.43 g oxygen per g of ammonia nitrogen.

2.2.2 Nitrite oxidation

Nitrite is oxidised to nitrate by NOB in one single step:

$$NO_2^- + 0.5O_2 \rightarrow NO_3^- \qquad (2.6)$$

This reaction is catalyzed by a nitrite oxidoreductase (NOR). Nitrite is the electron donor of NOB respiration, while oxygen is the final electron acceptor.

The free energy yield ($\Delta G^{\circ\prime}$) from the oxidation of nitrite is only -74 kJ $\cdot$ mole^{-1}. The consequence is a low growth yield, even when compared with AOB.

2.2.3 Equations including anabolism

Equations for synthetic-oxidation using a representative measurement of yield and oxygen consumption for AOB and NOB are as follows:

$$55NH_4^+ + 76O_2 + 109HCO_3^- \rightarrow C_5H_7O_2N + 54NO_2^- + 57H_2O \\ + 104H_2CO_3$$ (2.7)

$$400NO_2^- + NH_4^+ + 195O_2 + 4H_2CO_3 + HCO_3^- \rightarrow C_5H_7O_2N \\ + 400NO_3^- + 3H_2O$$ (2.8)

The overall equation for nitrification is as follows:

$$NH_4^+ + 1.83O_2 + 1.98HCO_3^- \rightarrow 0.021C_5H_7O_2N + 0.98NO_3^- \\ + 1.041H_2O + 1.88H_2CO_3$$ (2.9)

In these equations, growth yields for AOB and NOB are 0.15 mg cells $\cdot$ mg NH_4-N^{-1} oxidised and 0.02 mg cells $\cdot$ mg NO_2-N^{-1} oxidized, respectively. Oxygen consumption ratios in the equations are 3.16 mg $O_2 \cdot$ mg NH_4-N^{-1} oxidised and 1.11 mg $O_2 \cdot$ mg NO_2-N^{-1} oxidised, respectively. Also, it can be calculated that 7.07 mg alkalinity as $CaCO_3$ is required per mg ammonia nitrogen oxidised.

2.3 MICROBIOLOGY OF NITRIFICATION

2.3.1 Ammonia oxidizers (AOB)

AOB form two monophyletic groups, one within the beta- and one within the gamma-proteobacteria (Purkhold *et al.* 2000). Most of the AOB are beta-proteobacteria : *Nitrosomonas, Nitrosospira* and *Nitrosococcus mobilis* (that is related to *Nitrosomonas*) whereas the other *Nitrosococcus* species are gamma-proteobacteria (Schmidt *et al.* 2003). It is generally accepted that nitrosomonads (including *Nitrosococcus mobilis*) and not nitrosospiras (including the genera *Nitrosospira, Nitrosolobus* and *Nitrosovibrio*) are important for ammonia oxidation in wastewater treatment plants (Wagner *et al.* 2002). Moreover, Könneke *et al.* (2005) were the first in isolating an ammonia-oxidising archeon.

2.3.2 Nitrite oxidizers (NOB)

NOB include *Nitrobacter, Nitrococcus*, both being part of the alpha-proteobacteria, and *Nitrospira* that forms a separate division (Schmidt *et al.* 2003). A fourth genus, *Nitrospina*, has only been found in marine environments.

The use of molecular tools showed that uncultured *Nitrospira*-like bacteria are the most commonly found NOB in wastewater treatment plants (Daims *et al.* 2000; 2001). This predominance of *Nitrospira*-like bacteria over *Nitrobacter* could be due to their different survival strategies. *Nitrospira*-like NOB are K-strategists, which means they are well-adapted to low nitrite and oxygen concentrations because of their low Ks (high affinity for substrates) even if they may possess a low μ_{max}. On the other hand, *Nitrobacter* is supposed to be a fast-growing r-strategist with low affinities to nitrite and oxygen (Schramm 1999), which gives a competitive advantage in environments with high substrate concentrations. Nitrite concentrations in reactors are generally low, therefore *Nitrobacter* are outcompeted by *Nitrospira*-like bacteria.

In systems with temporally or spatially elevated nitrite concentrations, such as sequencing batch or biofilm reactors, both nitrite-oxidizers should be able to co-exist (Wagner *et al.* 2002). Thus, *Nitrobacter* and *Nitrospira*-like bacteria have been observed simultaneously by fluorescent *in situ* hybridization (FISH) in a nitrifying sequencing batch biofilm reactor (Daims 2001).

Figure 2.1 presents the phylogeny of autotrophic nitrifiers based on 16S rRNA gene sequences (Kowalchuk and Stephen 2001). Anaerobic ammonium oxidisers are also included.

2.4 FACTORS AFFECTING NITRIFICATION

As all microbial processes, nitrification is affected by environmental factors such as temperature, pH, substrates concentrations, and inhibiting factors. These factors will affect nitrifiers growth rate and, as a consequence, nitrification rate. However, AOB and NOB will not be affected similarly, NOB being considered as more sensitive to variations of the environmental conditions. The consequence of an environmental perturbation will often be nitrite build-up. Whereas uncontrolled nitrite accumulation is undesired, a possible way of nitrification-denitrification optimization is to carry out a partial nitrification or nitritation, consisting in stopping the oxidation of ammonia at the stage of nitrite, and then converting nitrite to gaseous nitrogen by denitrification. By this way, 25% of the consumed oxygen is saved, as well as 40% of the carbon required for heterotrophic denitrification (which becomes closed to $2\,gCOD \cdot gN_{removed}^{-1}$ with methanol) (Voets *et al.* 1975; Turk and Mavinic 1986). This is of particular

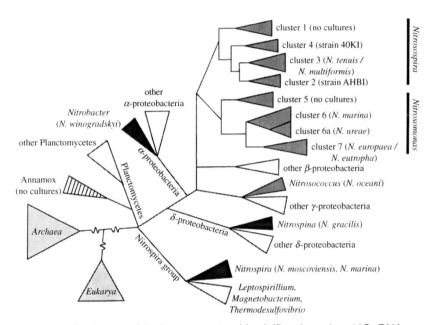

Figure 2.1. Phylogeny of the known autotrophic nitrifiers based on 16S rRNA gene sequences. Aerobic ammonia oxidisers are shown in dark gray, nitrite oxidisers in black, and anaerobic ammonia oxidisers are striped. Representative species (or isolates) are named for each group or sequence cluster (Kowalchuk and Stephen 2001)

interest when removing nitrogen from wastewaters with high nitrogen concentrations and an unbalanced COD:N ratio (Bernet and Spérandio 2006).

2.4.1 Temperature

Like all biological processes, nitrification is influenced by temperature. Nitrifying bacteria are in most cases mesophilic and have an optimum temperature for growth between 28 and 36°C. Therefore, growth rates of nitrifying bacteria will decrease with decreasing temperatures but nitrification is still possible at temperatures as low as 5°C. The effect of temperature on growth rate can be expressed by the following van't Hoff-Arrhenius equation:

$$\mu = \mu_{20} \cdot \theta^{T-20} \tag{2.10}$$

where μ = the rate coefficient (d^{-1}), μ_{20} = the value of μ at 20°C (d^{-1}), θ = temperature coefficient (dimensionless), T = temperature (°C).

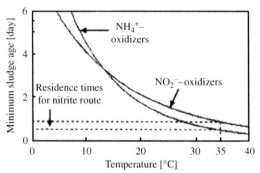

Figure 2.2. Influence of temperature on the minimum sludge age for maintaining ammonium and nitrite oxidisers (Hellinga *et al.* 1998)

The relationship between temperature and maximum growth rate is different between ammonia- and nitrite-oxidising bacteria. At elevated temperature (>15°C), ammonium-oxidisers have a higher growth rate than nitrite oxidisers. Carefully controlling the sludge age has been shown to be a good operating parameter for a stable partial nitrification (Hellinga *et al.* 1998): when the temperature is ~30–35°C, it is possible to select a sludge age which allows to maintain AOB and to wash out NOB (Figure 2.2). This is the basis of the nitritation application in chemostat process without sludge retention.

Temperature (35°C) combined with oxygen limitation and pH was shown to be efficient to maintain a stable nitrite accumulation in a biofilm reactor (Bougard *et al.* 2006).

2.4.2 pH

Optimum nitrification occurs under neutral to moderately alkaline conditions (pH 7.5 to 8.0) (Gieseke *et al.* 2006). However, nitrification has been shown to occur in acidic conditions like in acid soils (De Boer and Kowalchuk 2001) or, more recently, in biofilm and suspended biomass reactors (Tarre and Green 2004).

pH has also an indirect effect on nitrification through chemical equilibriums, the most important being the equilibrium ammonium/ammonia (NH_4^+/NH_3) and nitrite/nitrous acid (NO_2^-/HNO_2). Indeed, free ammonia (FA) and free nitrous acid (FNA) are known to inhibit nitrification and their concentrations depend on total ammonium and nitrite nitrogen concentrations, respectively, but also on pH.

2.4.3 Ammonium and FA concentrations

Ammonium and FA follow the next equation:

$$NH_4^+ + OH^- \leftrightarrow NH_3 + H_2O \qquad (2.11)$$

The FA nitrogen concentration can be calculated using the following formula derived from the above equation (Anthonisen et al. 1976):

$$S_{NH3} = \frac{S_{TAN} \times 10^{pH}}{1/K_a + 10^{pH}} \text{ with } 1/K_a = \exp[6334/(273 + T)] \qquad (2.12)$$

K_a : Ionisation constant for ammonium (at 25°C $K_a = 10^{-9.24}$)
T : Temperature in °C
Considering also the NO_2^-/NO_2H equilibrium:

$$NO_2^- + H^+ \leftrightarrow HNO_2 \qquad (2.13)$$

the concentration of free nitrous acid is:

$$S_{HNO2} = \frac{S_{NO2^-}}{K_n \times 10^{pH}} \text{ with } K_n = \exp[-2300/(273 + T)] \qquad (2.14)$$

K_n : Ionisation constant for nitrous acid (at 25°C $K_n = 10^{-3.4}$)
Anthonisen et al. (1976) were the first to study the effect of FA concentration on AOB and NOB. They established boundary conditions for FA inhibition of both AOB and NOB (Figure 2.3).

Due to possible acclimatisation, inhibition concentration level can significantly vary from a process to another (Abeling and Seyfried 1992, Anthonisen et al. 1976). The inhibition threshold is also dependant on the cell concentration. Overall it is now obvious that in all cases, NOB are more sensitive to inhibition than AOB.

2.4.4 Oxygen concentration

Nitrification is very sensitive to low oxygen concentrations. Nitrifying bacteria have very high affinities for oxygen (i.e. low Ks values), but since they are located in aggregates or biofilms in wastewater treatment processes, actual oxygen concentrations can be much lower than dissolved oxygen (DO) concentrations measured in the bulk.

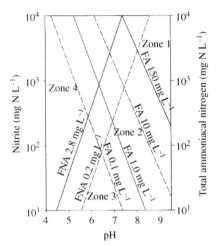

Figure 2.3. Relationship between concentrations of free ammonia (FA) and free nitrous acid (FNA) and inhibition to nitrifiers. The dashed lines mark the lower limit and the solid lines mark the upper limit of the range of boundary conditions of zones of nitrification inhibition. Zone 1 = Inhibition of nitritation and nitratation by FA; Zone 2 = Inhibition of nitratation by FA; Zone 3 = Complete nitrification Zone 4 = Inhibition of nitratation by FNA (Philips *et al.* 2002 (after Anthonisen *et al.* 1976))

   Pure nitrifying bacteria cultivation shows a difference in affinities for oxygen between *Nitrosomonas europaea* and *Nitrobacter winogradskyi* with a lower K_{O2} for *N. winograsdskyi* (Laanbroek and Gerards 1993). The same observation has been made in mixed culture (Jayamohan *et al.* 1988). Therefore, low DO concentration, around 0.2 to 0.5 mg L^{-1}, is another possible condition for limiting NOB growth (von Münch *et al.* 1996; Zeng *et al.* 2003). This solution leads to Simultaneous Nitrification-Denitrification (SND) in the same reactor and offers the potential to save the costs for a second anoxic-aerated tank. However, lower nitrification rates are consequently obtained, due to partial limitation of AOB, and nitrous oxide (N$_2$O) production is a possible problem (Zeng *et al.* 2003). It is suggested that nitrite is produced and directly consumed inside the flocs or aggregates and hence this process would be particularly adapted to granular and biofilm systems. Oxygen limitation has been shown to be efficient in biofilm system to favour nitrite build-up (Garrido *et al.* 1997; Bernet *et al.* 2000).

2.4.5 Inhibiting compounds

Due to their low growth rate, inhibition of nitrifying bacteria, even partial, can cause a complete stop of the nitrification process in activated sludge systems.

However, this complete stop only occurs after a progressive wash out of nitrifying organisms from the system over several weeks and nitrifying bacteria are generally not more sensitive to toxics than heterotrophs (Henze *et al.* 2002). Nitrification can be inhibited by organic compounds (especially sulphur components, aniline components phenols and cyanide (Henze *et al.* 2002) and metals (Cu, Ni, Cr, Zn, Co).

The modes of action for the specific inhibition of autotrophic nitrification have been reviewed by McCarty (1999) who focused on inhibitors that can be present in soils. AMO has a broad substrate range for catalytic oxidation and the inhibitory effects of many compounds are due to competition for the active site. Other compounds, such as acetylenes, are oxidised by AMO to highly reactive products which covalently bind the enzyme causing irreversible inhibition. Cu has been shown to have a significant role in the activity of AMO. A large class of compounds containing thiono-S also inhibits AMO activity by binding with Cu within the active site. Heterocyclic N compounds form another important class of nitrification inhibitors with little known about their mode of action, although evidence suggests that their inhibitory influence is closely related to the presence of ring N (McCarty 1999).

A very extensive study on inhibition of nitrification in the activated sludge process has been published by Tomlinson *et al.* (1966).

2.5 MODELLING

Nitrification kinetics is usually described by a conventional Monod expression, considering either a one-step process (ammonia conversion to nitrate) or a two step process (ammonia oxidation into nitrite, nitrite conversion into nitrate).

2.5.1 Conventional one-step nitrification model

The most commonly used model for nitrifying sludge modelling is presented in Table 2.1 and 2.2. Stoichiometric coefficients (n_{ij}) are given in the matrix and kinetic expressions (ρ_j) are presented in the last column. For each component, reaction rate r_i is obtained by the following equation:

$$r_i = \sum_j n_{ij} \cdot \rho_j \qquad (2.15)$$

Temperature effect can be considered for maximal growth and decay rates with the following expressions (Henze *et al.* 2002):

$$\mu_{mA} = \mu_{mA20°C} \cdot e^{(0.09(T°C-20))} \qquad (2.16)$$

Table 2.1. Conventional simplified model for nitrification (Henze *et al.* 2002)

Component i	1	2	3	4	5	6	
Process j	$X_{B,A}$	X_I	S_O	S_{NO3}	S_{NH}	S_{ALK}	Kinetic expression (ρ_j)
1 Growth	1		$-\dfrac{(4.57 - Y_A)}{Y_A}$	$\dfrac{1}{Y_A}$	$-i_{XB} - \dfrac{1}{Y_A}$	$\left[-i_{XB} - \dfrac{2}{Y_A}\right]/14$	$\mu_{mA}\dfrac{S_{NH}}{S_{NH}+K_{NH}}$ $\times \dfrac{S_O}{S_O+K_{OA}}X_{B,A}$
2 Decay	-1	f_{Xi}	$1-f_{Xi}$		$i_{XB}-f_{Xi}\,i_{Xi}$		$b_A X_{B,A}$

Table 2.2. Kinetic and stoichiometric parameters for nitrification (Henze *et al.* 2002)

	Symbol	Values at 20°C Typical (range)	Values at 10°C	Unit
Maximum specific growth rate for autotrophic bacteria	μ_{mA}	0.7 (0.6–1)	0.3	d^{-1}
Decay rate for autotrophic bacteria	b_A	0.1 (0.05–0.15)	0.05	d^{-1}
Yield coefficient for autotrophic bacteria	Y_A	0.24*	0.24*	$gCOD_x \cdot gN^{-1}$
Saturation constant for ammonia	K_{NH}	1 (0.25–1)	1	$mg\,N \cdot L^{-1}$
Saturation constant for oxygen	K_{OA}	0.5 (0.4-1)	0.5	$mg\,O_2 \cdot L^{-1}$
Nitrogen fraction in biomass	i_{XB}	0.086*	0.086*	$gN \cdot gCOD_x^{-1}$
Nitrogen fraction in inert matter	i_{Xi}	0.06	0.06	$gN \cdot gCOD_x^{-1}$
Fraction of inert matter from biomass decay	f_{Xi}	0.08	0.08	

*correspond to $Y_A = 0.17$ $gSS \cdot gN^{-1}$ and $i_{XB} = 0.12$ $gN \cdot gSS^{-1}$ assuming a conversion factor of 1.42 g $COD_x \cdot gSS^{-1}$ for biomass

$$b_A = b_{A20°C} \cdot e^{(0.09(T°C-20))} \tag{2.17}$$

As an example, the specific ammonium removal rate can be expressed:

$$r_{NH} = \left(-\frac{1}{Y_A} - i_{XB}\right) \cdot \mu_{mA} \frac{S_{NH}}{S_{NH}+K_{NH}} \frac{S_O}{S_O+K_{OA}} X_{B,A} \tag{2.18}$$

This expression does not include bicarbonate limitation which can occur in water with low alkalinity or wastewater with high ammonia to bicarbonate ratio. Wett and Rauch (2003) proposed a sigmoidal function to model this effect whereas Ganigué et al. (2007) included a Monod term (switching function).

2.5.2 Two-step nitrification model

Nitrification is described more accurately with a two step model, taking into account the differences between growth kinetics of AOB and NOB groups. Table 2.3 presents the corresponding growth and decay expressions. Table 2.4 shows an example of the values of kinetic and stoichiometric parameters obtained from literature, basically by means of model fitting to experimental data. It should be pointed out that these parameters can change significantly depending on environmental conditions and should not be considered as a unique set of values. Variability of AOB and NOB species naturally also leads to variation in those parameters.

Dissociation of AOB and NOB kinetics is necessary to consider nitrite accumulation which can be observed transiently or permanently in specific bioreactors such as SBR or processes fed with high strength effluents (Béline et al. 2008), or inside a dense matrix (biofilm or granule). Nitrite rising is most commonly caused by high temperature, inhibition by free ammonia or oxygen limitation.

It is observed in Table 2.4 that maximal growth rate of AOB is lower than those of NOB for low or conventional temperature, but becomes higher for high temperature ($\geq 30°C$). This could be included in the model with different Arrhenius constant for the two biomasses.

As previously explained, saturation constant for oxygen of AOB and NOB are different too. Due to a higher saturation constant, NOB are more sensitive to an oxygen limitation. It is the reason why nitrite accumulation can be observed when low DO concentration is maintained in a bioreactor. This phenomenon also occurs inside a biofilm if oxygen diffusion limits the process.

At last, inhibition of NOB by different organic or metallic species can be observed, at a concentration which does not disturb significantly AOB group. This leads also to nitrite accumulation and partial nitrification. This situation definitely needs a more complex model including inhibition expressions.

2.5.3 Advanced model including inhibition

Depending on concentration of wastewater, pH and operating conditions, inhibition of nitrification could be either neglected or considered for nitrification modelling (Figure 2.3).

Table 2.3. Process matrix: two-step model for nitrification

Component i / Process j	1 $X_{B,AOB}$	2 $X_{B,NOB}$	3 X_I	4 S_O	5 S_{NO2}	6 S_{NO3}	7 S_{NH}	8 S_{ALK}	Kinetic expression (ρ_j)
1 Growth of AOB	1			$-\dfrac{(3.43 - Y_{A,AOB})}{Y_{A,AOB}}$	$\dfrac{1}{Y_{A,AOB}}$		$-i_{XB} - \dfrac{1}{Y_{A,AOB}}$	$\left[-i_{XB} - \dfrac{2}{Y_{A,AOB}}\right]/14$	$\mu_{mAOB}\dfrac{S_{NH}}{S_{NH}+K_{NH,AOB}}\dfrac{S_O}{S_O+K_{O,AOB}}X_{B,AOB}$
2 Growth of NOB		1		$-\dfrac{(1.14 - Y_{A,NOB})}{Y_{A,NOB}}$	$-\dfrac{1}{Y_{A,NOB}}$	$\dfrac{1}{Y_{A,NOB}}$	$-i_{XB}$		$\mu_{mNOB}\dfrac{S_{NO2}}{S_{NO2}+K_{NO2,NOB}}\dfrac{S_O}{S_O+K_{O,AOB}}\dfrac{S_{NH}}{S_{NH}+K_{NH,NOB}}X_{B,NOB}$
3 Decay of AOB	-1		f_{XI}	$-(1 - f_{XI})$			$i_{XB} - f_{XI}\,i_{XI}$		$b_{A,AOB}X_{B,AOB}$
4 Decay of NOB		-1	f_{XI}	$-(1 - f_{XI})$			$i_{XB} - f_{XI}\,i_{XI}$		$b_{A,NOB}X_{B,NOB}$

Table 2.4. Example of set of kinetic and stoichiometric parameters for AOB and NOB (Henze *et al.* 2002, Pambrun *et al.* 2006)

	Symbol	Values at 15°C	Values at 30°C	Unit
Maximum specific growth rate of AOB	μ_{mAOB}	0.4	1.9	d^{-1}
Maximum specific growth rate of NOB	μ_{mNOB}	0.5	1	d^{-1}
Specific decay rate of AOB	$b_{A,AOB}$	0.05	0.4	d^{-1}
Specific decay rate of NOB	$b_{A,NOB}$	0.09	0.4	d^{-1}
Yield coefficient for AOB	$Y_{A,AOB}$	0.20	0.20	$gCOD_x \cdot gN^{-1}$
Yield coefficient for NOB	$Y_{A,NOB}$	0.04	0.04	$gCOD_x \cdot gN^{-1}$
Saturation constant for oxygen and AOB	$K_{O,AOB}$	0.5	0.5	$mgO_2 \cdot L^{-1}$
Saturation constant for oxygen and NOB	$K_{O,NOB}$	1	1	$mgO_2 \cdot L^{-1}$
Saturation constant for ammonia and AOB	$K_{NH,AOB}$	0.7*	0.7*	$mgN \cdot L^{-1}$
Saturation constant for nitrite and NOB	$K_{NO2,NOB}$	0.7*	0.7*	$mgN \cdot L^{-1}$
Saturation constant for ammonia and NOB	$K_{NH,NOB}$	0.05	0.05	$mgN \cdot L^{-1}$

*function of pH

Mathematical models were proposed for FA and FNA inhibition on nitritation and nitratation rates (Pambrun *et al.* 2006, Volcke *et al.* 2002, Ganigué *et al.* 2007). Expression for growth of AOB becomes:

$$\rho_1 = \mu_{m,\,AOB} \cdot \frac{S_{NH3}}{K_{NH3} + S_{NH3} + \frac{(S_{NH3})^2}{K_{iNH3}^{AOB}}} \cdot \left(\frac{S_O}{K_{O,AOB} + S_O}\right)$$
$$\cdot \left(\frac{K_{iHNO2}^{AOB}}{K_{iHNO2}^{AOB} + S_{HNO2}}\right) \cdot X_{B,AOB} \qquad (2.19)$$

Note that the species considered in this model are here the specific free ammonia (NH_3) and the nitrous acid form (HNO_2), both depending on pH. As presented on Figure 2.4, Haldane model gives a correct prediction of FA inhibition on nitritation rate (AOB). Inhibition constants $K_{i\,NH3}^{AOB}$ can vary very significantly due to acclimatising of nitrifying bacteria and depending on the type of reactor. It was estimated at 241 $mgNH_4$-N $\cdot L^{-1}$ with acclimatised biomass on the data shown in Figure 2.4. However, $K_{i\,NH3}^{AOB}$ ranged from 95

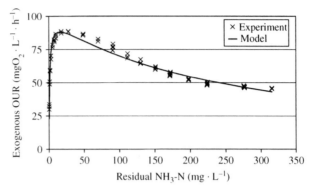

Figure 2.4. NH_3 inhibition of AOB activity measured by Oxygen Uptake Rate (OUR) measurement (model with $K_{NH3} = 1$ mgN$\cdot$L^{-1}, $K_{i\,NH3}^{AOB} = 241$ mgN$\cdot$L^{-1}) (Pambrun et al. 2006)

to 600 mgN$\cdot$L^{-1} in literature (Carrera et al. 2004; Guanigué et al. 2007; Magri et al. 2007; Pambrun et al. 2006).

The inhibition of AOB by HNO_2 is described by a switching function characteristic of non-competitive inhibition. Inhibition constant for FNA ($K_{i\,NHO3}^{AOB}$) was estimated in the range from 0.05 to 0.5 mgN$\cdot$L^{-1} (Carrera et al. 2004; Guanigué et al. 2007; Magri et al. 2007; Pambrun et al. 2006). However, this inhibition plays a secondary role in conventional pH range (6–8) due to the low amount of nitrogen present in the form of free nitrous acid.

Expression for growth of NOB can also include ammonia inhibition:

$$\rho_2 = \mu_{mNOB} \frac{S_{NO2}}{S_{NO2} + K_{NO2,NOB}} \frac{S_O}{S_O + K_{O,NOB}} \frac{K_{i,NH3}^{NOB}}{S_{NH3} + K_{i,NH3}^{NOB}} X_{B,NOB} \qquad (2.20)$$

The parameter $K_{i\,NH3}^{NOB}$ plays an essential role in modelling nitrification as it determines the concentration for which inhibition of NOB is 50% leading to nitrite rising. Anthonisen et al. (1976) observed NOB inhibition by NH_3 concentration from 0.1 to 1 mgN$\cdot$L^{-1}. Wong-Chong and Loehr (1975) have observed NOB inhibition for concentration of 40 mg NH_3-N$\cdot$L^{-1}, after an acclimation phase, whereas the non-acclimated microorganisms were inhibited by 3.5 mg NH_3-N$\cdot$L^{-1}. After a six months acclimatising period, Pambrun et al. (2006) estimated the parameter $K_{i\,NH3}^{NOB}$ at 11 mgN$\cdot$L^{-1}. A more accurate model (mixed inhibition) for free ammonia inhibition on AOB can be proposed to include the effect of FA on affinity constant.

Inhibition of nitrite oxidizers (NOB) by nitrous acid could be included that the model could further describe processes operating at pH lower than 7 or at high HNO_2 level. Anthonisen *et al.* (1976) observed NOB inhibition when HNO_2 concentration is >0.2–2.8 mg $HNO_2 \cdot L^{-1}$ or 0.06–0.83 mg $HNO_2\text{-}N \cdot L^{-1}$.

REFERENCES

Abeling, U., and Seyfried, C.F. (1992) Anaerobic-aerobic treatment of high-strength ammonium wastewater-nitrogen removal via nitrite. *Water Sci. Technol.* **26**(5–6), 1007–1015.

Anthonisen, A.C., Loehr, R.C., Prakasam, T.B.S. and Srinath, E.G. (1976) Inhibition of nitrification by ammonia and nitrous acid. *J. Water Pollut. Control Fed.* **48**(5), 835–852.

Béline, F., Boursier, H., Daumer, M.L., Guiziou, F., and Paul, E. (2007) Modelling of biological processes during aerobic treatment of piggery wastewater aiming at process optimisation. *Biores. Technol.* **98**, 3298–3308.

Bernet, N., Peng, D., Delgenès, J.P. and Moletta, R. (2000) Nitrification at low oxygen concentration in biofilm reactor. *J. Environ. Eng.-ASCE* **127**(3), 266–271.

Bernet, N. and Spérandio, M. (2006) Application of biological treatment systems for nitrogen-rich wastewaters. In *Advanced Biological Treatment Processes for Industrial Wastewaters*, F. Cervantes, S. Pavlostathis, and A. van Haandel (ed.), IWA Publishing, London, pp. 186–208.

Bougard, D., Bernet, N., Cheneby, D. and Delgenès, J.P. (2006) Nitrification of a high-strength wastewater in an inverse turbulent bed reactor: effect of temperature on nitrite accumulation. *Process Biochem.* **41**, 106–113.

Carrera, J., Jubany, I., Carvallo, L., Chamy, R. and Lafuente, J. (2004) Kinetic models for nitrification inhibition by ammonium and nitrite in a suspended and an immobilised biomass systems. *Process Biochem.* **39**, 1159–1165.

Daims, H., Nielsen, P.H., Nielsen, J.L., Juretschko, S. and Wagner, M. (2000) Novel *Nitrospira*-like bacteria as dominant nitrite-oxidizers in biofilms from wastewater treatment plants: diversity and in situ physiology. *Water Sci. Technol.* **41**(4–5), 85–90.

Daims, H., Nielsen, J.L., Nielsen, P.H., Schleifer, K.H. and Wagner, M. (2001) In situ characterization of *Nitrospira*-like nitrite oxidizing bacteria active in wastewater treatment plants. *Appl. Environ. Microbiol.* **67**, 5273–5284.

De Boer, W. and Kowalchuk, G.A. (2001) Nitrification in acid soils: micro-organisms and mechanisms. *Soil Biol. Biochem.* **33**, 853–866.

Guanigué, R., López, H., Balaguer, M.D. and Colprim, J. (2007) Partial ammonium oxidation to nitrite of high ammonium content urban landfill leachates. *Water Res.* **41**, 3317–3326.

Garrido, J.M., van Benthum, W.A.J., van Loosdrecht, M.C.M. and Heijnen, J.J. (1997) Influence of dissolved oxygen concentration on nitrite accumulation in a biofilm airlift suspension reactor. *Biotechnol. Bioeng.* **53**, 168–178.

Gieseke, A., Tarre, S., Green, M. and de Beer, D. (2006) Nitrification in a biofilm at low pH values: Role of in situ microenvironments and acid tolerance. *Appl. Environ. Microbiol.* **72**, 4283–4292.

Hellinga, C., Schellen, A.A.J.C., Mulder, J.W., van Loosdrecht, M.C.M. and Heijnen, J.J. (1998). The SHARON process: An innovative method for nitrogen removal from ammonium-rich waste water. *Water Sci. Technol.* **37**(9), 135–142.

Henze, M., Harremoës, P., la Cour Jansen, J., Arvin, A. (2002) *Wastewater treatment – Biological and Chemical Processes*. Springer-Verlag, Berlin, Heidelberg, New-York.

Jayamohan, S., Ohgaki, S. and Hanaki, K. (1988) Effect of DO on kinetics of nitrification. *Water Supply* **6**, 141–150.

Könneke, M., Bernhard, A.E., de la Torre, J.R., Walker, C.B., Waterbury, J.B. and Stahl, D.A. (2005) Isolation of an autotrophic ammonia-oxidizing marine archaeon. *Nature* **437**(7058), 543–546.

Kowalchuk, G.A. and Stephen, J.R. (2001) Ammonia-oxidizing bacteria: A model for molecular microbial ecology. *Ann. Rev. Microbiol.* **55**, 485–529.

Laanbroek, H.J. and Gerards, S. (1993) Competition for limiting amounts of oxygen between *Nitrosomonas europaea* and *Nitrobacter winogradskyi* grown in mixed continuous cultures. *Arch. Microbiol.* **159**, 453–459.

McCarty, G.W. (1999) Modes of action of nitrification inhibitors. *Biol. Fertil. Soils* **29**, 1–9.

Magri, A., Corominas, L.I., Lopez, H., Campos, E., Balaguer, M.D., Colprim, J. and Flotats, X. (2007) A model for the simulation of the SHARON process: pH as a key factor. *Environ. Technol.* **28**, 255–265.

Pambrun, V., Paul, E. and Spérandio, M. (2006) Modelling the partial nitrification in Sequencing Batch Reactor for biomass adapted to high ammonia concentration. *Biotechnol. Bioeng.* **95**, 120–131.

Philips, S., Laanbroek, H.J. and Verstraete, W. (2002) Origin, causes and effects of increased nitrite concentrations in aquatic environments. *Rev. Environ. Sci. Biotechnol.* **1**, 115–141.

Purkhold, U., Pommerening-Roser, A., Juretschko, S., Schmid, M.C., Koops, H.P. and Wagner, M. (2000) Phylogeny of all recognized species of ammonia oxidizers based on comparative 16S rRNA and amoA sequence analysis: Implications for molecular diversity surveys. *Appl. Environ. Microbiol.* **66**, 5368–5382.

Schmidt, I., Sliekers, O., Schmid, M., Bock, E., Fuerst, J., Kuenen, J.G., Jetten, M.S.M. and Strous, M. (2003) New concepts of microbial treatment processes for the nitrogen removal in wastewater. *Fems Microbiol. Rev.* **27**, 481–492.

Schramm, A., Santegoeds, C.M., Nielsen, H.K., Ploug, H., Wagner, M., Pribyl, M., Wanner, J., Amann, R. and De Beer, D. (1999) On the occurrence of anoxic microniches, denitrification, and sulfate reduction in aerated activated sludge. *Appl. Environ. Microbiol.* **65**, 4189–4196.

Tarre, S. and Green, M. (2004) High-rate nitrification at low pH in suspended- and attached-biomass reactors. *Appl. Environ. Microbiol.* **70**, 6481–6487.

Turk, O. and Mavinic, D.S. (1986) Preliminary assesment of a shortcut in nitrogen removal from wastewater. *Can J. Civil Eng.* **13**, 600–605.

Voets, J.P., Vanstaen, H. and Verstraete, W. (1975) Removal of nitrogen from highly nitrogenous wastewaters. *J. Water Pollut. Control Fed.* **47**, 394–397.

Volcke, E.I.P., Hellinga, C., van den Broeck, S., van Loosdrecht, M.C.M. and Vanrolleghem, P.A. (2002) Modelling the SHARON process in view of coupling with Anammox. In: Proceedings 1st IFAC International Scientific and Technical Conference on Technology, Automation and Control of Wastewater and Drinking Water Systems (TiASWiK'02). Gdansk-Sobieszewo, Poland, June 19–21, pp. 65–72.

Wagner, M., Loy, A., Nogueira, R., Purkhold, U., Lee N., and Daims, H. (2002) Microbial community composition and function in wastewater treatment plants. *Antonie Van Leeuwenhoek* **81**, 665–680.

Wett, B. and Rauch, W. (2003) The role of inorganic carbon limitation in biological nitrogen removal of extremely ammonia concentrated wastewater. *Water Res.* **37**, 1100–1110.

Winogradsky M.S. (1890) Recherches sur les organismes de la nitrification. *Ann. Inst. Pasteur* **5**, 257–275.

Wong-Chong, G.M. and Loehr, R.C. (1975) The kinetics of microbial nitrification. *Water Res.* **9**, 1099–1106.

3
Principles of denitrifying processes

*F. Cuervo-López, S. Martínez Hernández,
A.-C. Texier, and J. Gómez*

3.1 INTRODUCTION

Nitrogen is a basic constituent of the living cells since it is necessary to the synthesis of essential macromolecules such as nucleic acids and proteins. Nitrogen is a versatile element, which exists in several oxidation states ranging from -3 (for NH_4^+) to the most oxidized state of $+5$ (for NO_3^-). Conversions of nitrogen compounds in the biosphere are illustrated in Figure 3.1.

In general, nitrogen enters to the biosphere by the nitrogen fixation process and it is returned to the atmosphere as N_2 after denitrification takes place. Ammonium, that is the result of both nitrogen fixation or oxidative dissimilation of nitrate or nitrite, is oxidized to nitrate by nitrification process. Nitrate can be

© 2009 IWA Publishing. *Environmental Technologies to Treat Nitrogen Pollution: Principles and Engineering*, Edited by Francisco J. Cervantes. ISBN: 9781843392224. Published by IWA Publishing, London, UK.

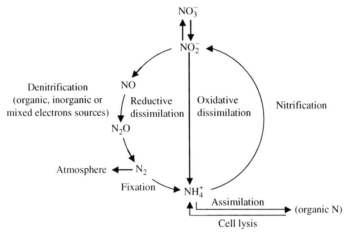

Figure 3.1. The nitrogen cycle

reduced both, in assimilation and dissimilation pathway. In case of assimilation, nitrate is reduced to ammonium for biomass synthesis. Traditional dissimilative pathways are two: 1) denitrification which implies reduction of nitrate as a part of respiratory metabolism for energy generation and the consequent reduced product released as N_2; and 2) ammonification or dissimilative nitrate reduction to ammonium (DNRA). For biological wastewater treatment the nitrogen removal should be conducted by nitrification and denitrification, in which both biological processes must be dissimilative and dissipative to prevent biomass generation. Nevertheless, further processes have been identified for the elimination of nitrogen compounds from polluted water. Among these processes are recognized anaerobic ammonium oxidation (ANAMMOX) to molecular nitrogen, using nitrite as electron acceptor (Strous et al. 1999); single reactor high activity ammonium removal over nitrite (SHARON), which was developed for the ammonium removal by a partial nitrification and a subsequent nitrite denitrification (Jetten et al. 2002); and the nitrogen removal over nitrite process (CANON) which involves the coupling of a partial nitrification and the ANNAMOX process with intermittent aeration (Schmidt et al. 2003). Although these three processes work acceptably, they also enclose some problems in their operation and control. Coupled with the foregoing, energy expenditures required represent a serious disadvantage, as well as the possible necessity to implement a combination of two or three of these processes, in order to obtain a complete nitrogen removal from wastewater.

3.2 THE DENITRIFYING PROCESS

3.2.1 Microbiology of denitrification

Denitrifying ability is distributed in a wide variety of bacterial groups; most of them being facultative. However, the capacity to denitrify has also been found in some Archaea and in Fungi (Thorndycroft et al. 2007). Denitrifying bacteria are widely spread in nature; they are in sea sludge, soils of industrial zones, pristine forests, etc. (Knowles 1982). These microorganisms involve numerous physiological and morphological characteristics. Some of the most common genera are shown in Table 3.1.

It is known that denitrifying microorganisms are able to grow in a wide temperature range from 4 to 45°C (Knowles 1982). For instance, *Pseudomonas stutzeri* (Lalucat et al. 2006) and *Comamonas denitrificans* sp. nov. grow between 20 and 37°C (Gumaelius et al. 2001). In general, acceptable grow up and denitrifying rates can be seen in the range from 30 and 35°C.

In terms of pH, studies show that optimal pH oscillate between 7 and 8.2 for many of the denitrifying microorganisms (Knowles 1982). For example, studies with *Pseudomonas denitrificans* cultures showed that optimal pH for growth and nitrate and nitrite reduction was between 7.2 and 7.6. *Comamonas denitrificans* cultures showed that most favorable pH was at 7.5 (Gumaelius et al. 2001).

Table 3.1. Some microbial genera carrying out the denitrifying process (Mateju et al. 1992)

Family	Genus	Morphology	Physiological characteristics
Alcaligenaceae	*Alcaligenes*	Rods	Organotrophic and lithotrophic
Pseudomonadaceae	*Pseudomonas*	Rods	Organotrophic and lithotrophic
Rhodobacteriaceae	*Paracoccus*	Cocci	Organotrophic and lithotrophic
Hydrogenophilaceae	*Thiobacillus*	Rods	Lithotrophic
Bradyrhizobiaceae	*Rhodopseudomonas*	Rods	Phototrophic
Bacillaceae	*Bacillus*	Rods	Fermentative
Rhizobiaceae	*Rhizobium*	Rods	Organotrophic N_2 fixers
Neisseriaceae	*Neisseria*	Cocci	Pathogenic organotrophic
Propionibacteriaceae	*Propionibacterium*	Rods	Organotrophic

Denitrifying microorganisms are able to consume numerous sources of carbon and energy for their growth and respiratory process. Some species are photoautotrophs, such as *Rhodopseudomonas sphaeroides*; others such as *Thiobacillus denitrificans*, are able to use hydrogen or reduced sulfur compounds as an energy source, and CO_2 as a carbon source (lithoautotrophs). Many other species can use several organic sources (organoheterotrophs) such as glycerol (Akuna *et al.* 1993), glucose (Cuervo-López *et al.* 1999), volatile fatty acids (Rijn *et al.* 1996), polyvinyl alcohol (Watanabe *et al.* 1995), and even recalcitrant compounds such as aromatic petroleum products (Peña-Calva *et al.* 2004; Martínez *et al.* 2007).

There are a few bacteria, which only accomplish some stages of the denitrifying pathway, resulting in the formation of denitrifying intermediaries, mainly NO_2^- or N_2O. In order to consider a microbial strain as a denitrifying microorganism, anoxic respiration is obligatory (anaerobic respiration using either nitrate or nitrite as electron acceptor), as well as the presence of nitrite reductase enzyme. Besides this, high yields of N_2O or N_2 should be obtained. Nevertheless, attention on this must be paid, because production of N_2O and N_2 could be also determined by the availability of reductive source in the medium. In fact, denitrification involves nitrate or nitrite reduction to molecular nitrogen, without accumulation of N_2O, which should only be considered as an intermediate of the respiratory process.

3.2.2 Biochemistry of denitrification

The denitrifying process might be described as a modular organization, where every reaction is catalyzed by a specific enzyme. The enzymes associated to this process are reductases which are synthesized when the environmental conditions become anoxic. Enzymes are shown in Table 3.2, as well as some of their characteristics (Moura and Moura 2001; Tavares *et al.* 2006).

Studies related to characterization of enzymes have been performed in *Paracoccus denitrificans*, *Pseudomonas denitrificans*, *Rhodobacter capsulatus* and *Rhodobacter sphaeroides*, from which it has accomplished interpolations to other biological systems. In general, denitrification takes place as follows: (*i*) nitrate is reduced to nitrite by nitrate reductase (*Nar*) which may be coupled to either the cell membrane or in the periplasmic space. Nitrite is produced in the cytoplasm, since the catalytic site of *Nar* is linked to the membrane (Alefounder and Ferguson 1980); (*ii*) a subsequent reduction of nitrite to nitric oxide is carried out by nitrite reductase (*Nir*); (*iii*) afterwards, nitric oxide is reduced to nitrous oxide by the enzyme nitric oxide reductase (*Nor*), which is located in the cytoplasmic membrane; (*iv*) finally, nitrous oxide is reduced to N_2 by the enzyme nitrous oxide reductase (*Nos*).

Table 3.2. Enzymes of the denitrifying process and their location (Jetten *et al.* 1997)

Enzyme	Location in the cell	Reaction	Prosthetic group
Nitrate reductase (*Nar*)	Cell membrane and periplasmic space	$NO_3^- + UQH_2 \rightarrow$ $NO_2^- + UQ + H_2O$	Molybdenum
Nitrite reductase (*Nir*)	Periplasmic space	$NO_2^- + Cu^{1+} + 2H^+ \rightarrow$ $NO + H_2O + Cu^{2+}$ (A) $NO_2^- + c^+ + 2H^+ \rightarrow$ $NO + H_2O + c^{3+}$ (B)	Copper and Hem
Nitric oxide reductase (*Nor*)	Cell membrane	$2NO + 2c^{2+} + 2H^+ \rightarrow$ $N_2O + H_2O + 2c^{3+}$	Hem Group
Nitrous oxide reductase (*Nos*)	Periplasmic space	$N_2O + 2c^{2+} + 2H^+ \rightarrow$ $N_2 + H_2O + 2c^{3+}$	Copper

UQH_2: reduced ubiquinone, UQ: ubiquinone, c^{2+}: reduced cytochrome, c^{3+}: oxidized cytochrome.

3.2.2.1 Nitrate reduction (NO_3^-) to nitrite (NO_2^-)

Four types of *Nar* have been described, two of them being respiratory enzymes, one is coupled to the membrane (*Nar*) and the other is located in the periplasmic space (*Nap*) (Moura and Moura 2001). Reaction mechanism of *Nar* in the denitrifying respiratory process is summarized in Table 3.2. Catalyzed reaction involves a reduced quinone by an electron source with molybdenum as cofactor (Hille 1996). *Nar* is composed of different subunits: a) subunit α (104–150 KDa) where the active site is containing molybdenum; b) subunit β (118–150 KDa) that mediates the electron transfer between the two others subunits; and c) subunit γ (19–20 KDa) that is linked to the membrane. It has been proposed that this reaction mechanism takes place in the core of molybdenum where the redox state changes from Mo(IV) to Mo(VI) (Jetten *et al.* 1997). The enzyme in its reduced stage (Mo(IV)) might coordinate the interaction between nitrate and molybdenum (Moura and Moura 2001).

Nar enzyme has been characterized in axenic batch cultures for several bacteria. Michaelis-Menten affinity constant (Km) of nitrate has been estimated between 0.3 and 3.8 mM (Ketchum *et al.* 1991). It is necessary to emphasize that Km value is influenced by the environmental conditions, such as the pH value. Thus, consumption rate is also affected. Wu and Knowles (1994) working with *Flexibacter canadensis*, observed nitrate consumption when glycerol, lactose, ethanol and glucose are present. Glucose showed to be the best electron source. Nitrate consumption was dependent on glucose concentration, approaching to a Michaelis-Menten kinetic with an apparent Km of 0.021 mM for glucose. The same authors observed that nitrate consumption rate was related with the intracellular reduction of nitrate and nitrite, and that nitrate consumption is dependent on the presence of an active nitrate reductase (Wu and Knowles 1994). Activity of *Nar* is inhibited by azide, thiocyanate, cyanide, chloramphenicol, dinitrophenol, nigericin (Wu and Knowles 1995) and other agents such as tungsten, which causes a chelation or competition with molybdenum (Stewart 1988).

3.2.2.2 Nitrite reduction (NO_2^-) to nitric oxide (NO)

Studies have shown two different *Nir* with different structure and prosthetic group which are mutually excluding; therefore, it is possible to find only one into the denitrifying microorganism (Berks *et al.* 1995). One of them is a protein that contains copper as prosthetic group (CuNIR) (reaction A, Table 3.2) and the other is the cytochrome *cd₁* with the Hem group as cofactor (reaction B). Activity of CuNIR is inhibited by chelating action of diethyldithiocarbamate on copper (Shapleigh and Payne 1985). CuNIR is a trimeric enzyme with molecular weight close to 40 KDa for each subunit (Dodd *et al.* 1997). Numerous studies have shown that this enzyme has a great versatility, since not only reduces nitrite to nitric oxide, but it is also capable of generating other products such as ammonium and nitrous oxide (Fee *et al.* 1995). At first, nitrite reductase cytochrome *cd₁* was described as a cytochrome oxidase in *P. aeruginosa* (Horio *et al.* 1961). The biosynthesis of this enzyme depends on the presence of nitrate. This enzyme is a tetrameric protein made up by two subunits of 60 to 67 KDa each one (Ohshima *et al.* 1993). The prosthetic groups are Hem C and Hem D_1, both are cytochrome *cd₁*. The proposed mechanism of nitrite reductase *cd₁* involves the union of nitrite to the ferrous Hem D_1, formation of nitric oxide and displacement as a product of the ferric Hem (Fulop *et al.* 1995). It is noteworthy that although this enzyme is also capable to reduce oxygen, its rate is much lower than the nitrite reduction.

Activity of *Nir* could strongly be inhibited by its own product, the NO. Nitrous oxide inhibits activity of cytochrome *cd₁* in *Paracoccus denitrificans* and *Pseudomonas aeruginosa* (Carr and Ferguson 1990). NO inhibits its own consumption (Goretski and Hollocher 1990) and the reduction of nitrite and N_2O in *Paracoccus denitrificans* (Carr *et al.* 1989) and *P. perfectomarina* (Frunzke and Zumft 1986). In the case of *Flexibacter canadensis* and *P. aeruginosa*, nitric oxide formation from nitrite was dependent on pH, probably due to structure modification of *Nir* in acidic pH values, because of nitrous acid production when nitrite is present or due to the disconnection of the enzymatic complex nitrite reductase-nitric oxide reductase (*Nir-Nor*) (Glass *et al.* 1997).

3.2.2.3 Nitric oxide reduction (NO) to nitrous oxide (N₂O)

Nor enzyme (Table 3.2), also known as cytochrome bc_1 has been purified from *Pseudomonas stutzeri* (Schäfer and Conrad 1993), *Paracoccus denitrificans* (Dermastia *et al.* 1991), *P. aeruginosa* (Voβwinkel *et al.* 1991), *Flexibacter canadensis* (Wu *et al.* 1995a) and *Achromobacter cycloclastes* (Jones and Hollocher 1993). *Nor* is associated to the membrane and is generated only under anaerobic conditions. Researches with *P. stutzeri* have shown that *Nor* consists of cytochrome *c* (NorC subunit) with 17 KDa and cytochrome *b* (NorB subunit) with 53 KDa (Zumft *et al.* 1994). Prosthetic groups for this enzyme are Hem C, Hem B and Fe no-Hem. Some evidences suggest that the active site of *Nor* is found at the binuclear site and is composed of Hem B and iron no-Hem (Berks *et al.* 1995). It is estimated that Km of *Nor* for NO in *P. stutzeri* ranges from 1.2 to 1.4 nM (Remde and Conrad 1991), while in *Azospirillum brasilense*, *P. aeruginosa*, and *P. fluorescens*, Km oscillates between 0.5 and 6 nM (Remde and Conrad 1991). The *Nor* activity might be decreased by the action of some detergents as Triton X-100, volatile fatty acids, nigericin, cyanide, azide, chloramphenicol, nitrite and nitric oxide as a substrate (Wu *et al.* 1995a; Wu and Knowles 1995).

3.2.2.4 Nitrous oxide reduction (N₂O) to molecular nitrogen (N₂)

Reduction of N_2O to N_2 is the final step in denitrifying pathway and this reaction is catalyzed by the nitrous oxide reductase enzyme (*Nos*) (Table 3.2). *Nos* is soluble and is located at the periplasmic space with a molecular weight of 70 KDa (Boogerd *et al.* 1981). This enzyme is composed by subunits, each subunit containing four copper atoms which are organized in two different binuclear centers, CuA and CuZ. Studies about the three-dimensional structure of CuZ have shown an atom of sulfur in the center that is coupled with four copper atoms (Cu I to Cu IV). It is proposed that N_2O is linked with Cu IV through the oxygen and in turn external nitrogen of N_2O might bonds with a single histidine and a single lysine to stabilize the complex (Prudencio *et al.* 2000). In order to have an active *Nos* and avoid accumulation of N_2O, it is required that copper participates in the denitrifying process. Experimental evidences have shown that this enzyme has a Km for N_2O in the range of micromoles. Activity of the *Nos* is inhibited specifically by acetylene and is known that this reaction is reversible (Kristjansson and Hollocher 1980). Other inhibitors could be azide, thiocyanate, and cyanide, presumably forming a ligand with copper (Frunzke and Zumft 1986). In axenic cultures of *Flexibacter canadensis* and *P. perfectomarina*, NO concentrations at 67 nM diminish quickly the nitrous oxide consumption (Frunzke and Zumft 1986; Wu *et al.* 1995b). Nitrite also could inhibit *Nos* enzyme in *F. canadensis* and *P. denitrificans* (Wu *et al.* 1995a).

3.2.2.5 Genetic control for denitrifying enzymes

In general, the genetic expression of the four denitrifying enzymes is strongly regulated. The effects caused by oxygen and nitrogen oxides are among the most important factors which regulate the enzymatic transcription (Knowles 1982). In this regard, a set of regulatory genes, which controls denitrification has been recognized, as well as a number of important elements controlling denitrification (Table 3.3). These elements are proteins known as factors similar to fumarate and nitrate reduction (FNR), or transcription factors which operate as sensors at the level of oxygen and as activators of the denitrifying regulatory genes under anaerobic conditions (Guest *et al.* 1996). That is, in the absence of O_2, genetic expression of denitrifying enzymes are activated.

Table 3.3. Denitrification regulatory genes, functions and transcription factors

Gen or locus	Function	Source: Regulatory factors	References
Anr	Denitrifying gen expression of *nar, nir, nor* and *nos*	*P. aeruginosa*: factor type FNR	Galimand *et al.* 1991
dnr, frnD	Gen expression of *nir* and *nor*	*Pseudomonas*: factor type FNR	Arai *et al.* 1995
fixK₂	Gen expression of *nar*	*B. japonicum*: factor type FNR	Fischer 1994
fnrP	Gen expression of *nar*	*Paracoccus*: factor type FNR	Van Spanning *et al.* 1997
narL	Responding to nitrate	*Pseudomonas*: Transcription factor NarXL	Härtig and Zumft 1996
nnr, nnrR	Gen expression of *nir* and *nor*	*Paracoccus* and *Rhodobacter*: factor type FNR	Van Spanning *et al.* 1995; Tosques *et al.* 1996

Nar, gene that codifies nitrate-oxide reductase; *Nir*, gene that codifies nitrite-oxide reductase; *Nor*, gene that codifies nitric-oxide reductase; *Nos*, gene that codifies nitrous oxide reductase; FNR, fumarate and nitrate reduction transcription factor; NarXL, nitrate and nitrite transcription factor.

Nitrogen oxides are required as an inductor of denitrification. Usually, nitrate is an inductor to every one of genes for denitrifying enzymes. It has also been observed in *P. stutzeri* that nitrite and nitric oxide are inducers of their own reductase (Körner and Zumft 1989). In terms of nitrous oxide, it operates as inducer of the genes in the nitrous oxide reductase and nitrate reductase. It is not known a sensor for N_2O; nonetheless, a protein identified as *NosR* could do this. In *P. aeruginosa* cultures, it has been compared the inducer effect of the oxygen and nitrate in the gene transcription. The effect was stronger with nitrate (Zennaro *et al.* 1993).

Reduction of NO_2^- and NO to N_2O represent the strongest regulator if compared to the rest of the denitrifying reactions. Actually, there is interdependence between nitrite and nitric-oxide reductase, because reduction of nitrite takes place only if it is assured a subsequent reduction of NO (Ye *et al.* 1992). Likewise, the inactivation of the nitric-oxide reductase results in low levels of nitrite reductase and electron flux for that enzyme (Braun and Zumft 1991).

Some studies in chemostat have shown that genes expression of denitrifying enzymes occurs in a coordinated manner, since in *P. denitrificans* the gene of N_2O reductase appears simultaneously to the gene of nitrate reductase (Baumann *et al.* 1996). In case of *P. stutzeri* the expression of the gene of NO reductase is given in combination with further genes from other denitrifying enzymes. In *E. coli*, two transcription factors have been identified and they have been designated as NarX and NarL. These enzymes operate as sensor and regulator, respectively, and they respond to the external levels of nitrate and nitrite, inducing the gene expression to nitrate reduction (Stewart *et al.* 1989).

The biochemical evidences clearly indicate that denitrifying respiratory pathway is a complex process subjected to various types of inhibition and control. Therefore, for a successful wastewater treatment a stable enzymatic activity should be attained.

3.2.3 Physiology of denitrification

Nitrate can be used for obtaining energy in prokaryote cells and for biosynthetic purposes. In either case, the first reaction implies nitrate reduction to nitrite. However, both processes are essentially different (Figure 3.2). Denitrification is a dissimilative pathway where nitrate could be used by the denitrifying microorganisms to obtain energy and to form N_2.

In general, some of the factors that induce and control the denitrifying system could be a low pressure of oxygen, presence of nitrogen oxide and controls, which permit expression of the entire denitrifying pathway. Thus, it must also be considered the effect caused by different environmental factors such as pH,

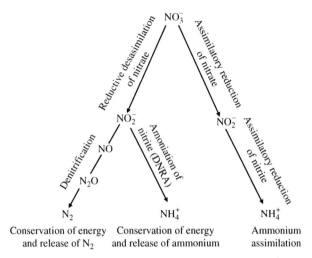

Figure 3.2. Nitrate assimilative and dissimilative pathways

temperature, stoichiometry of the process, carbon/nitrogen ratio and type of carbon source. Here are a few details for every one of these factors.

3.2.3.1 Effect of oxygen

Denitrifying activity is inhibited in reversible manner under aerobic conditions. There is evidence suggesting that control of oxygen on the nitrate reduction seems to reside in the nitrate transport through the cytoplasmic membrane (Hernández and Rowe 1988). It has also been suggested that the sensibility to oxygen is related to the microbial genus. Working with axenic *P. aeruginosa* cultures, it was found that oxygen effect was minimum with 0.4 mg/L of dissolved O_2, while working with *F. canadensis*, nitrate consumption was inhibited using 0.15 mg/L of dissolved O_2 (Wu and Knowles 1994). Oh and Silverstein (1999) found that oxygen concentrations of 0.09 mg/L resulted in a decrease of 35% in denitrifying rate and that denitrifying activity was practically zero at 5.9 mg O_2/L. In general, there are reports indicating that nitrate consumption could be affected by oxygen at concentrations ranging from 0.08 to 7.7 mg/L (Wilson and Bouwer 1997). It is also possible that the end product and its concentration were different in each case. Heterogeneity of these results may be due to differences in cell concentration, nitrate concentration and other experimental conditions. Oxygen could also inhibit nitrite reduction in *P. denitrificans* because of competence of electrons between the oxidase and nitrite reductase, which is still active when oxygen is present (Alefounder *et al.* 1983). Thus, the evidences show that presence of oxygen will have an effect on conversion of nitrate to N_2 (decreasing the denitrifying yield, Y_{N2}), although possibly consumption efficiency of the reducing source (organic and inorganic) might not be affected.

3.2.3.2 Effect of nitrogen oxides

Among nitrogen oxides, the effect caused by NO_3^- is probably the best studied (Knowles 1982). It has been observed that denitrifying rate is affected by this compound. Studies with *Alcaligenes odorans* show that for nitrate concentrations between 0.04 and 6.2 g NO_3^-/L, the process follows a first-order kinetic, without accumulation of nitrogen intermediaries (Knowles 1982). Respect to NO_2^- effect, a number of studies shows that denitrifying activity is diminished at concentrations of 0.27 g NO_2^-/L (Shimizu *et al.* 1978). Nevertheless, in continuous culture is showed that *P. denitrificans* can accomplish denitrification even if nitrite concentration is 1.5 g NO_2^-/L (Baumann *et al.* 1997). Some authors indicate that inhibitory effect should not be primarily attributed to the NO_2^- itself, but rather due to HNO_2, its non-dissociated form, which generation is directly related with pH decrease in the medium (Almeida *et al.* 1995). However, no information on biomass concentration is given and the sensibility of culture is closely related to this parameter. Contrasting to nitrite, HNO_2 can be transported quickly through the cell membrane (Kroneck and Zumft 1990; Baumann *et al.* 1997).

3.2.3.3 Effect of pH and temperature

Organotrophic denitrification generates bicarbonate ions, which tend to alkalinize the culture medium. Literature reports a wide range for pH values (6–9) where denitrification operates properly. pH will vary depending on both, nitrate and electrons source concentration which are related to consumption efficiency (Delwiche and Bryan 1976). Information points out that pH presents a major effect on the activity for two of the enzymes of the denitrifying pathway, *Nir* and *Nor* (Voßwinkel *et al.* 1991; Wu *et al.* 1995a).

In the majority of the works, it is observed a decrease in the efficiency and denitrifying rate when the pH value is acidic. This could be associated to the nitrous acid formation under these conditions, which seems to be more toxic than nitrite. Moreover, it was shown that pH value affects gas releasing during denitrification (Knowles 1982). Thus, when pH is lower than 5, NO is obtained as main product. For pH between 6 and 7, less NO is generated and over pH 7, N_2 is produced (Wicht 1996). Works with different sources of carbon, such as lactate, acetate and ethanol, have shown that at pH values in the range from 6 to 9, denitrification achieved high efficiency without accumulation of intermediaries, and the Y_{N2} value was higher than 0.95. However, when pH was 5, efficiencies and yields decreased remarkably. Thus, the acidic pH values stimulate changes in the consumption efficiency and affect the yield of the denitrifying process. Likewise, pH effect on the electron source and nitrate cannot be ignored, as their ionization states changes will result in modifications in their Ks values and thus, their consumption rate.

Denitrification could be carried out in a temperature range between 5 and 35°C (Lalucat et al. 2006). Temperature has an effect on the consumption rate of nitrate and growth rate of the microorganisms. It is recommendable to keep a constant temperature between 20 and 35°C to obtain a constant and acceptable denitrifying rate. Effects of low and high temperatures are mainly related with the physicochemical changes in the cell membrane structures, either for lipids and proteins. The temperature can also induce changes in the genetic expression processes affecting the consumption efficiency of substrates and possibly the yields.

3.2.3.4 Effect of C/N ratio

One of the problems that could be present in denitrifying microbial consortia is when denitrification and DNRA are performed at the same time (Tiedje 1988). As it was illustrated in Figure 3.2, the first step in DNRA is the reduction of nitrate to nitrite, which despite being necessary is not the critical step to the reaction, because conversion of nitrite to ammonium is the key step. DNRA might function as: a) mechanism to eliminate toxic accumulation of nitrite; b) an electrons source that permits reoxidation of NADH and c) producing process of energy by means of phosphorylation in the transport of electrons (Tiedje 1988).

It seems that the most applicable factor to determine whether a dissimilative pathway directs either denitrification or DNRA is the C/N ratio (Cuervo et al. 1999; Cervantes et al. 2001; Peña-Calva et al. 2004). At high C/N ratio, nitrate will probably be reduced to ammonia, due to excess of reducing power, consequently DNRA will be the pathway and ammonia will be the main product. If C/N ratio is close to the stoichiometric value, then the main final product will be N_2 (Tiedje et al. 1982; Martínez et al. 2007). Likewise, changes in Gibbs free energy ($\Delta G^{\circ\prime}$) are less favorable at higher C/N ratios as shown in the following reactions:

$$CH_3COOH + 1.6NO_3^- \rightarrow 2CO_2 + 0.8N_2 + 1.6OH^- + 1.2H_2O$$
$$\Delta G^{\circ\prime} = -843\,\mathrm{kJ/reaction}$$

(3.1)

$$CH_3COOH + NO_3^- \rightarrow 2CO_2 + NH_3 + OH^-$$
$$\Delta G^{\circ\prime} = -533 \, \text{kJ/reaction}$$
(3.2)

It has been observed experimentally that using a defined medium with nitrate and glucose at C/N ratio between 1.38 and 5.64, ammonium production by DNRA diminishes (Akunna et al. 1994a). Rustrian et al. (1997) showed that using glucose as carbon source at high C/N ratios (51, 87 and 138) the DNRA was the dominant metabolic pathway. Under these conditions, it was quantified about 50% of nitrogen was assimilated for biomass production. When C/N ratios were 4, 7, 17 and 34, denitrification was the predominant pathway; nevertheless, for the last two C/N ratios, there was a significant presence of residual carbon. When acetate is used at stoichiometric C/N ratio in batch culture or sequenced batch culture, it has been found that it is possible to achieve consumption efficiencies for nitrate close to 100% and yields close to 1, since there is no accumulation of nitrite or ammonium (Peña-Calva et al. 2004; Martínez et al. 2007; Hernández et al. 2008). On the other hand, in continuous culture using acetate, lactate or glucose at C/N ratios near the stoichiometric value, there is also a total elimination of substrates without ammonium or intermediates formation and the yields obtained are almost 1, even at loading rates of 2.5 g NO_3^--N/Ld (Cuervo-López et al. 1999).

There are evidences indicating that the type of electron source has an important influence also in the dissimilative pathway. Akunna et al. (1993) evaluated the effect of different electron donors (glucose, glycerol, acetate, lactate and methanol) in a denitrifying process. They found that the first two promote fermentation and ammonium production instead of molecular nitrogen formation, while nitrite and nitrate were converted completely to N_2 with acetate and lactate. In many denitrifying studies it is not specified if nitrate is reduced to molecular nitrogen; mass balances are not included and it is not feasible to determine yields and specific velocities, which are necessaries to evaluate the respiratory process.

Some research studies mention that when the C/N ratio is very high, methanogenesis might compete with denitrification. However, it has been observed that methanogenesis could be affected by the presence of nitrate and nitrite in batch culture (Chen and Lin 1993). Nonetheless, from the thermodynamic point of view this behavior is not understandable. The changes in $\Delta G^{\circ\prime}$ for the two processes are:

$$CH_3COOH + 1.6NO_3^- \rightarrow 2CO_2 + 0.8N_2 + 1.6OH^- + 1.2H_2O$$
$$\Delta G^{\circ\prime} = -843 \, \text{kJ/reaction}$$
(3.3)

$$CH_3COOH + H_2O \rightarrow H_2CO_3 + CH_4$$
$$\Delta G^{\circ\prime} = -67\,kJ/\text{reaction} \tag{3.4}$$

The change in free energy during methanogenesis is significantly less exergonic than in denitrification. Similarly, from the kinetic point of view, the rates values for the two processes are significantly different. Nevertheless, Hanaki and Polprasert (1989) and Akunna et al. (1994a) found that in continuous denitrifying reactors, denitrification and methanogenesis may be simultaneously observed.

3.2.3.5 Alternative electron sources

Methanol, ethanol, glucose, aspartate and volatile fatty acids are the most commonly employed reducing sources for denitrification. Methane is another electron donor, which represents a significant potential to perform denitrification (Islas-Lima et al. 2004; Raghoebarsing et al. 2006). Depending on the type of organic substrate, both consumption rate and changes in $\Delta G^{\circ\prime}$ will be different. The type of substrate will determine the general behavior of the microbial denitrifying population. Table 3.4 shows denitrification in presence of different electron sources and changes in Gibbs free energy.

It is possible to observe that in any case the processes are clearly exergonic and that the size of $\Delta G^{\circ\prime}$ is depending on the type of electron source. If denitrification is accomplished in presence of acetate or sulfide, differences in $\Delta G^{\circ\prime}$ are not significant, but if p-cresol is the electron donor, the difference will be very significant. In kinetic terms, the differences are also dependent on the electron

Table 3.4. $\Delta G^{\circ\prime}$ values of denitrification in presence of different electron sources

Compound	Equation	$\Delta G^{\circ\prime}$ (kJ/reaction)
Acetic acid	$CH_3COOH + 1.6NO_3^- \rightarrow 2CO_2 + 0.8N_2 + 1.6OH^- + 1.2H_2O$	−843
Ethanol	$C_2H_6O + 2.4NO_3^- + 0.4H^+ \rightarrow 1.2N_2 + 2HCO_3^- + 2.2H_2O$	−1230
Lactic acid	$C_3H_6O_3 + 2.4NO_3^- \rightarrow 1.2N_2 + 3HCO_3^- + 1.2H_2O + 0.6H^+$	−1260
Glucose	$C_6H_{12}O_6 + 4.8NO_3^- \rightarrow 2.4N_2 + 6HCO_3^- + 1.2H^+ + 2.4H_2O$	−2686
Phenol	$C_6H_6O + 5.6NO_3^- + 0.2H_2O \rightarrow 2.8N_2 + 6HCO_3^- + 0.4H^+$	−2818
p-Cresol	$C_7H_8O + 6.8NO_3^- \rightarrow 3.4N_2 + 7HCO_3^- + 0.2H^+ + 0.4H_2O$	−3422
Benzene	$C_6H_6 + 6NO_3^- \rightarrow 3N_2 + 6HCO_3^-$	−2977
Toluene	$C_7H_8 + 7.2NO_3^- + 0.2H^+ \rightarrow 3.6N_2 + 7HCO_3^- + 0.6H_2O$	−3524
Xylene	$C_8H_{10} + 8.4NO_3^- + 0.4H^+ \rightarrow 4.2N_2 + 8HCO_3^- + 1.2H_2O$	−4136
Sulphide	$S^{-2} + 2NO_3^- + 4H^+ \rightarrow SO_4^{-2} + N_2 + 2H_2O$	−922

source and might be understood if it is considered that specific consumption rate (q_s) of denitrifying process not only is associated to the concentration of substrate, but also is associated to the type of substrate. Each denitrifying consortia has a specific Ks constant for every electron donor.

When organic compounds are not easily assimilated, such as aromatic hydrocarbons, their consumption rates may represent the limiting step in denitrifying process. Taking into account this situation, studies about the elimination of mixtures of readily oxidable organic compounds with recalcitrant organic matter have been made. In this regard, there are evidences of microbial consortia capable of accomplish denitrification with aromatic compounds such as benzene, toluene and m-xylene (BTX), which are considered as recalcitrant. Peña-Calva *et al.* (2004) and Martínez *et al.* (2007) achieved elimination efficiencies of toluene and nitrate close to 100%, as well as yields formation of HCO_3^- and N_2 close to 1. Denitrifying process was clearly dissimilative, as the biomass formation was not detected. In order to increase the denitrifying rate of toluene, Hernández *et al.* (2008) used a sequencing batch reactor (SBR). It was observed an increase in denitrifying activity while the number of cycles increases.

There is evidence that phenolic compounds could be consumed both, in continuous and batch cultures by microbial consortia under denitrifying conditions (Khoury *et al.* 1992). O'Connor and Young (1996) studying a consortium under denitrifying conditions, and considering only kinetic aspects concluded that the pattern of consumption of cresols was: p-cresol > m-cresol >> o-cresol. Meza-Escalante *et al.* (2008) achieved high efficiencies of elimination for p-cresol with yields of 0.75 for bicarbonate and 0.85 for N_2. In this work, it was evidenced that specific oxidation rate for intermediaries of p-cresol was 8-fold lower than oxidation rate of p-cresol itself; thus, accumulation of carbonated intermediaries in the denitrifying process was observed. Information related to denitrification with chloro-phenol compounds are still limited (Chang *et al.* 2004). Bae *et al.* (2002) found that 2-chlorophenol and 3-chlorophenol could be mineralized under denitrifying conditions with high efficiency, but the process was very slow.

On the other hand, contaminants which contain sulfur have a great relevancy due to their adverse environmental impact to ecosystems. It is said that sulfide is toxic, corrosive, besides it produces extremely unpleasant odors. Under denitrifying lithoautotrophic conditions and depending on stoichiometry of the reaction, sulfide might be oxidized to sulfate (SO_4^{2-}) or elemental sulfur ($S°$). Either in batch studies or in continuous cultures, reduced sulfurous compounds ($S°$, S^{2-} and $S_2O_3^{2-}$) can be used for nitrate reduction to N_2 (Sublette and Sylvester 1987; Beristain-Cardoso *et al.* 2006; Sierra-Alvarez *et al.* 2007).

Organotrophic and lithotrophic denitrification have been widely and individually studied. However, there are evidences with microbial consortia showing that denitrification is a process that allows simultaneous elimination of nitrate, organic carbon, sulfide and obtaining N_2, CO_2 and $S°$ or (SO_4^2), respectively (Kim and Son 2000; Reyes-Ávila *et al.* 2004; Beristain-Cardoso *et al.* 2008). Elimination of sulfide and nitrate via denitrification could operate with high efficiencies and yields, even when recalcitrant organic matter is present (Sierra-Alvarez *et al.* 2005). Meza-Escalante *et al.* (2008) through kinetic studies pointed that addition of sulfide in batch denitrifying cultures in presence of *p*-cresol, resulted in an apparent inhibitory effect in denitrification, decreasing consumption specific rates for nitrate and nitrite, as well as for *p*-cresol oxidation. These authors propose that this effect was related with accumulation of $S°$, due to its lower oxidation rate. This is in agreement with the results of Reyes *et al.* (2004). Beristain-Cardoso *et al.* (2009b) showed in denitrifying batch cultures that a total oxidation of sulfide and phenol could take place coupled to the nitrate reduction to N_2, only when nitrate is at stoichiometric concentrations. Beristain-Cardoso *et al.* (2009a) showed that it is possible to efficiently remove phenol, sulfide and nitrate into an inverse fluidized bed reactor obtaining CO_2, SO_4^{2-} and N_2 as main final products. Specific consumption rates for nitrate and sulfide during the simultaneous oxidation of aromatic compounds seem to be restricted by the accumulation of $S°$, which is very slowly oxidized, but not due to toxicity effects, as it was suggested by Reyes *et al.* (2004). In order to understand the simultaneous oxidation of organic matter and sulfide, further research on the microbiological and pathways aspects is still needed.

3.2.4 Mathematical model of denitrification

The consumption rate equation for the denitrification process can be obtained from the mass balance equation:

$$S = Scx + Scp + Scm + Scit + Sr \qquad (3.5)$$

Where S is the nutrient and energy source, Scx and Scp are the fraction of substrate consumed for biomass and product formation, respectively; Scm is the substrate used for maintenance, $Scit$ is the substrate consumed for intermediates formation and Sr is the residual substrate. Equation (3.5) can be expressed as a function of time:

$$-\frac{dS}{dt} = \frac{dScx}{dt} + \frac{dScp}{dt} + \frac{dScm}{dt} + \frac{dScit}{dt} + \frac{dSr}{dt} \qquad (3.6)$$

When the substrate consumption is efficient and the environmental conditions are well defined, the Sr might be considered close to zero. Likewise, if no intermediates are accumulated $Scit$ can also be negligible and the Scm value will have the lowest value. Thus, the equation could mainly be expressed in terms of substrate used for biomass and products formed. Then

$$-\frac{dS}{dt} = \frac{dScx}{dt} + \frac{dScp}{dt} \tag{3.7}$$

In many biological processes the volumetric consumption rate dS/dt is a first order kinetic, then the rate can be defined as follows:

$$\text{Volumetric rate (r)} = -\frac{dS}{dt} = q_s S \tag{3.8}$$

Where q_s is the specific consumption rate of substrate, a constant of first order and its value is also dependent on the environmental culture conditions. The q_s can also be expressed as a function of the biomass concentration (X). It is known that the substrate conversion or yield for biomass and products respect to the consumed substrate ($Y_{x/Scx}$, $Y_{p/Scp}$, respectively) can be expressed by:

$$Y_{x/Scx} = \frac{dX}{dScx} \tag{3.9}$$

where

$$dScx = \frac{dX}{Y_{x/Scx}} \tag{3.10}$$

for product

$$Y_{p/Scp} = \frac{dP}{dScp} \tag{3.11}$$

and

$$dScp = \frac{dP}{Y_{p/Scp}} \tag{3.12}$$

Where Scx and Scp can be obtained from the general equation:

$$Sc = S - Sr \tag{3.13}$$

where S is the initial concentration.

If equations 3.10 and 3.12 are substituted in equation 3.7, then:

$$q_s S = \left(\frac{dX}{dt} \cdot \frac{1}{Y_{x/scx}} \right) + \left(\frac{dP}{dt} \cdot \frac{1}{Y_{p/scp}} \right) \tag{3.14}$$

where

$$\frac{dX}{dt} = \mu X \tag{3.15}$$

Equation 3.15 is the general microbial growth equation. The product rate formation can be expressed as follows:

$$\frac{dP}{dt} = q_p P \tag{3.16}$$

where q_p is the specific product formation rate.

Rearranging equations 3.14, 3.15 and 3.16 results in:

$$q_s S = \left(\frac{\mu}{Y_{x/Scx}} \right) X + \left(\frac{q_p}{Y_{p/Scp}} \right) P \tag{3.17}$$

If S, X and P are expressed under the same base (carbon or nitrogen, for instance) the equation can experimentally be simplified to

$$q_s = \frac{\mu}{Y_{x/Scx}} + \frac{q_p}{Y_{p/Scp}} \tag{3.18}$$

The equation summarizes the metabolism of denitrifying cultures, where $\mu/Y_{X/Scx}$ express the anabolism (or biosynthetic process) and $q_p/Y_{P/Scp}$ the catabolism (or the respiratory process). In all wastewater treatment processes, it is desirable that biomass has a very long doubling time, thus μ will be very small and the process becomes clearly dissimilative. Under these conditions, $Y_{X/Scx}$ will not be significant, as the biomass production will be negligible. Regarding this, Cuervo-Lopez et al. (1999), Peña-Calva et al. (2004), Reyes et al. (2004),

Hernández *et al.* (2008) and Martínez *et al.* (2007) have shown that denitrification process could be clearly dissimilative. Namely, biomass production was very low (less than 5% of consumed substrate was used for biomass). In fact, the yield values for molecular nitrogen and bicarbonate were very close to 1. In these cases, the $\mu/Y_{X/Scx}$ term of equation 3.18 becomes negligible. In other words, the main process was the respiration, while biosynthesis was negligible. Many processes for wastewater treatment are modeled considering the microbial growth rate and many attempts have been made to optimize it. However, this might be a mistake, as these processes are (or must be) essentially dissimilative, where biomass formation and its rate are negligible.

Equation 3.7 does not consider the chemical properties of the substrate. For instance, glucose, ethanol and acetate do not have the same physicochemical properties (e.g. their Ks values are different) and it is well known that the rate for any denitrifying culture is also dependent on the kind of substrate. In order to characterize kinetically the process, taking in account the substrate concentration and the chemical properties of the substrate, the modified Monod's equation in terms for substrate consumption rate can be useful. When the process is dissimilative, q_s is the better choice:

$$q_s = \frac{q_{max}S}{K_s + S} \tag{3.19}$$

where q_{max} is the maximal consumption rate of substrate. Each kind of substrate has a specific K_s value. It must be considered that, in some cases, the size of Ks might also vary when the environmental conditions change.

It was pointed out that any process must be characterized in thermodynamic and kinetic terms. Thermodynamics and kinetics are not always correlated in denitrifying cultures. In fact, Meza-Escalante *et al.* (2008) observed that sulfide is consumed faster than *p*-cresol under denitrifying conditions, but the $\Delta G^{\circ\prime}$ value is much higher for the latter process. Thus, for wastewater treatment processes, both thermodynamics and kinetics must be taken into account. The efficiencies and yields are also closely related to both aspects.

REFERENCES

Akunna, J.C., Bizeau, C. and Moletta, R. (1993) Nitrate and nitrite reductions with anaerobic sludge using various carbon sources: glucose, glycerol, acetic acid, lactic acid and methanol. *Water Res.* **27**, 1303–1312.

Akunna, J.C., Bizeau, C. and Molleta, R. (1994a) Nitrate reduction by anaerobic sludge using glucose at various nitrate concentrations: ammonification, denitrification and methanogenic activities. *Environ. Technol.* **15**, 41–49.

Alefounder, P.R. and Ferguson, S.J. (1980) The location of the dissimilatory nitrite reductase and the control of dissimilatory nitrate reductase in *Paracoccus denitrificans. Biochem. J.* **192**, 231–240.

Alefounder, P.R., Greenfield, A.J., McCarthy, J.E.G. and Ferguson, S.J. (1983) Selection and organization of denitrifying electron transfer pathways in *Paracoccus denitrificans. Biochim. Biophys. Acta.* **724**, 20–39.

Almeida, J.S., Julio, S.M., Reis, M.A.M., and Larrondo, J.T. (1995) Nitrite Inhibition of Denitrification by *Pseudomas fluorescens. Biotechnol. Bioeng.* **46**, 94–201.

Arai, H., Igarashi, Y. and Kodama T. (1995) Expression of the *nir* and *nor* genes for denitrification of *Pseudomonas aeruginosa* requires a nocel CRP/FNR-related transcriptional regulator, DNR in addition to ANR. *FEBS Lett.* **371**, 73–76.

Bae, H.-S., Yamagishi, T. and Suwa, Y. (2002) Evidence for degradation of 2-chlorophenol by enrichment cultures under denitrifying conditions. *Microbiology* **148**, 221–227.

Baumann, B., Snozzi, M, Zehnder, A.J.B. and van de Meer, J.R. (1996) Dynamics of denitrification actcivity of *Paracoccus denitrificans* in continuous culture during aerobic-anaerobic changes. *J. Bacteriol.* **178**, 4367–4374.

Baumann, B., van der Meer, J.R., Snozzi, M. and Zehnder, A.J.B. (1997) Inhibition of denitrification activity but not of mRNA induction in *Paracoccus denitrificans* by nitrite at a suboptimal pH. *Antonie van Leeuwenhoek.* **72**, 183–189.

Beristain-Cardoso, R., Sierra-Alvarez, R., Rowlette, P., Razo-Flores, E., Gómez, J. and Field, J.A. (2006) Sulfide oxidation under chemolithoautotrophic denitrifying conditions. *Biotechnol. Bioeng.* **95**, 1148–1157.

Beristain-Cardoso, R., Texier, A.-C., Alpuche-Solís, A., Gómez, J. and Razo-Flores, E. (2009a) Phenol and sulfide oxidation in a denitrifying biofilm reactor and its microbial community analysis. *Process Biochem.* Doi:10.1016/j.procbio.2008.09.002.

Beristain-Cardoso, R., Texier, A.-C., Sierra-Álvarez, R., Field, J.A., Razo-Flores, E. and Gómez, J. (2008) Simultaneous sulfide and acetate oxidation under denitrifying conditions using an inverse fluidized bed reactor. *J. Chem. Technol. Biotechnol.* **83**, 1197–1203.

Beristain-Cardoso, R., Texier, A.-C., Sierra-Álvarez, R., Razo-Flores, E., Field, J.A. and Gómez, J. (2009b) Effect of initial sulfide concentration on sulfide and phenol oxidation under denitrifying conditions. *Chemosphere,* **74**, 200–2005.

Berks, B.C., Ferguson, S.J., Moir, J.W.B. and Richardson, D.J. (1995) Enzymes and associated electron transport systems that catalyse the respiratory reduction of nitrogen oxides and oxyanions. *Biochim. Biophys. Acta.* **1232,** 97–173.

Boogerd, F.C., van Verseveld, H.W. and Stouthamer, A.H. (1981) Respiration-driven proton translocation with nitrite and nitrous oxide in *Paracoccus denitrificans. Biochim. Biophys. Acta.* **638**, 181–191.

Braun, C. and Zumft, W.G. (1991) Marker exchange of the structural genes for nitric oxide reeductase blocks the denitrification pathway of *Pseudomonas stutzeri. J. Biol. Chem.* **266**, 22785–22788.

Carr, G.J. and Ferguson, S.J. (1990) Nitric oxide formed by nitrite reductase of *Paracoccus denitrificans* is sufficiently stable to inhibit cytochrome oxidase activity and is reduced by its reductase under aerobic conditions. *Biochim. Biophys.Acta.* **1017**, 57–62.

Carr, G.J., Page, M.D. and Ferguson, S.J. (1989) The energy-conserving nitric-oxide-reductase system in *Paracoccus denitrificans*: distinction of the nitrite reductase that catalyses synthesis of nitric oxide and evidence from trapping experiments for nitric oxide as a free intermediate during denitrification. *Eur. J. Biochem.* **179**, 683–692.

Cervantes F. J., De la Rosa D. and Gomez J. (2001). Nitrogen removal from wastewaters at low C/N ratios with ammonium and acetate as electron donors. *Biores. Technol.* **79**, 165–170.

Chang, C.-C., Tseng, S.-K., Chang, C.-C. and Ho, C.-M. (2004) Degradation of 2-chlorophenol via a hydrogenotrophic biofilm under different reductive conditions. *Chemosphere.* **56**, 989–997.

Chen, K.C. and Lin, Y.F. (1993) The relationship between denitrifying bacteria and methanogenic bacteria in a mixed culture system of acclimated sludge. *Water Res.* **27**, 1749–1759.

Cuervo-López, F.M., Martinez, F., Gutiérrez-Rojas, M., Noyola, R.A. and Gomez, J. (1999) Effect of nitrogen loading rate and carbon source on denitrification and sludge settleability in UASB reactors. *Water Sci. Technol.* **40**(8), 123–130.

Delwiche, C.C. and Bryan, B.A. (1976) Denitrification. *Ann. Rev. Microbiol.* **30**, 241–262.

Dermastia, M., Turk, T. and Holocher, T.C. (1991) Nitric oxide reductase. Purification from *Paracoccus denitrificans* with use of a single column and some characteristics. *J. Biol. Chem.* **266**, 10899–10905.

Dodd, F.E., Hasnain, S.S., Abraham, Z.H.L., Ready, R.R. and Smith, B.E. (1997) Structure of a blue-copper nitrite reductase and its substrate-bound complex. *Acta Crystallog. Sect. D.* **53**, 406–418.

Fee, J.A., Sanders, D., Slutter, C.E., Doan, P.E., Aasa, R., Karpefors, M. and Vänngard, T. (1995) Multi-frequency EPR evidence for a binuclear Cu_A center in citochrome c oxidase: studies with a ^{63}Cu and ^{65}Cu enriched soluble domaine of the cytochrome ba_3 subunit II from *Thermus thermophilus. Biochem. Biophys. Res. Commun.* **212**, 77–83.

Firth, J.R. and Edwards, C. (1999) Effects of cultural conditions on denitrification by *Pseudomonas stutzeri* measured by Membrane Inlet Mass Spectrometry. *J. Appl. Microbiol.* **87**, 353–358.

Fischer, H.M. (1994) Genetic regulation of nitrogen fixation in rhizobia. *Microbiol. Rev.* **58**, 352–386.

Frunzke, K. and Zumft, W.G. (1986) Inhibition of nitrous-oxide respiration by nitric oxide in the denitrifying bacterium *Pseudomonas perfoectomarina. Biochim. Biophys. Acta.* **852**, 119–125.

Fulop, V., Moir, J.W.B., Ferguson, S. and Hajdu, J. (1995) The anatomy of a bifunctional enzyme: structural basis for reduction of oxygen to water and synthesis of nitric oxide by cytochrome cd_1. *Cell.* **81**, 369–377.

Galimand, M., Gamper, M., Zimmerman, A. and Hass, D. (1991) Positive FNR-like control of anaerobic arginine degradation and nitrate respiration in *Pseudomonas aeruginosa. J. Bacteriol.* **173**, 1598–1606.

Glass, C., Silverstein, J. and Oh, J. (1997) Inhibition of denitrification in activated sludge by nitrite. *Wat. Environ. Res.* **69**(6), 1086–1093.

Goretski, J. and Hollocher, T.C. (1990) The kinetic and isotopic competence of nitric oxide as an intermediate in denitrification. *J. Biol. Chem.* **265**, 889–895.

Guest, J.R., Green, J., Irvine, A.S. and Spiro, S. (1996) The FNR modulon and FNR-regulated gene expression, p. 317–342. In E.C.C. Lin and A.S. Lynch (ed) Regulation of gene expression in *Escherichia coli.* Chapman and Hall, New York, N.Y.

Gumaelius, L., Magnusson, G., Pettersson, B. and Dalhammar, G. (2001) *Comamonas denitrificans* sp. nov., an efficient denitrifying bacterium isolated from activated Sludge. *Int. J. Syst. Evol. Micr.* **51**, 999–1006.

Hanaki, K., and Polprasert, C. (1989) Contribution of methanogenesis to denitrification with an upflow filter. *J. Wat. Poll. Control Fed.* **61**(9), 1604–1611.

Härtig, E. and Zumft, W.G. (1996) Nitrate and nitrite response of denitrification genes studied by mRNA analysis. *BIOspektrum.* **2**, 88.

Hernández L., Buitrón, G., Gómez, J., Cuervo-López F. M. (2008). Denitrification of toluene and sludge settleability. 4th Sequencing Batch Reactor Conference, Rome, Italy, 7–10 april.

Hernández, D. and Rowe, J.J. (1988) Oxigen inhibition of nitrate uptake is a general regulatory mechanism in nitrate respiration. *J. Biol. Chem.* **263**, 7937–7939.

Hille, R. (1996) The mononuclear molybdenum enzymes. *Chem. Rev.* **96**, 2757–2816.

Horio, T., Higashi, T., Yamanaka, T., Matsubara, H. and Okuniki, K. (1961) Purification and properties of cytochrome oxidase from *Pseudomonas aeruginosa. J. Biol. Chem.* **236**, 944–951.

Islas-Lima S., Thalasso, F. and Gómez, J. (2004) Evidence of anoxic methane oxidation coupled to denitrification. *Water Res.* **38**, 13–16.

Jetten, M.S.M., Logemann, S., Muyzer, G., Robertson, L.A., Vries, S., VanLoosdrecht, M.C.M. and Kuenen, J.G. (1997) Novel principles in the microbial conversion of nitrogen compounds. *Antonie van Leeuwenhock.* **71**, 75–93.

Jetten, M.S.M., Schmidt, M., Schmidt, I., Wubben, M., Van Dongen, U. and Abma, W. (2002) Improved nitrogen removal by application of new nitrogen-cycle *bacteria. Rev. Environ. Sci. Bio/Technol.* **1**, 51–63.

Jones, A.M. and Hollocher, T.C. (1993) Nitric oxide reductase of *Achromobacter cycloclastes. Biochim. Biophys. Acta.* **1144**, 359–366.

Ketchum, P.A., Denariaz, G., LeGall, J. and Payne, W.J. (1991) Menaquinol-nitrate oxidoreductase of *Bacillus halodenitrificans. J. Bacteriol.* **173**, 2498–2505.

Khoury, N., Dott, W. and Kampfer, P. (1992) Anaerobic degradation of phenol in batch and continuous cultures by a denitrifying bacterial consortium. *Appl. Microbiol. Biotechnol.* **37**, 524–528.

Kim, I. and Son, J. (2000) Impact of COD/N/S ratio on denitrification by the mixed cultures of sulfate reducing bacteria and sulfur denitrifying bacteria. *Water Sci. Technol.* **42**(34), 69–76.

Knowles, R. (1982) Denitrification. *Microbiol. Rev.* **46**(1): 43–70.

Körner, H. and Zumft, W.G. (1989) Expression of denitrification enzymes in response to the dissolved oxigen level and respiratory substrate in continuous culture of *Pseudomonas stutzeri. Appl. Environ. Microbiol.* **55**, 1670–1676.

Kristjansson, J.K. and Hollocher, T.C. (1980) First practical assay for soluble nitrous oxide reductase of denitrifying bacteria and a partial kinetic characterization. *J Biol. Chem.* **255**, 704–707.

Kroneck, P.M.H. and Zumft, W.G. (1990) Bio-inorganic aspects of denitrification: Structures and reactions of NxOy compounds and their interaction with iron and copper proteins. In: Revsbech NP & Sørensen J (Eds) Denitrification in Soil and Sediment (pp 1–20). Plenum Press, New York

Lalucat, J., Bennasar, A., Bosch, R., García-Valdés, E., and Palleroni, N.J. (2006) Biology of *Pseudomonas stutzeri. Microbiology and Molecular Biology Reviews.* **70**(2), 510–547.

Martínez, S., Cuervo-López, F.M. and Gómez, J. (2007) Toluene mineralization by denitrification in an up flow anaerobic sludge (UASB) reactor. *Biores. Technol.* **98**, 1717–1723.

Mateju, V., Cizinska, S., Krejei, J. and Janoch, T. (1992) Biological water denitrification. A review. *Enzyme Microb. Technol.* **14**, 170–183.

Meza-Escalante, E., Texier, A.-C., Cuervo-López, F., Gómez, J. and Cervantes, F.J. (2008) Inhibition of sulphide on the simultaneous removal of nitrate and *p*-cresol by a denitrifying sludge. *J. Chem. Technol. Biotechnol.* **83**, 372–377.

Moura, I. and Moura, J.J.G. (2001) Structural aspects of denitrifying enzymes. *Curr. Opin. Chem. Biol.* **5**, 168–175.

O'Connor, O. and Young, L.Y. (1996) Effects of six different functional groups and their position on the bacterial metabolism of monosubstituted phenols under anaerobic conditions. *Environ. Sci. Technol.* **30**(5), 1419–1428.

Oh, J. and Silverstein, J. (1999) Acetate limitation and nitrite accumulation during denitrification. *J. Environ. Eng.* **125**(3), 234–242.

Ohshima, T., Sugiyama, M., Ouzumi, N., Iijima, S. and Kobayashi, T. (1993) Cloning and sequencencing of a gene encoding nitrite reductase from *Paracoccus denitrificans* and expresion of the gene in *Escherichia coli*. *J. Ferment. Bioeng.* **76**, 82–88.

Peña-Calva, A., Olmos, D.A., Viniegra, G.G., Cuervo-López, F.M. and Gómez, J. (2004) Denitrification in presence of benzene, toluene, and *m*-xylene. *Appl. Biochem. Biotech.* **119**, 195–208.

Prudencio, M., Pereira, A.S., Tavares, P., Besson, S., Cabrito, I., Brown, K., Samyn, B., Devreese, B., Van Beeumen, J. and Rusnak, F. (2000) Purification, characterization and preliminary crystallographic study of copper containing nitrous oxide reductase from *Pseudomonas nautica*. *Biochemistry.* **39**, 3899–3907.

Raghoebarsing, A.A., Pol, A., van de Pas-Schoonen, K.T., Smolders, A.J.P., Ettwig, K.F., Rijpstra, W.I.C., Schouten, S., Sinninghe Damste, J.S., Op den Camp, H.J.M., Jetten, M.S.M. and Strous, M. (2006) A microbial consortium couples anaerobic methane oxidation to denitrification. *Nature.* **440**, 918–921.

Remde, A. and Conrad, R. (1991) Metabolism of nitric oxide in soil and denitrifying bacteria. *FEMS Microbiol. Ecol.* **85**, 81–93.

Reyes-Ávila, J., Razo-Flores, E. and Gómez, J. (2004) Simultaneous biological removal of nitrogen, carbon and sulfur by denitrification. *Water Res.* **38**, 3313–3321.

Rijn, J.V., Tal, Y. and Barak, Y. (1996) Influence of volatile fatty acids on nitrite accumulation by a *Pseudomonas Stutzeri* strain isolated from a denitrifying fluidized bed reactor. *Appl. Environ. Microbiol.* **62**(7), 2615–2620.

Rustrian, E., Delgenes, J.P., Bernet, N. and Moletta, R. (1997) Nitrate reduction in acidogenic reactor: influence of wastewater COD/N-NO3 ratio on denitrification and acidogenic activity. *Environ. Technol.* **18**, 309–315.

Schäfer, F. and Conrad R. (1993) Metabolism of nitric oxide by *Pseudomonas stutzeri* in culture and in soil. *FEMS Microbiol. Ecol.* **102**, 119–127.

Schmidt, I., Sliekers, O., Schmid, M., Bock, E., Fuerst, J., Kuenen, J.G., Jetten, M.S.M. and Strous, M. (2003) New concepts of microbial treatment processes for the nitrogen removal in wastewater. *FEMS Microbiol. Rev.* **27**, 481–492.

Shapleigh, J.P. and Payne, W.J. (1985) Differentiation of cd1cytochrome and cupper nitrite reductase production in denitrifiers. *FEMS Microbiol. Lett.* **26**, 275–279.

Shimizu, T, Furuki, T, Waki, T. and Ichikawa, K. (1978) Metabolic characteristics of denitrification by *Paracoccus denitrificans*. *J. Ferment. Technol.* **56**, 207–213.

Sierra-Alvarez, R., Beristain-Cardoso, R., Salazar M., Gómez, J., Razo-Flores, E. and Field, J.A. (2007) Chemolithotrophic denitrification with elemental sulfur for groundwater treatment. *Water Res.* **41**, 1253–1262.

Sierra-Alvarez, R., Guerrero, F., Rowlette, P., Freeman, S. and Field, J.A. (2005) Comparison of chemo-, hetero- and mixotrophic denitrification in laboratory-scale UASBs. *Water Sci. Technol.* **52**(1–2), 337–342.

Stewart, V. (1988) Nitrate respiration in relation to facultative metabolism in enterobacteria. *Microbiol. Rev.* **52**, 190–232.

Stewart, V., Parales, J.Jr. and Merkel, S.M. (1989) Structure of genes NarL and NarX of the nar (nitrate reductase) locus in *Escherichia coli* K-12. *J. Bacteriol.* **171**, 2229–2234.

Strous, M., Kuenen, J.G. and Jetten, M.S.M. (1999) Key physiology of anaerobic ammonium oxidation. *Appl. Environ. Microbiol.* **65**, 3248–3250.

Sublette, K.L. and Sylvester, N.D. (1987) Oxidation of hydrogen sulfide by mixed cultures of *Thiobacillus denitrificans* and heterotrophs. *Biotechnol. Bioeng.* **29**, 759–761.

Tavares, P., Pereira, A.S., Moura, J.J.G. and Moura, I. (2006) Metalloenzymes of the denitrification pathway. *J. Inorg. Biochem.* **100**, 2087–2100.

Thorndycroft, F.H., Butland, G., Richardson, D.J. and Watmough, N.J. (2007) A new assay for nitric oxide reductase reveals two conserved glutamate residues form the entrance to a proton-conducting channel in the bacterial enzyme. *Biochem. J.* **401**, 111–119.

Tiedje J.M. (1988) In *Biology of anaerobic microorganisms*. Zehnder A.J.B. (Ed.). Chap. 4th Wiley, NY.

Tiedje J.M., Sextone, A.J. Myrold, D.D. and Robinson, J.A. (1982) Denitrification: ecological niches, competition and survival. *Antonie van Leeuwenhoek. J. Microbiol. Serol.* **48**, 569–583.

Tosques, I.E., Kwiatkowski, A.V., Shi, J. and Shapleigh, J.P. (1996) Characterization and regulation of the gene encoding nitrite reductase in *Rhodobacter sphaeroides* 2.4.3. *J. Bacteriol.* **179**, 1090–1095.

Van Spanning, R.J.M., de Boer, A.P.N., Reijnders, W.N.M., Spiro, S., Westerhoff, H.V., Stouthamer, A.H., and van der Oost, J. (1995) Nitrite and nitric oxide reduction in *Paracoccus denitrificans* is under the control of NNR, a regulatory protein that belong to the FNR family of transcriptional activators. *FEBS Lett.* **360**, 151–154.

Van Spanning, R.J.M., de Boer, A.P.N., Reijnders, W.N.M., Westerhoff, H.V., Stouthamer, A.H., and van der Oost, J. (1997) FnrP and NNR of *Paracoccus denitrificans* are both members of the FNR family of transcriptional activators but have distinct roles in respiratory adaptation in response to oxygen limitation. *Mol. Microbiol.* **23**, 893–907.

Voβwinkel, R., Neidt, I. and Bothe, H. (1991) The production and utilization of nitric oxide by a new denitrifying strain of *Pseudomonas aeruginosa*. *Arch. Microbiol.* **156**, 62–69.

Watanabe, Y., Okabe, S., Hirata, K. and Masuda, S. (1995) Simultaneous removal of organic materials and nitrogen by micro-aerobic biofilms. *Water Sci. Technol.* **31**(1), 195–203.

Wicht, H. (1996) A model for predicting nitrous oxide production during denitrification in activated sludge. *Water Sci. Technol.* **34**(5–6), 99–106.

Wilson, P. and Bouwer, E.J. (1997) Biodegradation of aromatic compounds under mixed oxygen/denitrifying conditions: a review. *J. Ind. Microbiol. Biotechnol.* **18**, 118–130.

Wu, Q. and Knowles, R. (1994) Cellular regulation of nitrate uptake in denitrifying *Flexibacter canadensis. Can. J. Microbiol.* **40**, 576–582.

Wu, Q. and Knowles, R. (1995) Effect of chloranphenicol on denitrification in *Flexibacter canadensis* and *"Pseudomonas denitrificans". Appl. Environ. Microbiol.* **61**(2), 434–437.

Wu, Q., Knowles, R. and Chan, Y.K. (1995a) Production and consumption of nitric oxide by denitrifying *Flexibacter canadensis. Can. J. Microbiol.* **41**, 585–591.

Wu, Q., Knowles, R. and Niven, D.F. (1995b) Effect of ionophores on denitrification in *Flexibacter canadensis. Can. J. Microbiol.* **41**, 227–234.

Ye, R.W., Arunakumari, A., Averill, B.A., and Tiedje, J.M. (1992) Mutants of *Pseudomonas fluorescens* deficient in dissimilatory nitrate reduction are also altered in nitric oxide reduction. *J. Bacteriol.* **174**, 2560–2564.

Zennaro, E.I., Ciabatti, F., Cutruzzola, F., D'Alessandro, R. and Silvestrini, M.C. (1993) The nitrite reductase gene of *Pseudomonas aeruginosa*: effect of growth conditions on the expression and construction of a mutant by gene disruption. *FEMS Microbiol. Lett.* **109**, 243–250.

Zumft, W.G., Braun, C. and Cuypers, H. (1994) Nitric oxide reductase from *Pseudomonas stutzeri*: primary structure and gene organization of a novel bacterial cytochrome *bc* complex. *Eur. J. Biochem.* **219**, 481–490.

4

The ANAMMOX process

J.R. Vázquez-Padín, I. Fernández,
A. Mosquera-Corral, J.L. Campos,
and R. Méndez

4.1 DISCOVERY AND STOICHIOMETRY

For a long time, it was thought that ammonium oxidation could only take place aerobically. Broda (1977) predicted, using thermodynamic calculations, the existence of chemolitoautotrophic bacteria capable to oxidize ammonium using nitrite as electron acceptor (Equation 4.1):

$$NH_4^+ + NO_2^- \rightarrow N_2 + 2H_2O \qquad -\Delta G^\circ = 335\,kJ/(mol\ NH_4^+) \qquad (4.1)$$

That prediction would be experimentally confirmed two decades later by Mulder (1992) in a denitrifying pilot plant, treating wastewaters from a yeast factory. That reactor removed an ammonium loading rate of 0.4 kg NH_4^+-N/(m$^3 \cdot$ d) by this process (Mulder *et al.* 1995). These authors called the process ANAMMOX (ANaerobic AMMonium OXidation) and by nitrogen balances, hypothesized the following stoichiometry (Equation 4.2, Jetten *et al.* 1999):

$$5NH_4^+ + 3NO_3^- + 3e^- \rightarrow$$
$$4N_2 + 9H_2O + 2H^+ \qquad -\Delta G^\circ = 297\,kJ/(mol\ NH_4^+) \quad (4.2)$$

Van de Graaf *et al.* (1995) proved that the anaerobic ammonium oxidation was a biological process by means of activity assays with and without sludge and using sterilized sludge as inoculum. The ammonium removal rate was proportional to the biomass concentration. They also showed, by experiments with tracers, that the electron acceptor in the reaction was actually nitrite and not nitrate as initially thought. Strous *et al.* (1998) optimized the process conditions and found the global equation of the process, based on a mass balance performed on anammox enriched cultures as follows (Equation 4.3, Van de Graaf *et al.* (2000)):

$$NH_4^+ + 1.3NO_2^- + 0.066HCO_3^- + 0.13H^+ \rightarrow$$
$$1.02N_2 + 0.26NO_3^- + 0.066CH_2O_{0.5}N_{0.15} + 2H_2O$$
$$-\Delta G^\circ = 357\,kJ/(mol\ NH_4^+) \qquad (4.3)$$

Nitrogen losses in biological nitrogen removal systems were registered in the late nineties when treating highly nitrogen loaded wastewaters without presence of organic matter. These processes were given different names and were later discovered to be due to the anammox activity. Hippen *et al.* (1997) observed, in biofilm contactors, that working under oxygen limited conditions, a considerable part of the nitrogen load was eliminated without the corresponding consumption of organic matter for denitrification. They called the process aerobic deammonification. Kuai and Verstraete (1998) registered nitrogen losses in a sequencing batch reactor and named the process OLAND (oxygen-limited autotrophic nitrification-denitrification). These nitrogen losses were initially thought to be catalyzed by ammonium oxidizers under micro-aerophilic conditions. It was finally demonstrated that anammox bacteria were responsible for nitrogen removal in both systems (Helmer *et al.* 2001; Pynaert *et al.* 2003).

4.2 THERMODYNAMIC AND KINETIC PARAMETERS

The anammox process is chemolitotrophic, which generally implies micro-organisms characterized by low growth rates and yields due to the low Gibbs free energy involved in the reaction. In this case, the Gibbs free energy and the activation energy calculated by Van de Graaf et al. (2000) are -357 kJ/(mol NH_4^+) and 70 kJ/(mol NH_4^+), respectively.

The growing of the anammox microorganisms can be described by a Monod equation that indicates the dependency of the bacteria growing on a limiting substrate, including a bacteria decay coefficient (Equation 4.4):

$$\mu = \mu_{max} \frac{C_S}{K_S + C_S} - k_d \tag{4.4}$$

Where:
μ = microorganisms growing rate (1/d)
μ_{max} = maximum microorganisms growing rate (1/d)
K_s = substrate saturation constant (g/L)
C_S = substrate concentration (g/L)
k_d = decay constant (g/(g·d))

If ammonium is the limiting substrate, the equation (4.5) would be written as:

$$\mu = \mu_{max} \frac{C_{NH_4}}{K_{NH_4} + C_{NH_4}} - k_d \tag{4.5}$$

If the limiting substrate is nitrite, the employed equation (4.6) would be:

$$\mu = \mu_{max} \frac{C_{NO_2}}{K_{NO_2} + C_{NO_2}} - k_d \tag{4.6}$$

The kinetic parameters calculated by Strous et al. (1998) for an enriched anammox culture, of the type Candidatus "Brocadia anammoxidans" are included in Table 4.1. The low growing rates of these microorganisms imply, on one hand, long start-up periods of time for this kind of systems, and on the other hand, low amounts of sludge produced, solving the problem of the disposal of sludge in excess. These facts will suppose the need of systems with a good biomass retention but also savings of money. Some of the most applied systems to improve the retention of anammox biomass are the Sequencing Batch Reactors (SBRs; Strous et al. 1998; Dapena-Mora et al. 2004a) and biofilm reactors (Fernández et al. 2008a; Cema et al. 2007).

Table 4.1. Kinetic parameters of *Candidatus* "Brocadia anammoxidans" (Strous *et al.* 1998)

Anaerobic rate	45 nmol NH_4^+/(mg protein·min)
Maximum growing rate	0.003 1/h
Duplication time	10.6 d
Ks (ammonium)	5 µM
Ks (nitrite)	<5 µM
Biomass yield	0.066 mol C/(mol NH_4^+)
Oxygen	Reversible inhibition
Free energy	−357 kJ/(mol NH_4^+)
Biomass yield	0.07 mol C/mol C

Strous *et al.* (1998) analyzed the final products of the process, observing that the main product was N_2, while N_2O, NO and NO_2 were produced only in trace quantities.

4.3 BIOCHEMISTRY

Van de Graaf *et al.* (1997) realized studies with markers to research the reaction mechanisms. These authors employed different combinations of nitrogen compounds to test the formation and consumption of possible intermediate products. They found that, when hydrazine (N_2H_4) was added to the anammox culture, this was immediately consumed, while it accumulated if hydroxylamine (NH_2OH) was also added. The concentration of hydrazine decreased as soon as hydroxylamine was consumed. Addition of $^{15}NH_2OH$, $^{14}NH_4^+$ and $^{15}NO_2^-$ yielded $^{14,15}N_2$ as the main product, while in experiments with $^{15}NH_2OH$, $^{15}NH_4^+$ and $^{14}NO_2^-$ the quantity of $^{15,15}N_2$ increased. These results permitted to hypothesize the mechanisms showed in Figure 4.1. Ammonium and hydroxylamine would be combined to form hydrazine, which then would be oxidized to N_2. This oxidation would generate 4 electrons used to reduce nitrite to hydroxylamine. During the course of ammonium oxidation, low quantities of nitrate are produced from nitrite. This oxidation of nitrite into nitrate generates the electrons needed for the CO_2 fixation.

Schalk *et al.* (1998) confirmed that hydrazine was an intermediate of the anammox process. These authors observed that, if hydrazine was added to an anammox culture, the obtained products were ammonium and N_2 in a 4:1 ratio (Equation 4.7), while with the addition of hydrazine and nitrite as electron acceptor the final product was N_2 (Equation 4.8). These authors observed that hydrazine was toxic in long-term periods and that it was not possible to grow an ammox directly from hydrazine and nitrite.

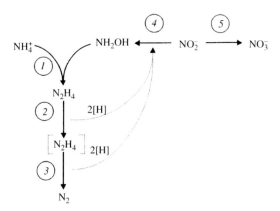

Figure 4.1. Proposed metabolic route of the anammox process (Schalk *et al.* 1998)

$$3N_2H_4 + 4H^+ \rightarrow 4NH_4^+ + N_2 \qquad (4.7)$$

$$3N_2H_4 + 4NO_2^- + 4H^+ \rightarrow 5N_2 + 8H_2O \qquad (4.8)$$

The anammox pathway proceeds via hydroxylamine and hydrazine as intermediates (Figure 4.1). From cell extracts of *Candidatus* "Brocadia anammoxidans", a new type of hydroxylamine oxidoreductase (HAO) was purified to homogeneity (Schalk *et al.* 2000). The enzyme was able to oxidize hydroxylamine and hydrazine. Inhibition experiments with H_2O_2 showed that hydroxylamine binds to a P-468 cytochrome, which is assumed to be the putative substrate binding site. The amino acid sequences of several peptide fragments of HAO from *Candidatus* "Brocadia anammoxidans" showed clear differences with the deduced amino acid sequence of HAO from the aerobic ammonia-oxidizing bacterium *Nitrosomonas europaea* (Schalk *et al.* 2000).

4.4 MICROBIOLOGY

Since the discovery of the anammox process in Delft (The Netherlands), evidence for anammox activity has been obtained in a variety of laboratory and engineered systems (Schmid *et al.* 2005), pilot plants for ammonium removal (van Dongen *et al.* 2001; Fux 2002), and even marine sediments (Thamdrup and Dalsgaard 2002) as well as the sub-oxic zone of the Black Sea (Kuypers *et al.* 2003).

Thereby, other anammox organisms different to *Candidatus* "Brocadia Anammoxidans" were identified: *Candidatus* "Kuenenia stuttgartiensis", *Candidatus* "Scalindua sorokinii", *Candidatus* "Scalindua brodae", *Candidatus* "Scalindua wagneri", *Candidatus* "Brocadia fulgida" and *Candidatus*

"Anammoxoglobus propionicus" were discovered and their 16S rRNA sequences were determined (Schmid *et al.* 2000; Schmid *et al.* 2003; Kuypers *et al.* 2003; Fujii *et al.* 2002; Kartal *et al.* 2007; Kartal *et al.* 2008).

The studies about enrichment or operation of the anammox process at laboratory or pilot scale reactors up to date were performed mainly with bacteria of the genus *C.* "Brocadia" or *C.* "Kuenenia", so the major part of the available information on anammox is about these types of bacteria.

The main difference between *Candidatus* "K. stuttgartiensis" and *Candidatus* "B. anammoxidans" is their anammox activity: 26.5 nmol N_2/(mg protein min) at pH 8 and 37°C for *C.* "Kuenenia" (Egli *et al.* 2001) and 55 nmol N_2/(mg protein·min) at pH 8 and 40°C for "Brocadia" (Jetten *et al.* 1999). However, *Candidatus* "K. Stuttgartiensis" has higher tolerance to nitrite, is more active in low cell density cultures and is less inhibited by phosphate compared to *Candidatus* "B. anammoxidans" (Egli *et al.* 2001).

The anammox organisms resemble each other in the phylogenetic analyses of their 16S rRNA sequences, which show they form a monophyletic branch, which consists of five distinct genera with about 90% sequence similarity to each other, within the phylum *Planctomycetes* (Figure 4.2.).

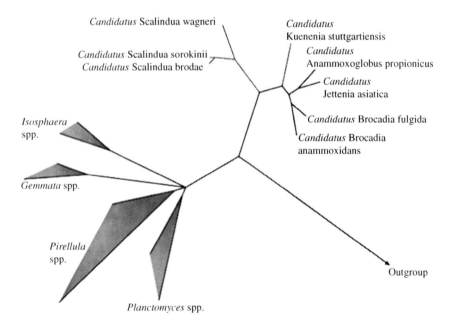

Figure 4.2. Anammox phylogenetic tree (Kuenen 2008)

As the rest of the species within the order *Planctomycetales*, they lack of peptidoglycan, an almost universal polymer found within the domain Bacteria. Instead, protein is the major constituent of their cell walls. Among the domain Bacteria, this lack of peptidoglycan is a characteristic shared only with the *Chlamydiae* and the cell-wall-free *Mycoplasms* (Lindsay *et al.* 2001).

As all known *Planctomycetes*, Anammox bacteria have an ultra-structure atypical for bacteria, with a compartmentalized cytoplasm (Fuerst 2005). Electron microscopy analysis revealed that anammox bacteria have a separate specific membrane bound compartment: the anammoxosome. HAO enzymes are present exclusively inside this anammoxosome, an organelle-like body that made up more than 30% of the cell volume (Van Niftrik *et al.* 2004). This dedicated intra-cytoplasmic compartment has been found to be surrounded by a membrane nearly exclusively composed of unconventional membrane lipids: ladderane lipids (Sinninghe-Damste *et al.* 2002; Schmid *et al.* 2003). The structure of ladderane lipids is unique in nature and has only been found in anammox bacteria so far. These lipids are composed of pentacycloanammoxic acids, which contain five linearly concatenated cyclobutane rings (Mascitti and Corey 2004). Due to the very slow metabolism of anammox bacteria, a very dense and impermeable membrane is required to maintain concentration gradients during the anammox reaction. Such a membrane also protects the cell from the toxic intermediates.

The genome of *Candidatus* "K. stuttgartiensis" has been sequenced (Strous *et al.* 2006). The genome data illuminate the evolutionary history of the Planctomycetes and anammox bacteria in particular found to be related to a clade of intracellular parasites known as the *Chlamydiae*. Candidate genes responsible for ladderane biosynthesis and biological hydrazine metabolism were identified. Besides, an unexpected metabolic versatility was discovered, this versatility could justify why although slow and specialized, anammox bacteria are global players in the biological nitrogen cycle (Arrigo 2005).

4.5 FACTORS THAT AFFECT THE ANAMMOX PROCESS

4.5.1 Effect of temperature

The specific substrate (ammonium and nitrite) consumption rates and nitrate production rate were estimated by Strous *et al.* (1999) at different temperatures. These authors observed that the optimum temperature for the anammox biomass was $40 \pm 3°C$. Between $20°C$ and $37°C$ the anammox activity depends on the temperature according to the Arrhenius law. The activation energy was estimated as 70 kJ/mol. At a temperature of $10°C$ or lower the anammox activity was null.

Toh *et al.* (2002) tried to select and enrich an autotrophic anaerobic ammonium oxidation consortium, from sludge collected from a municipal treatment plant in Sydney, at two different temperatures, 37°C and 55°C. While mesophilic anammox activities were successfully obtained in batch and continuous cultures, thermophilic anammox organisms could not be selected at 55°C.

Isaka *et al.* (2007) have reported that they could efficiently operate an anammox reactor at 20°C with a relatively high biomass concentration of 20 g SS/L. Dosta *et al.* (2008) obtained a dependency between temperature and specific anammox activity measured by batch tests (Figure 4.3). Their results agree with those obtained by Strous *et al.* (1999), reporting an exponential behaviour in a relatively wide temperature range, between 10°C and 40°C, with the maximum specific activity at 35–40°C. Dosta *et al.* (2008) have also

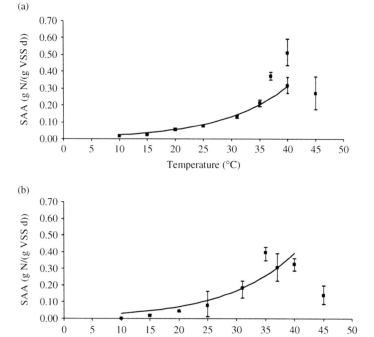

Figure 4.3. Influence of the temperature on the maximum specific anammox activity (SAA): Dependency profiles for granular (a) and biofilm (b) anammox biomass (experimental SAA (■); modified Arrhenius model (—)). (Adapted from Dosta *et al.* 2008)

reported that it was possible to operate an anammox SBR at 18°C by slowly
adapting the biomass to the low temperatures.

4.5.2 Effect of pH

Strous *et al.* (1999) studied the anammox activity dependence on the pH, finding
a physiological interval of pH for the anammox biomass of 6.7–8.3 (Figure 4.4).
The optimum value for operation was around 8. Egli *et al.* (2001) observed
anammox activity even at a pH value of 8.5–9.0. The difference in the range
of operation found by these authors could also be due to the fact that Strous
et al. (1999) worked with *Candidatus* "Brocadia anammoxidans" and Egli *et al.*
(2001) with *Candidatus* "Kuenenia stuttgartiensis".

Ahn *et al.* (2004) studied the anaerobic ammonium removal from piggery
waste characterized by its high loads (56 g COD/L and 5 g TKN-N/L) using a
lab-scale up-flow anaerobic sludge bed reactor under mesophilic conditions. The
pH of the effluent remained at 9.3–9.5 and the anaerobic ammonium removal
in the reactor was not inhibited in spite of the pH values which were out of the
optimal operation range.

4.5.3 Effect of substrate and product concentrations

Strous *et al.* (1999) exposed anammox biomass to high ammonium and nitrate
concentrations (up to 70 mM during one week) in a sequencing bath reactor
(SBR) and did not observe negative effects on the activity.

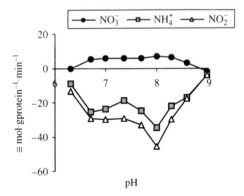

Figure 4.4. Influence of the pH in the maximum ammonium and nitrite consumption
rate and nitrate production rate (Adapted from Strous (2000))

Nevertheless, the same authors found that the biomass completely lost its activity at concentrations of nitrite of 98 mg NO_2^--N/L (7 mM). Furthermore, if the biomass was subjected at nitrite concentrations higher than 5 mM (70 mg N/L) for a period of 12 h, the process (for the type *Brocadia anammoxidans*) was totally inhibited. The activity was recovered by the addition of traces (50 µM) of hydrazine or hydroxylamine (intermediate products of the anammox process).

Strous *et al.* (1999) adjusted the obtained experimental data to the Luong model for processes with inhibition (Equation 4.9):

$$q = q_{max} \cdot \frac{C_S(1 - C_S/K_i)^\alpha}{K_S + C_S} \qquad (4.9)$$

Where:

q = specific substrate consumption rate (mmol substrate/(g protein· min))
$q_{máx}$ = maximum specific substrate consumption rate (mmol substrate/
        (g protein· min))
K_S = affinity constant (mM)
K_i = inhibition constant (mM)
C_S = substrate concentration (mM)
α = curve shape factor

The values of the coefficients found by Strous *et al.* (1999) for the nitrite inhibition were K_i = 57 mM, α = 0.8 for the ammonium oxidation process and K_i = 70 mM, α = 0.7 for the nitrite reduction one.

Egli *et al.* (2001) showed that *Candidatus* "Kuenenia stuttgartiensis" has a relatively important tolerance to nitrite concentrations up to 180 mg NO_2^--N/L.

Fux *et al.* (2002) also reported an inhibition of the anammox process in an anammox pilot plant fed with the effluent of a Sharon (Single reactor High Ammonia Removal Over Nitrite, which was previously described by Hellinga *et al.* (1998) and Van Dongen *et al.* (2001). The reactor achieved 50% of NH_4^+ oxidation to NO_2^-. The Sharon reactor is a chemostate where partial nitritation occurs and it is based in the selection of ammonia oxidizers by control of the hydraulic retention time and the temperature. The case of the anammox reactor operated by Fux *et al.* (2002), the anammox organisms of the type *Candidatus* "Kuenenia stuttgartiensis" were strongly inhibited by a nitrite surplus of 60 mg NO_2^--N/L. The anammox activity was slowly restored two weeks after the influent nitrogen load was reduced to nearly 50%.

A more recent work studied the short-term inhibitory effects of both substrates and nitrate on the anammox activity (Dapena-Mora *et al.* 2007).

Table 4.2. IC50 of ammonia, nitrite and nitrate
(Dapena-Mora *et al.* 2007)

Compound	IC50 (mg N/L)
Ammonia	770
Nitrite	350
Nitrate	630

These authors estimated the 50% inhibitory concentration (IC50) for each compound (Table 4.2).

As in the case of nitrifying microorganisms, the unionized form of substrates (NH_3 and HNO_2) were considered by Fernández *et al.* (2008b) as the true inhibitors of anammox microorganisms. For this reason, they studied the maximum levels of NH_3 and HNO_2, which allow maintaining a stable operation of anammox systems and found that concentrations higher than 20–25 mg NH_3-N/L and 0.5 µg HNO_2-N/L should be avoided to guarantee the stable operation of the process.

4.5.4 Effect of oxygen

Van de Graaf *et al.* (1996) demonstrated, using batch assays, that oxygen inhibited completely the anammox process, but this inhibition was reversible (Strous *et al.* 1997). These authors did not observe anammox activity in micro-aerobic conditions (2.0, 1.0 and 0.5% of air saturation). Nevertheless, Sliekers *et al.* (2002; 2003) operated an anammox SBR and a gas-lift under oxygen-limited conditions (<0.1% O_2 saturation and 0.5 mg O_2/L, respectively) and, although the maximum anammox activity decreased, a stable coexistence of anammox and aerobic ammonium oxidizers permitted a completely autotrophic removal of nitrogen in both systems. In this case, the existence of anammox in presence of oxygen can be attributed to the oxygen gradients existing through the biofilm. Oxygen would be consumed by the nitrifiers (in the outer zones of the biofilm) and anammox would grow inside the biofilm in an anoxic environment (Egli 2003; Cema *et al.* 2007). In those systems, nitrite oxidizing bacteria (NOB) growth (and subsequent nitrate production) is prevented due to the lower affinity for oxygen compared to ammonia oxidizing bacteria (AOB) and for the lower affinity for nitrite compared to anammox bacteria.

The autotrophic removal of nitrogen in a single reactor was called by different names: CANON, Completely Autotrophic Nitrogen removal Over Nitrite process (Strous 2000; Helmer *et al.* 2001); OLAND, Oxygen-Limited Autotrophic Nitrification-Denitrification (Kuai and Verstraete 1998, Pynaert *et al.* 2004) or aerobic/anoxic deammonification (Helmer *et al.* 2001).

4.5.5 Effect of inhibitors

Van de Graaf *et al.* (1995) studied the effect of antibiotics and other possible inhibitors on the anammox activity in batch cultures (Table 4.3). These authors showed that the addition of 800 mg/L ampicillin inhibited ammonium removal almost completely, while a concentration of 400 mg/L of this compound reduced the activity by 71%. They attributed these observed inhibitory effects to the antibiotic and not to the presence of organic matter.

Other tested compounds, such as 2,4-dinitrophenol, CCCP and $HgCl_2$, resulted to be strong inhibitors of the anaerobic ammonium oxidation (inhibition of 95% of the activity at concentrations of 1 mM or lower, which corresponded approximately to 0.21 mmol/g VSS).These authors also showed that 20 mg/L of chloramphenicol caused partial inhibition of ammonium removal, reducing the rate to 64% of the control level. Nevertheless, Toh and Ashbolt (2002) reported that anammox activity was greatly affected by the addition of phenol but not by chloramphenicol.

Fernández *et al.* (2006) reported a decrease of 75% of the specific anammox activity when 20 mg/L of chloramphenicol were fed to an anammox SBR. These authors also observed that the presence of 10 mg/L of tetracycline hydrochloride during continuous experiments caused a substantial loss of reactor capacity.

Table 4.3. Inhibitory effects on the anammox process of different compounds, using ammonium and nitrate as substrates (Van de Graaf *et al.* 1995)

Compound	Concentration (mg/l)	% Ammonium removal
Penicillin	0	100 ± 13
	1	83 ± 13
	100	64 ± 10
Chloramphenicol	0	100 ± 5
	20	64 ± 5
	200	2 ± 2
Ampicillin	0	100 ± 3
	400	29 ± 3
	800	6 ± 4
$HgCl_2$	0	100 ± 4
	271	0
2,4-Dinitrophenol	0	100 ± 5
	37	47 ± 7
	368	1 ± 2
CCCP	0	100 ± 3
	41	0

To obtain more information about specific requirements for anammox, Van de Graaf *et al.* (1996) started up a Fluidized Bed Reactor (FBR) fed with ammonium and nitrite and using sand as support material. With the enriched sludge they carried out a series of batch experiments, adding different chemicals to the medium. These authors found that both organic (acetate, propionate, glucose, fructose and lactose) and inorganic (sulphide, sulphur, sulphite, thiosulphate) electron donors had no effect or increased the ammonium oxidation rate when added at low concentrations (1–5 mM) but, in continuous assays, the organic compounds had a negative effect because they favor the growth of heterotrophic bacteria (Table 4.4). Nevertheless, Güven *et al.* (2005) showed that anammox bacteria could compete successfully with heterotrophic denitrifiers for propionate because they can carry out propionate oxidation

Table 4.4. Effects of diverse compounds on the anammox activity, using ammonium and nitrite as substrates (Van de Graaf *et al.* 1996)

Compound	Concentration	Effect
Batch assays		
Acetate	1–5 mM	Activity increase
Propionate	1 mM	No effect
Glucose	1 mM	Activity increase
Fructose	1 mM	Activity increase
Lactose	1 mM	No effect
Casamino acids	50 mg/L	No effect
Sulfide	1–5 mM	Activity increase
Sulfur	1–5 mM	Activity increase
Sulphite	1 mM	Activity increase
Thiosulfate	1 mM	Activity increase
KCl	50 mM	No effect
$KHCO_3$	20–40 mM	No effect
EDTA	100 mg/L	No effect
KH_2PO_4	5–50 mM	Activity loss
Continuous assays		
Acetate	2 mM	−28%
Glucose	1 mM	−12%
Pyruvate	1 mM	−20%
Formate	5 mM	−10%
Cysteine	0.8 mM	−11%
Sulphide	2 mM	+20–60%
Thiosulphate	2 mM	+47%
Hydroxilamine	2 mM	−28%
N_2O	0.7 mM	No effect
Nitrite	2.4–6 mM	+11, 24, or 51%

simultaneously with anaerobic ammonium oxidation. The addition in batch experiments of 5 or 50 mM of phosphate (commonly used as a pH buffer in batch experiments) caused complete inhibition of the anammox process, while addition of 50 mM KCl or 40 mM KHCO₃ had no effect. These authors also observed that light caused an inhibition of the process of 30–50%. Dapena-Mora *et al.* (2007) found that the presence of salts, phosphate or flocculant at relatively high concentrations does not represent a problem for the process. However, to guarantee the proper operation of the anammox process organic matter and sulphide must be absent.

4.5.6 Effect of shear stress

Arrojo *et al.* (2006) have reported that shear stress can cause a decrease in the activity of anammox biomass. According to their work, despite anammox sludge is highly resistant to mechanical stress, it would not be advisable to operate a SBR with specific mechanical stirring power higher than 0.09 kW/m³. Moreover, Arrojo *et al.* (2008) found that anammox biomass is less resistant to shear stress caused by gas-flow mixing. In this case, the authors advised not to operate with specific input power higher than 0.0172 kW/m³.

4.6 ANAMMOX SPECIFIC ANALYTICAL TECHNIQUES

4.6.1 Detection

For starting up an anammox system it is necessary the use of analytical methods that permit to detect the presence of anammox microorganisms in very low concentrations. In this case, the objective is to detect rather than quantify the anammox microorganisms and with this aim, the following methods can be used:

- Fluorescence in situ hybridization (FISH) is a rather fast method, but might not be sensitive enough to detect anammox organisms if they appear in relatively low numbers (below 1000–10,000 cells/mL) or if the sample is highly autofluorescent. As more anammox microorganisms are discovered, the designed probes to detect these microorganisms increase, permitting to detect all the anammox bacteria or distinguish between the existent different types (Schmid *et al.* 2005).
- The PCR (Polymerase Chain Reaction) approach, in which anammox specific FISH probes are used as primers, enables the detection of anammox organisms down to the genus level even faster and with higher sensitivity than the usual FISH approach (Schmid *et al.* 2005).

- Detection of the intergenic spacer region (ISR), between the 16S rRNA gene and the 23S rRNA gene with specific ISR probes, gives good indications about the activity status of the anammox cells, because the amount of ISR is directly linked to the ribosome turnover rate in the cells.
- The conversion of hydroxylamine to hydrazine is a unique reaction of the anammox process (Van de Graaf *et al.* 1997) that can be used specifically to detect anammox activity in environmental samples (Schmid *et al.* 2003).
- Much more sensitive are tracer experiments with ^{15}N labelled ammonium. Under anoxic conditions labelled ^{15}N-ammonium reacts uniquely, in a 1:1 ratio with unlabelled ^{14}N-nitrite, to ^{29}N$_2$ (^{14}N^{15}N) via the anammox reaction. This method was successfully used to assess the contribution of anammox to the nitrogen conversion in marine and estuarine environments, where the cell count of anammox is low (Thamdrup and Dalsgaard 2002; Dalsgaard *et al.* 2003; Kuypers *et al.* 2003; Trimmer *et al.* 2003).
- Anammox bacteria have lipids with unique properties (Sinninghe-Damste *et al.* 2002; Van Niftrik *et al.* 2004) that can be used as biomarkers for the presence of these cells in an environmental sample (Kuypers *et al.* 2003; Schmid *et al.* 2003).

4.6.2 Anammox activity measurements

Application of the anammox process to the treatment of wastewaters requires the utilization of batch activity tests to assess biomass activities and control the operation, measure kinetic parameters or know the influence of possible toxic compounds.

The methods commonly used to measure microorganisms activities can be divided in two groups:

4.6.2.1 Methods based in measurements in the liquid phase

Some of these methods are based on the measurement of the substrate (NH_4^+, NO_2^-) consumption rate by means of monitoring the concentration of these compounds in the liquid phase along time.

For the application of this method to the anammox process, some preliminary studies have been made to develop biosensors for the determination of NO_x^- and NO_3^- (Larsen *et al.* 1997; Revsbech *et al.* 2000). The determination of the evaluated compound is performed using a biological culture that oxidizes or reduces the measured compound, and measuring the redox potential of the reaction. When totally developed, the application of these biosensors to the anammox process would allow its continuous following and control.

In the meanwhile, concentrations of nitrogenous compounds in the liquid phase must be followed by spectrophotometric or chromatographic methods, requiring a high number of analyses, meaning time and reagents consumption. Up to date, this has been the most widely used method to measure anammox activity (Strous 2000; Sliekers *et al.* 2002; Helmer *et al.* 2001; Toh *et al.* 2002).

Other methods are based on the alkalinity production or consumption. ANITA (Ammonium NITrification Analyser) is an example of this kind of methods. For a nitrifying culture, the ammonium oxidizing rate is measured by the flow rate of the alkaline solution needed to neutralize the acidity produced by the metabolic activity of the biomass (Ficara *et al.* 2000). The difficulty to apply this method for measuring anammox activity relies on the low production of protons by the reaction (only 0.13 mol H^+ for each mole of ammonium consumed).

4.6.2.2 *Methods based in measurements in the gas phase*

Some methods are based on measuring the volume and composition of the gas phase.

Anaerobic batch tests are often performed by adding a large amount of substrate to the sludge and monitoring the resulting gas production with a Mariotte flask system (Soto *et al.* 1993). This system is not sensitive enough to measure small amounts of gas produced accurately and therefore demands the addition of large amounts of substrate. This is not possible in the case of anammox process, as it suffers substrates inhibition.

Another method to follow the gas production is to measure the pressure increase during the experiment, in a sealed vial, as well as the gas composition (Buys *et al.* 2000). By using a pressure sensor, small pressure increments may be measured accurately, so that the experiments can be performed at relatively low substrate concentrations. This method was extensively described and applied by Dapena-Mora *et al.* (2007). They utilized a standardized temperature of 30°C, mixing speed of 150 rpm and a buffered pH of 7.8. These authors reported that this method was very accurate and neither the initial biomass concentrations nor the substrate to biomass ratio had significant influence on the measured activity.

4.7 MODELLING

Mathematical models are useful and represent an alternative in order to test the response of a system to perturbations and to test different operational scenarios. The use of models allows saving time and operational costs, especially when working with slow growing organisms like anammox bacteria. The computing power nowadays allows calculating several days, even several years of operation even when dealing with complicated differential equations.

Different types of microorganisms are involved in anammox or CANON reactors since AOB, NOB, heterotrophs and anammox bacteria among others can coexist. The kinetic and stoichiometric basis used for AOB, NOB and heterotrophs is normally the Activated Sludge Models (ASM) (Henze *et al.* 2000). Kinetic and stoichiometric parameters for anammox bacteria were taken mainly from the work carried out by Strous *et al.* (1998) who studied deeply these microorganisms developing a culture enriched at 80%. Generally, the process of growth of anammox bacteria is modelled with two Monod terms for ammonium and nitrite and one inhibition term for oxygen (Equation 4.10). No inhibition by nitrite is considered due to the low concentrations registered during stable operation and also probably due to a lack of information and agreement about the inhibition mechanism.

$$\text{Anammox}_{growth} = \mu_{max} \frac{C_{NH_4}}{K_{NH_4} + C_{NH_4}} \frac{C_{NO_2}}{K_{NO_2} + C_{NO_2}} \frac{K_{O_2}}{K_{O_2} + C_{O_2}} \quad (4.10)$$

Where:

μ_{max} = maximum microorganisms growing rate (1/d)
K_{NH_4} = substrate saturation constant for ammonium (g/L)
K_{NO_2} = substrate saturation constant for nitrite (g/L)
K_{O_2} = substrate inhibition constant for oxygen (g/L)
C_{NH_4} = ammonium concentration (g/L)
C_{NO_2} = nitrite concentration (g/L)
C_{O_2} = oxygen concentration (g/L)

However, for the decay or endogenous respiration of bacteria, there is not a uniform criterion. Some authors use the endogenous respiration on aerobic and anoxic conditions, according to the ASM3 (a more realistic approach under a microbiological point of view, Van Loosdrecht and Henze 1998) and other authors consider the death-regeneration process used in the ASM1. A resume of different works about simulation of reactors carrying out the anammox process is shown in Table 4.5. It is indicated the ASM followed and the presence or absence of considering heterotrophs as a part of the biofilm, since anammox process is mainly used to treat wastewater with low COD/N ratio and some models avoid including heterotrophic growth.

Van der Star *et al.* (2008) grew free cells in a membrane bioreactor and therefore a simple model could be developed avoiding the inclusion of mass

Table 4.5. Models simulating anammox reactors

Reactor[a]	Software	ASM	Heterotrophs	Reference[b]
RBC	Aquasim	ASM3	No	[1]
Biofilm	Aquasim	ASM3	No	[2]; [3]
Biofilm	Aquasim	ASM3	Yes	[4]; [5]
MABR	Aquasim	ASM3	No	[6]
MABR	Aquasim	ASM1	Yes	[7]
SBR	West	ASM1	Yes	[8]

[a] RBC: Rotating biological contactor; MABR: Membrane Aerated Biofilm Reactor.
[b] [1] Koch *et al.* 2000; [2] Hao *et al.* 2002a; [3] Hao *et al.* 2002b; [4] Hao and Van Loosdrecht 2004; [5] Hao *et al.* 2005; [6] Terada *et al.* 2007; [7] Lackner *et al.* 2007; [8] Dapena-Mora *et al.* 2004b.

transfer phenomena. However, anammox bacteria grow normally forming biofilms or granules due to their low growth rate (Fernández *et al.* 2008a). To simulate the operation of such biofilms (granules can be considered as biofilms without carrier material), two different compartments are defined, a homogeneous bulk liquid compartment and a biofilm compartment separated by a boundary layer (Figure 4.5). For simple and more or less smooth biofilms systems, 1-D (one dimension) models represent a good compromise between required modelling output and effort involved in solving these models (Morgenroth *et al.* 2004) and are therefore widely used. The main assumption made by working with 1-D models is that the composition of particulate and dissolved compounds in the plan parallel to the biofilm surface is homogeneous. The biofilm is discretised in different slices and the partial equation is solved for each slice, therefore, the more discretisation steps, the more computational time is required.

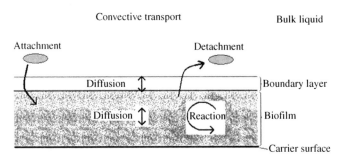

Figure 4.5. Processes involved in the biofilm compartment

4.7.1. Simulation of experimental data

The first model including anammox bacteria was developed by Koch *et al.* (2000) in Aquasim to describe the operation of a RBC where nitrogen losses were registered (Siegrist *et al.* 1998). These authors considered AOB, NOB and anammox bacteria, although the NOB were almost inexistent. Good agreement between simulation and experimental data for steady state and short term experiments was possible with the developed model. The presence of an external aerobic zone with AOB and the presence of anammox in a more internal anoxic zone were observed.

Dapena-Mora *et al.* (2004b) extended ASM1 including growth and decay of anammox bacteria. They simulated the start up and operation of an anammox SBR inoculated with sludge from a wastewater treatment plant (WWTP). After the start-up, AOB and NOB were washed out due to anoxic conditions but heterotrophs still remained in the biofilm probably because they were able to live on cell lysis products (however, their presence had no influence in reactor performance). Simulated concentrations of ammonium, nitrite and nitrate in the effluent fitted to the experimental data.

4.7.2. Simulations without experimental data

Models have also been developed without comparing with experimental data. These models were used to optimize the performance of CANON reactors and to test the effects of different perturbations on temperature, dissolved oxygen concentration (DO), ammonium load, COD concentration, etc. The time required for studying the effects of all the parameters involved in the process through experimentation would not be feasible due to the low growth rate of anammox bacteria.

4.7.2.1. Key parameters

Hao *et al.* (2002a; b) performed simulations to optimize the performance of a CANON reactor. The influence of ammonium surface load, temperature, biofilm thickness and DO was studied. It was observed that for each temperature and ammonium load there was an optimal thickness and DO concentration. The nitrogen removal performance increases by increasing the biofilm thickness, until reaching an optimal value. Thicker biofilms do not improve nitrogen removal and would require higher aeration. From simulations it was inferred that, the operation at low temperatures required thicker biofilms if the same ammonium load was applied, needing therefore a higher DO concentration. The DO was determined as the key parameter and an unbalance between applied and required DO is responsible for the decreased nitrogen removal efficiency.

Terada *et al.* (2007) also confirmed that the ratio between the superficial fluxes of oxygen and ammonium was the key parameter. Under oxygen excess conditions, ammonium is oxidized in excess and undesired nitrite oxidation takes place. The feasibility of a counter-diffusion reactor and its comparison with a co-diffusion one was carried out. In a MABR, the oxygen and the ammonium diffuse in opposite directions creating a biomass profile different from co-diffusion reactors since the aerobic zone is located on the support material and the anoxic zone close to the biofilm-bulk interface. The simulation results shown that counter-diffusion biofilm has a wider range for high autotrophic nitrogen removal than the co-diffusion biofilm.

4.7.2.2. Effect of COD

Although anammox processes are normally suitable to treat wastewater with low COD/N ratio, some COD could be present in the influent and therefore heterotrophic biomass cannot be neglected due to their possible fast growth rate.

Hao and Van Loosdrecht (2004) determined that increasing COD loading requires a higher DO concentration for optimal nitrogen removal and that the COD removal in the biofilm is mainly accomplished by oxidation, not by denitrification. However, if the biofilm is thick enough and the DO concentration is optimized, even with a COD/N ratio of 10, the 90% of nitrogen can be removed mainly by anammox activity.

Lackner *et al.* (2007) studied with a model the effect of the COD concentrations on the operation with perturbations consisting of introducing COD in the influent in co- and counter-diffusion biofilms. They conclude that counter-diffusion biofilms are more sensitive since perturbations in the influent COD/N ratio of 2 or higher for more than 50 d caused a loss of anammox capacity, which cannot be recovered within a reasonable time frame. The reason was that the heterotrophs physically displace anammox from the outer zone of the biofilm, whereas in co-diffusion heterotrophs grow more in the aerobic zone.

4.7.2.3. Sloughing

A punctual increase in shear stress (e.g. caused by a change in hydrodynamic conditions) in the biofilm-bulk interface can cause sloughing events. Anammox activity in a biofilm requires a very stable biofilm process. If regular sloughing of the biofilm occurs, slow growing anammox bacteria have no chance to establish in a biofilm system making either impossible or too long the process start-up. Once the biofilm developed, simulations performed by Lackner *et al.* (2007) indicated that the recovery after a sloughing event is possible even with 90–95% of biofilm loss. However, the recovery is significantly faster in a co-diffusion biofilm than in counter-diffusion.

Although quite an amount of research has been performed in the field of the anammox process further work is necessary in order to establish the optimal conditions for the start up and operation of industrial scale anammox reactors.

REFERENCES

Ahn, Y.H., Hwang, I.S. and Min, K.S. (2004) Anammox and partial denitritification in anaerobic nitrogen removal from piggery waste. *Water Sci. Technol.* **49**(5–6), 145–153.

Arrigo, K.R. (2005) Marine microorganisms and global nutrient cycles. *Nature* **437**, 349–355.

Arrojo, B., Mosquera-Corral, A., Campos, J.L. and Méndez, R. (2006) Effects of mechanical stress on Anammox granules in a sequencing batch reactor (SBR). *J. Biotechnol.* **123**, 453–463.

Arrojo, B., Figueroa M., Mosquera-Corral, A., Campos, J.L. and Méndez, R. (2008) Influence of gas flow-induced shear stress on the operation of the Anammox process in a SBR. *Chemosphere* **72**, 1687–1693.

Broda, E. (1977) Two kinds of lithotrophs missing in nature. *Z. Allg. Mikrobiol.* **17**, 123–143.

Buys, B.R., Mosquera-Corral, A., Sánchez, M. and Méndez, R. (2000) Development and application of a denitrification test based on gas production. *Water Sci. Technol.* **41**(12), 113–120.

Cema, G., Wiszniowski, J., Zabczyński, S., Zabłocka-Godlewska, E., Raszka, A. and Surmacz-Górska, J. (2007) Biological nitrogen removal from landfill leachate by deammonification assisted by heterotrophic denitrification in a rotating biological contactor (RBC). *Water Sci. Technol.* **55**(8–9), 35–42.

Dalsgaard, T., Canfield, D.E., Petersen, J., Thamdrup, B. and Acuña-González, J. (2003) N2 production by the Anammox reaction in the anoxic water column of Golfo Dulce, Costa Rica. *Nature* **422**, 606–608.

Dapena-Mora, A., Arrojo, B., Campos, J.L., Mosquera-Corral, A. and Méndez, R. (2004a) Improvement of the settling properties of Anammox sludge in an SBR. *J. Chem. Technol. Biotechnol.* **79**, 1417–1420.

Dapena-Mora, A., Van Hulle, S.W.H., Campos, J.L., Mendez, R., Vanrolleghem, P.A. and Jetten, M.S.M. (2004b) Enrichment of Anammox biomass from municipal activated sludge: experimental and modelling results. *J. Chem. Technol. Biotechnol.* **79**, 1421–1428.

Dapena-Mora, A., Fernández, I., Campos, J.L., Mosquera-Corral, A., Méndez, R. and Jetten, M.S.M. (2007) Evaluation of activity and inhibition effects on Anammox process by batch tests based on the nitrogen gas production. *Enzyme Microb. Technol.* **40**, 859–865.

Dosta, J., Fernández, I., Vázquez-Padín, J.R., Mosquera-Corral, A., Campos, J.L., Mata-Álvarez J. and Méndez R. (2008) Short- and long-term effects of temperature on the Anammox process. *J. Hazard. Mater.* **154**, 688–693.

Egli, K., Franger, U., Álvarez, P.J.J., Siegrist, H., Vandermeer, J.R. and Zehnder, A.J.B. (2001) Enrichment and characterization of an anammox bacterium from a rotating biological contractor treating ammonium-rich leachate. *Arch. Microbiol.* **175**, 198–207.

Egli, K. (2003) On the use of anammox in treating ammonium-rich wastewater. PhD Thesis, Swiss Federal Institute of Technology, Zurich, DISS. ETH No. 14886.

Fernández, I., Mosquera-Corral, A., Campos, J.L. and Méndez, R. (2006) Effect of broad-spectrum antibiotics on Anammox granular biomass. In Proc. of the International Water Conference – IWC 2006, Porto, 12–14 June.

Fernández, I., Vázquez-Padín, J.R., Mosquera-Corral, A., Campos, J.L. and Méndez, R. (2008a) Biofilm and granular systems to improve Anammox biomass retention. *Biochem. Eng. J.*, in press. (doi: 10.1016/j.bej.2008.07.011).

Fernández, I., Dosta, J., Campos, J.L., Mosquera-Corral, A. and Méndez, R. (2008b) Short- and long-term effects of ammonia and nitrite on the Anammox process. In Proc. of the Third International Meeting on Environmental Biotechnology and Engineering, Palma de Mallorca, 21–25 September.

Ficara, E., Rocco, A. and Rozzi, A. (2000) Determination of nitrification kinetics by the ANITA-Dostat biosensor. *Water Sci. Technol.* **41**(12), 121–128.

Fuerst, J.A. (2005) Intracellular compartmentation in planctomycetes. *Annu. Rev. Microbiol.* **59**, 299–328.

Fujii, T., Sugino, H., Rouse, J.D. and Furukawa, K. (2002) Characterization of the Microbial Community in an Anaerobic Ammonium-Oxidizing Biofilm Cultured on a Nonwoven Biomass Carrier. *J. Biosci. Bioeng.* **94**, 412–418.

Fux, C., Böhler, M., Huber, P., Bruner, I. and Siegrist, H. (2002) Biological Treatment of ammonium-rich wastewater by partial nitritation and subsequent anaerobic ammonium oxidation (anammox) in a pilot plant. *J. Biotechnol.* **99**, 295–306.

Güven D., Dapena A., Kartal B., Schmid M., Maas B., Méndez R., Op den Camp H.J.M., Jetten M.S.M., Strous M. and Schmid I. (2005) Propionate Oxidation by and methanol inhibition of anaerobic ammonium-oxidizing bacteria. *Appl. Environ. Microbiol.* **71**, 1066–1071.

Hao, X., Heijnen, J.J. and van Loosdrecht, M.C.M. (2002a) Model-based evaluation of temperature and inflow variations on a partial nitrification-ANAMMOX biofilm process. *Water Res.* **36**, 4839–4849.

Hao, X., Heijnen, J.J. and van Loosdrecht, M.C.M. (2002b) Sensitivity analysis of a biofilm model describing a one-stage completely autotrophic nitrogen removal (CANON) process. *Biotechnol. Bioeng.* **77**, 266–277.

Hao, X. and van Loosdrecht M.C.M. (2004) Model-based evaluation of the influence of COD on a partial nitrification-Anammox biofilm (CANON) process. *Water Sci. Technol.* **49**(11–12), 83–90.

Hao, X.D., Cao, X.Q., Picioreanu, C. and van Loosdrecht, M.C.M. (2005) Model-based evaluation of oxygen consumption in a partial nitrification–Anammox biofilm process. *Water Sci. Technol.* **52**(7), 155–160.

Hellinga, C., Schellen, A.A.J.C., Mulder, J.W., van Loosdrecht, M.C.M. and Heijnen, J.J. (1998) The SHARON process: an innovate method for nitrogen removal from ammonium-rich waste water. *Water Sci. Technol.* **37**(9), 135–142.

Helmer, C., Tromm, C., Hippen, A., Rosenwickel, K.H., Seyfried, C.F. and Kunst, S. (2001) Single stage biological nitrogen removal by nitritification and anerobic ammonium oxidation in biofilm systems. *Water Sci. Technol.* **43**(1), 311–320.

Henze, M., Gujer, W., Mino, T. and van Loosdrecht, M.C.M. (2000) Activated sludge models; ASM1, ASM2, ASM2d and ASM3, IWA Publishing, London.

Hippen, A., Rosenwickel, K.H., Baumgarten, G. and Seyfried C.F. (1997) Aerobic deammonification: a new experience in the treatment of wastewaters. *Water Sci. Technol.* **35**(10), 111–120.

Isaka, K., Sumino, T. and Tsuneda, S. (2007) High nitrogen removal performance at moderately low temperature utilizing anaerobic ammonium oxidation reactions. *J. Biosci. Bioeng.* **103**(5), 486–490.

Jetten, M.S.M., Strous, M., van de Pas-Schoonen, K.T., Schalk, J., van Dongen, L., van de Graaf, A.A., Logemann, S., Muyzer, G., van Loosdrecht, M.C.M. and Kuenen, J.G., (1999) The anaerobic oxidation of ammonium. *FEMS Microbiol. Rev.* **22**, 421–437.

Kartal, B., Rattray, J., van Niftrik, L.A., van de Vossenberg, J., Schmid, M.C., Webb, R.I., Schouten, S., Fuerst, J.A., Damste, J.S., Jetten, M.S.M. and Strous, M. (2007) *Candidatus* Anammoxoglobus propionicus a new propionate oxidizing species of anaerobic ammonium oxidizing bacteria. *Syst. Appl. Microbiol.* **30**, 39–49.

Kartal, B., Van Niftrik, L., Rattray, J., Van De Vossenberg, J.L.C.M., Schmid, M.C., Sinninghe-Damsté, J., Jetten, M.S.M. and Strous, M. (2008) *Candidatus* 'Brocadia fulgida': An autofluorescent anaerobic ammonium oxidizing bacterium *FEMS Microbiol. Ecol.* **63**, 46–55.

Koch, G., Egli, K., Van der Meer, J.R. and Siegriest, H. (2000) Mathematical modeling of autotrophic denitrification in a nitrifying biofilm of a rotating biological contactor. *Water Sci. Technol.* **41**(4–5), 191–198.

Kuai, L. and Verstraete, W. (1998) Ammonium removal by the oxygen-limited auto-trophic nitrification-denitrification system. *Appl. Environ. Microbiol.* **64**, 4500–4506.

Kuenen, G. (2008) Anammox bacteria: from discovery to application. *Nat. Rev. Microbiol.* **6**, 320–326.

Kuypers, M.M.M., Sliekers, A.O., Lavik, G., Schmid, M., Jørgensen, B.B., Kuenen, J.G., Sinninghe-Damste, J.S., Strous, M. and Jetten, M.S.M. (2003) Anaerobic ammonium oxidation by Anammox bacteria in the Black Sea. *Nature* **422**, 608–611.

Lackner, S., Terada, A. and Smets, B.F. (2007) Heterotrophic activity compromises autotrophic nitrogen removal in membrane-aerated biofilms: Results of a modeling study. *Water Res.* **42**, 1102–1112.

Larsen, L.H., Kjær, T. and Revsbech, N.P. (1997) A microscale NO_3^- biosensor for environmental applications. *Anal. Chem.* **69**, 3527–3531.

Lindsay, M.R., Webb, R.I., Strous, M., Jetten, M.S., Butler, M.K., Forde, R.J. and Fuerst, J.A. (2001) Cell compartmentalisation in planctomycetes: novel types of structural organisation for the bacterial cell. *Arch. Microbiol.* **175**, 413–429.

Mascitti, V. and Corey, E.J. (2004) Total synthesis of ($\pm$)-pentacycloanammoxic acid. *J. Am. Chem. Soc.* **126**, 15664–15665.

Morgenroth, E., Eberl, H.J., van Loosdrecht, M.C.M., Noguera, D.R., Pizarro, G.E., Picioreanu,, C., Rittmann, B.E., Schwarz, A.O. and Wanner, O. (2004) Comparing biofilm models for a single species biofilm system. *Water Sci. Technol.* **49**(11–12), 145–154.

Mulder, A. (1992) Anoxic Ammonium Oxidation US patent 427849 (5078884).

Mulder, A., Van de Graff, A.A., Robertson, L.A. and Kuenen, J.G. (1995) Anaerobic ammonium oxidation discovered in a denitrifying fluidized bed reactor. *FEMS Microbiol. Ecol.* **16**, 177–184.

Pynaert, K., Smets, B.F., Wyffels, S., Beheydt, D., Siciliano, S.D. and Verstraete, W. (2003) Characterization of an autotrophic nitrogen removing biofilm from a highly loaded lab-scale rotating biological contactor. *Appl. Environ. Microbiol.* **69**, 3626–3635.

Pynaert, K., Smets, B.F., Beheydt, D. and Verstraete, W. (2004). Start-up of autotrophic nitrogen removal reactors via sequential biocatalyst addition. *Environ. Sci. Technol.* **38**, 1228–1235.

Revsbech, N.P., Kjær, T., Damgaard, L. and Larsen, L.H. (2000). Biosensors for analysis of water, sludge, and sediments with emphasis on microscale biosensors. In J. Buffle & G. Horvai (eds.) In situ monitoring of aquatic systems: Chemical analysis and speciation. Wiley.

Schalk, J., Oustad, H., Kuenen, J.G. and Jetten, M.S.M. (1998) The anaerobic oxidation of hydrazine: a novel reaction in microbial nitrogen metabolism. *FEMS Microbiol. Lett.* **158**, 61–67.

Schalk, J., De Vries, S., Kuenen, J.G. and Jetten, M.S.M. (2000) Involvement of a novel hydroxylamine oxidoreductase in anaerobic ammonium oxidation. *Biochemistry* **39**, 5405–5412.

Schmid, M., Twachtmann, U., Klein, M., Strous, M., Juretschko, S., Jetten, M., Metzger, J, Schleifer, K.H. and Wagner, M. (2000) Molecular evidence for genus level diversity of bacteria capable of catalyzing anaerobic ammonium oxidation. *Sys. Appl. Microbiol.* **23**, 93–106.

Schmid, M., Walsh, K., Webb, R., Rijpstra, W.I.C., van de Pas-Schoonen, K.T., Verbruggen, M.J., Hill, T., Moffett, B., Fuerst, J., Schouten, S., Sinninge-Damste, J.S., Harris, J., Shaw, P., Jetten, M.S.M. and Strous, M. (2003) Candidatus "Scalindua brodae", sp. nov., Candidatus "Scalindua wagneri", sp. nov., two new species of anaerobic ammonium oxidizing bacteria. *Syst. Appl. Microbiol.* **26**, 529–538.

Schmid, M.C., Maas, B., Dapena, A., van de Pas-Schoonen, K., van de Vossenberg, J., Kartal, B., van Niftrik, L., Schmidt, I., Cirpus, I., Kuenen, G., Wagner, M., Damste, J.S.S., Kuypers, M., Revsbech, N.P., Mendez, R., Jetten, M.S.M. and Strous, M. (2005) Biomarkers for In Situ Detection of Anaerobic Ammonium-Oxidizing (Anammox) Bacteria. *Appl. Environ. Microbiol.* **71**, 1677–1684.

Siegrist, H., Reithaar, S., Koch, G. and Lais, P. (1998) Nitrogen loss in a nitrifying rotating contactor treating ammonium-rich wastewater without organic carbon. *Water Sci. Technol.* **38**(8–9), 241–248.

Sinninghe-Damste, J.S., Strous, M., Rijpstra, W.I.C., Hopmans, E.C., Geenevasen, J.A.J., van Duin, A.C.T., van Niftrik, L.A. and Jetten, M.S.M. (2002) Linearly concatenated cyclobutane lipids form a dense bacterial membrane. *Nature* **419**, 708–712.

Sliekers, O., Derwort, N., Campos-Gomez, J.L., Strous, M., Kuenen, J.G. and Jetten, M.S.M. (2002) Completely autotrophic nitrogen removal over nitrite in a single reactor. *Water Res.* **36**, 2475–2482.

Sliekers, A.O., Third, K.A., Abma, W., Kuenen, J.G. and Jetten, M.S.M. (2003) CANON and Anammox in a gas-lift reactor. *FEMS Microbiol. Lett.* **218**, 339–344.

Soto, M., Méndez, R. and Lema, J.M. (1993) Methanogenic and non-methanogenic activity test. Theoretical basis and experimental set up. *Water Res.* **27**, 1361–1376.

Strous, M., van Gerven, E., Kuenen, J.G. and Jetten, M. (1997) Effects of aerobic and microaerobic conditions an anaerobic ammonium-oxidizing (Anammox) sludge. *Appl. Environ. Microbiol.* **63**(6), 2446–2448.

Strous, M., Heijnen, J.J., Kuenen, J.G. and Jetten, M. (1998) The sequencing batch reactor as a powerful tool for the study of slowly growing anaerobic ammonium-oxidizing microorganisms. *Appl. Microbiol. Biotechnol.* **50**, 589–596.

Strous, M., Kuenen, J.G. and Jetten M.S.M. (1999) Key Physiology of Anaerobic Ammonium Oxidation. *Appl. Environment. Microbiol.* **65**(7), 3248–3250.

Strous, M. (2000) Microbiology of anaerobic ammonium oxidation. PhD Thesis, Technical University of Delft, The Netherlands.

Strous, M., Pelletier, E., Mangenot, S., Rattei, T., Lehner, A., Taylor, M.W., et al. (2006) Deciphering the evolution and metabolism of an anammox bacterium from a community genome. *Nature* **440**, 790–794.

Terada, A., Lackner, S., Tsuneda S. and Smets, B.F. (2007) Redox-stratification controlled biofilm (ReSCoBi) for completely autotrophic nitrogen removal: the effect of co-versus counter-diffusion on reactor performance. *Biotechnol. Bioeng.* **97**, 40–51.

Thamdrup, B. and Dalsgaard, T. (2002) Production of N2 through anaerobic and ammonium oxidation coupled to nitrate reduction in marine sediments. *Appl. Environ. Microbiol.* **68**, 1312–1318.

Toh, S.K., Webb, R.I. and Ashbolt, N.J. (2002) Enrichment of autotrophic anaerobic ammonium oxidizing consortia from various wastewaters. *Microb. Ecol.* **43**, 154–167.

Toh, S.K. and Ashbolt, N.J. (2002) Adaptation of anaerobic ammonium-oxidising consortium to synthetic coke-ovens wastewater. *Appl. Microbiol. Biotechnol.* **59**, 344–352.

Trimmer, M., Nicholls, J.C. and Deflandre, B. (2003) Anaerobic ammonium oxidation measured in sediments along the Thames estuary, United Kingdom. *Appl. Environ. Microbiol.*, **69**, 6447–6454.

Van de Graaf, A.A., Mulder, P., de Bruijn, P., Jetten, M.S.M., Robertson L.A. and Kuenen J.G. (1995) Anaerobic oxidation of ammonium is a biologically mediated process. *Appl. Environ. Microbiol.* **61**, 1246–1251.

Van de Graaf, A.A., de Bruijn P., Robertson L.A., Jetten M.S.M. and Kuenen J.G. (1996) Autotrophic growth of anaerobic ammonium-oxidizing micro-organisms in a fluidized bed reactor. *Microbiology* (UK) **142**, 2187–2196.

Van de Graaf A.A., de Bruijn, P., Robertson, L.A., Jetten, M.S.M. and Kuenen, J.G. (1997) Metabolic pathway of anaerobic ammonium oxidation on basis of 15N-studies in a fluidized bed reactor. *Microbiology* (UK) **143**, 2415–2421.

Van der Star, W.R.L., Miclea, A.I., van Dongen, U.G.J.M., Muyzer, G., Picioreanu, C. and van Loosdrecht, M.C.M. (2008) The Membrane Bioreactor: A Novel Tool to Grow Anammox Bacteria as Free Cells. *Biotechnol. Bioeng.* **101**, 286–294.

Van Dongen, U., Jetten, M.S.M. and van Loosdrecht, M.C.M. (2001) The SHARON®-Anammox® process for treatment of ammonium rich wastewater. *Water Sci. Technol.* **44**(1), 153–160.

Van Loosdrecht, M.C.M. and Henze, M. (1999) Maintenance, endogenous respiration, lysis, decay and predation, *Water Sci. Technol.* **39**(1) 107–117.

Van Niftrik, L.A., Fuerst, J.A., Sinninghe-Damsté, J.S., Kuenen, J.G., Jetten, M.S.M. and Strous, M. (2004) The anammoxosome: an intracytoplasmic compartment in anammox bacteria. *FEMS Microbiol. Lett.* **233**, 7–13.

5

Environmental technologies to remove nitrogen from municipal wastewaters

M. Spérandio and N. Bernet

5.1 INTRODUCTION

Domestic wastewater contains a large variety of organic pollutants and nutrients. Nitrogen is mainly present as ammonia form (60–70%), soluble and particulate organic nitrogen (urea, proteins) and seldom as nitrate and nitrite (Table 5.1). Conventional biological systems aim to remove nitrogen by means of hydrolysis of organic nitrogen, nitrification of ammonia, and denitrification of nitrate. In the following sections, the main biological treatment processes applied for nitrogen removal from domestic wastewater will be described.

Table 5.1. Characteristics of typical municipal wastewaters (concentrations in mg/L) (Henze 2002)

	Concentrated	Moderate	Diluted	Very diluted
COD	740	530	320	210
BOD$_5$	350	250	150	100
TKN	80	50	30	20
N-NH$_4^+$	50	30	18	12
N-NO$_3^-$	0.5	0.5	0.5	0.5
N-NO$_2^-$	0.1	0.1	0.1	0.1
SS (COD)	440	320	190	130
Total phosphorus	14	10	6	4

TKN: Total Kjeldhal Nitrogen, N-NH$_4^+$: Ammonium nitrogen, N-NO$_x$: Nitrate and nitrite nitrogen, COD: Chemical Oxygen Demand, SS: Suspended Solids, TP: Total Phosphorus, BOD: Biochemical Oxygen Demand.

5.2 SUSPENDED BIOMASS PROCESSES

5.2.1 Reactor configurations and treatment concepts

5.2.1.1 Multi tank activated sludge process

Activated sludge process is commonly designed and operated to achieve nitrification and denitrification in a single sludge system with mixed autotrophic and heterotrophic bacteria. Performing nitrification and denitrification with the same biomass needs to alternate aerobic and anoxic conditions, in single or multi-reactor systems (Figure 5.1).

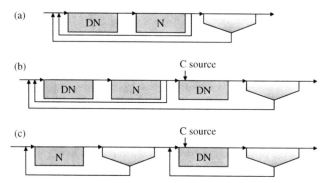

Figure 5.1. Configurations of activated sludge process for nitrification/denitrification (N/DN). a: Modified Ludzack Ettinger system, b: pre- and post-denitrification system, c: separated sludge system

The main configuration is the two stage activated sludge process (Modified Ludzack Ettinger) which combines an anoxic bioreactor followed by an aerobic bioreactor and a secondary clarifier (Figure 5.1). Sludge is pumped from the bottom of the clarifier to the anoxic stage with a flow of 50% to 100% of the influent flow. In addition, an internal recirculation of mixed liquor is imposed from the aerobic to the anoxic reactor for recycling nitrate produced by nitrification to the anoxic pre-denitrification zone. Basically, this system is optimal if organic matter for denitrification is provided by the wastewater as for example in domestic sewage.

However, this configuration leads to a residual concentration of nitrate in the effluent function of the ratio of internal nitrate recycling.

For reaching a very low nitrate concentration, an extra post-denitrification zone can be used. In this case denitrification takes place at a low endogenous rate. A secondary readily biodegradable carbon source (e.g., methanol) can be used for obtaining higher denitrification rate and reaching the desired low nitrate concentrations (Figure 5.1b).

In some cases, separated sludge systems can be chosen with a nitrifying activated sludge process dissociated from the denitrifying process. This last process is especially adapted to wastewater with very low COD/N ratio (Figure 5.1c).

Basically, selection of a biological nutrient removal (BNR) process configuration depends on the wastewater characteristics (nitrogen forms, biodegradability of organic matter, COD/N ratio), the effluent requirement and the volume constraints. Pre-denitrification reactor is preferable if the biodegradable COD of wastewater is valuable for denitrification of nitrate produced by nitrification. It means that the denitrification rate obtained with the biodegradable organic matter of influent should be significantly higher than an endogenous denitrification rate. When effluent COD is unfavorable for pre-denitrification, or if the COD/N ratio is too low for a complete denitrification, post-denitrification reactor is necessary.

5.2.1.2 Alternated aerated activated sludge

A single reactor with alternated aeration by switching on/off aerators is a convenient technique for nitrogen removal in activated sludge process. This process is commonly applied for nitrification/denitrification of domestic sewage designed with a very low loading rate (or extended aeration system).

Optimal control of aerobic and anoxic phase duration is possible with basic sensors for measuring DO, pH, or redox potential. Both phases can be controlled in order to maximize the nitrogen removal rate of the nitrification and denitrification. pH gives a clear indication of nitrification when decreasing and indicates denitrification when increasing (pH signal also increases with CO_2 stripping when ammonia is depleted). It is then possible to switch from aerobic to anoxic period when a minimal threshold value is reached ("valley" detection). This phenomenon is especially significant in the case of bicarbonate limited wastewater. DO and redox potential also show rapid drops and bending points when ammonia and nitrate are depleted, and then are commonly used for optimal control of nitrification and denitrification (Wareham *et al.* 1993; Mauret *et al.* 2001). Adaptation of aerobic and anoxic period time allows warranting maximal N removal despite instability of nitrification and denitrification rate as well as wastewater characteristics.

5.2.1.3 Sequencing batch reactors

Sequencing batch reactor for treating domestic wastewater is applied for nitrification, nitrification/denitrification or only denitrification. The major advantage of this process is the design simplicity, due to the absence of a separate clarifier. This process generally allows reaching low nitrogen concentration in the effluent and is known to produce sludge with very good settling properties.

A normal SBR cycle consists in the following phases: wastewater feeding, an anoxic phase, an aerobic phase, a settling period and effluent withdraw (Figure 5.2). Operation of three or four reactors in parallel allows avoiding hydraulic equalizing tank. In certain case, SBR processes can be operated in step-feed mode with a given number of feeding period, and successive anoxic and aerobic period, followed by the settling and the withdrawing. Sludge is generally wasted after settling period. Total cycle time is variable, commonly about 5 to 10 h, depending on the hydraulic retention time, the concentration of influent and the settling capacities.

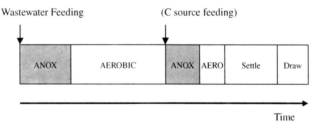

Figure 5.2. Example of sequential phases for SBR operation with advanced nitrogen removal. Second anoxic phase can be improved by a secondary organic carbon source. Last aerobic phase allows nitrogen gas to be stripped before settling

5.2.2 Processes and design criteria

5.2.2.1 Nitrification and aerobic volume

The principal design criteria for nitrogen removal in activated sludge process is the sludge retention time (SRT) and the aerobic volume fraction of the process ($f_{aerobic}$). The product of these parameters is named the aerobic sludge age ($SRT_{aerobic}$). This parameter must be chosen properly in order to maintain aerobic autotrophic bacteria responsible for nitrification, considering the minimal annual temperature of the process.

Aerobic sludge retention time is defined the same for total biomass (X in SS) and for a specific functional group as autotrophic nitrifiers:

$$SRT_{aerobic} = \frac{V_{aerobic} \cdot X}{Q_W \cdot X_W} = \frac{V_{aerobic} \cdot X_{B,A}}{Q_W \cdot X_{B,A,W}} \qquad (5.1)$$

$V_{aerobic}$ = aerobic volume of bioreactor (m^3)
X = total biomass concentration in bioreactor (Mixed Liquor Suspended Solids) (kg/m^3)
Q_W = wasted sludge flow rate (m^3/d)
X_W = wasted sludge concentration (SS) (kg/m^3)
$X_{B,A}$ = autotrophic nitrifying biomass concentration in reactor (kg/m^3)
$X_{B,A,W}$ = wasted sludge autotrophic biomass concentration (kg/m^3)

From a mass balance on autotrophic biomass (neglecting influent biomass), it comes that the aerobic sludge retention time (SRT) is directly linked to the observed growth rate ($\mu_{A,obs}$) by the following expression:

$$SRT_{aerobic} = \frac{X_{B,A}}{dX_{B,A}/dt} = \frac{1}{\mu_{A,obs}} \qquad (d) \qquad (5.2)$$

From Table 2.1, observed autotrophic growth rate is expressed:

$$\mu_{A,obs} = \mu_{mA} \frac{S_{NH}}{S_{NH} + K_{NH}} \frac{S_O}{S_O + K_{OA}} - b_A \qquad (1/d) \qquad (5.3)$$

Knowing the maximal growth rate (μ_{mA}) for a given temperature, it is possible to find the necessary sludge age and consequently the aerobic volume for maintaining nitrifying biomass in the process.

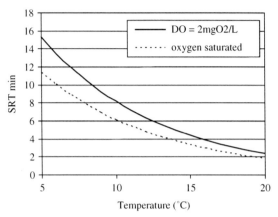

Figure 5.3. Theoretical minimal aerobic sludge age as a function of temperature for conventional nitrifying biomass

$$SRT_{aerobic,MIN} = \frac{1}{\mu_{mA} - b_A} \qquad (d) \qquad\qquad (5.4)$$

This value can be estimated including the effect of dissolved oxygen concentration maintained in the bioreactor (Figure 5.3).

Obviously, the maximal growth rate of a nitrifying biomass can be reduced by the presence of inhibiting compounds and this aspect should be evaluated for industrial wastewaters.

The minimal aerobic SRT is temperature dependent (Figure 5.3), and the chosen SRT is generally at least 2 or 3 times higher than the minimal value (Metcalf and Eddy 1991). This leads to ensuring a good stabilization of nitrifying biomass despite variation of influent load. Moreover, it allows reaching a low ammonia concentration in the effluent of the process. In steady state condition, the effluent concentration in ammonia is obtained from (Equation 5.3).

Example 5.1: *A domestic wastewater treatment plant designed for nitrification is operated at 12°C. The minimal aerobic sludge age for maintaining nitrifying biomass is 4.5 days (equation 5.4 with parameters of Table 2.2). If the system is totally aerobic with a dissolved oxygen concentration of 2 mgO₂/L and designed at a sludge retention time of 10 days, the mean ammonia concentration in the treated effluent is estimated at 1.7 mgN-NH₃/L (equations 5.2 and 5.3 with parameters from Table 2.2).*

Then the size of the reactor for activated sludge system treating domestic wastewater is generally deduced from the relation between sludge age, and sludge loading (B_X).

$$1/SRT = Y_H \cdot B_X - k_d \qquad (5.5)$$

Y_H = biomass yield (kgVSS/kgCOD)
k_d = decay rate (d^{-1})
B_X = sludge loading (kgCOD/kgVSS.d)

The parameters Y_H and k_d used in the expression 5.5 are global parameters for the total suspended solids production and decay. They are attributed to heterotrophic biomass because it composes the major fraction of the sludge.

In the previous example the sludge age of 10 days commonly results in a design at a sludge loading of 0.32–0.35 kgCOD/kgVSS.d (0.1–0.14 kgBOD$_5$/kgSS.d). Volume of the process is obtained from the following expression:

$$V = \frac{Q \cdot (S_{IN} - S_{OUT})}{B_X \cdot X} \qquad (5.6)$$

S_{IN} = inlet COD concentration
S_{OUT} = effluent COD concentration

This method is subjected to inaccuracy if the wastewater characteristics are not conventional. For example for very low COD/N ratio, it could be necessary to estimate the fraction of nitrifying biomass in the total mixed liquor (Metcalf and Eddy, 1991).

Example 5.2: *Design of an aerobic nitrifying activated sludge treating a conventional domestic wastewater (COD = 750 mgO$_2$/L, TKN = 80 mgN/L) at flow rate of Q = 1000 m^3/day, leads to the following results: V$_{aerobic}$ = 610 m^3*

With a chosen SRT of 12 days, assuming COD removal close to 90% (commonly observed for low loaded processes) with a steady state MLSS chosen at 3.5 kgVSS/m^3 (with Y$_H$ = 0.44 kgVSS/kgCOD and biomass decay rate k$_d$ = 0.056 d^{-1}).

5.2.2.2 Oxygen demand

Calculation of oxygen demand (F_{O2}) for nitrification is based on the estimated nitrified nitrogen (F_N) flux over the process and the amount of oxygen needed to oxidize 1 kg ammonia nitrogen (4.57 kgO$_2$/kg NH$_3$-N):

$$F_{O2} = 4.57 \cdot F_N \qquad (5.7)$$

Nitrified nitrogen is obtained with the influent nitrogen after retaining biomass assimilation (12% of the volatile suspended solids):

$$F_N = Q \cdot S_{N,IN} - Q \cdot S_{N,OUT} - 0.12 P_X \qquad (5.8)$$

F_N = total nitrified nitrogen flux (kg/d)
$S_{N,IN}$ = concentration of Total Kjeldahl nitrogen in the wastewater (kg/m^3)
$S_{N,OUT}$ = ammonia concentration in the effluent (kg/m^3)
P_X = total sludge production (kgVSS/d)
The sludge production is commonly estimated from the COD (or BOD) load and the observed yield which is linked to the sludge age:

$$P_X = Y_{OBS} \cdot Q \cdot (S_{IN} - S_{OUT}) \qquad (5.9)$$

Y_{OBS} = heterotrophic observed sludge yield (kgVSS/kgCOD)

5.2.2.3 Effluent denitrification

The main parameters, which should be considered for the design of denitrifying system are the COD/N ratio (stoichiometry) and the specific denitrification rate (kinetic). Because denitrification is catalyzed by heterotrophic bacteria, organic carbon is necessary as an electron donor. In anoxic bioreactor, organic carbon will be used by heterotrophic biomass with simultaneous respiration on nitrate or nitrite. A fraction of organic matter is converted to biomass (growth) resulting in a sludge production, whereas another fraction is oxidized to carbon dioxide, giving electrons for the reduction of nitrate to dinitrogen gas.

Increasing the sludge age leads to increase in anoxic endogenous activity and reduces sludge production. This contributes to denitrification (at a low rate) without external carbon source. In other words, increase of sludge age allows decreasing the necessary organic matter consumption for denitrifying a given amount of nitrate.

It is then possible to evaluate the stoichiometry for denitrification over the plant considering the observed sludge production yield (for a purely anoxic system):

$$\Delta COD / \Delta N = \frac{2.86}{1 - 1.42 Y_{OBS}} \qquad (5.10)$$

2.86 = oxygen equivalent reduced for one mass of nitrogen as nitrate reduced to dinitrogen (gO$_2$/gN)
1.42 = COD equivalent for a mass of biomass (gO$_2$/gVSS)
In single sludge process one should consider the fraction of COD removed aerobically, which is a loss of organic matter for denitrification. As a consequence in practice, the COD/N ratio is always higher than value given by the equation 5.10.

Example 5.3: *For a domestic wastewater treatment plant operated at a low-loaded condition ($Y_{OBS} = 0.28$ kgVSS/kgCOD), theoretical ratio needed for total denitrification is 4.8 kgCOD/kgN-NO_3 from equation 5.10, but the real ratio becomes 9.6 kgCOD/kgN-NO_3 if the bioreactor is alternatively aerobic and anoxic 50% of the time.*

Moreover, a fraction of COD should be considered as not biodegradable (or inert) in domestic wastewater (around 20%) with the conventional retention time of activated sludge plant. Consequently, for this case, total denitrification is feasible for wastewater with COD/N ratio higher than 12.

This calculation illustrates the importance of maximizing the anoxic use of organic matter in a nitrification-denitrification. This is reinforced by the fact that nitrate use for COD removal allows oxygen economy. The reduction of oxygen consumption due to denitrification is directly obtained from oxidation-reduction equivalent: 2.86 kg of oxygen/kg N-NO_3.

Calculation of anoxic volume for denitrification (V_{DN}) is obtained from mass balance introducing a specific denitrification rate (q_{DN}). Denitrification rate has been measured for different organic carbon sources (Table 5.2).

Nitrate balance over the anoxic reactor in steady state condition gives:

$$Q \cdot S_{NO3,IN} = q_{DN} \cdot X \cdot V_{DN} + Q \cdot S_{NO3,OUT} \tag{5.11}$$

$S_{NO3,IN}$ = influent nitrate concentration
q_{DN} = specific denitrification rate
$S_{NO3,OUT}$ = desired effluent nitrate concentration
$S_{NO3,IN}$ = influent nitrate concentration
X = mixed liquor volatile suspended solid concentration

Example of MLE process: Design of an anoxic/aerobic activated sludge process with two separate zones is proposed (Figure 5.4). Aerobic volume was obtained from previous calculation with a nitrifying biomass at 12°C.

Table 5.2. Heterotrophic denitrification rate with different carbon sources

Carbon source	Specific denitrification rate q_{DN} (20°C) mgN $- NO_3^-$/gVSS $\cdot$ h
Easily biodegradable (methanol, ethanol, VFA, simple carbohydrates)	10 ± 5
Organic matter of domestic wastewater	3 ± 1
Endogenous activity	$0.2 - 1$

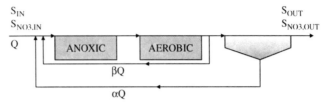

Figure 5.4. Flow sheet of a conventional two zone nitrification/denitrification activated sludge process

Firstly, the recirculation ratio $(\alpha + \beta)$ should be determined based on the nitrified nitrogen and considering the desired effluent nitrate concentration:

$$F_N = (\alpha + \beta + 1)QS_{NO3,OUT} \qquad (5.12)$$

Secondly, calculation of the anoxic reactor is given by a nitrate balance on the first reactor:

$$(\alpha + \beta)QS_{NO3,OUT} + QS_{NO3,IN} = q_{DN}V_{ANOXIC}X \qquad (5.13)$$

5.3 BIOFILM PROCESSES

Biofilm processes are increasingly being favored instead of activated sludge processes for several reasons, such as (Ødegaard 2006):

- Less space is required.
- Biomass separation has less influence on the final treatment result
- The attached biomass becomes more specialized at a given point in the process train because there is no biomass return.

5.3.1 Reactor configurations and treatment concepts

The treatment concepts are the same as those applied in suspended biomass processes: nitrification and denitrification. However, some of the biofilm processes are specifically aerobic (trickling filters) and cannot be used alone for an efficient nitrogen removal.

5.3.1.1 Trickling filters

This is a very old low technology process since the first trickling filters were built for sewage treatment in the late 1800. The reactor is filled with a mineral or plastic medium. The wastewater to be treated is distributed over the top of the filter, trickles down over the media and is collected at the bottom. There is no active aeration of the filter which is passively ventilated thanks to the variation in temperature between filter medium, wastewater and the surrounding air (Harremoës and Henze 2002).

Traditionally used for the oxidation of biodegradable organic matter, interest in the use of trickling filters for both carbon oxidation and nitrification has increased for about 20 years (Daigger *et al.* 1994). To achieve complete nitrification, effluent recirculation and/or the use of two filters in series is usually required (Pearce 2004). To achieve a complete nitrogen removal, recirculation of trickling filter effluent via a denitrification step is needed. However, significant amounts of nitrogen removal (30–50%) can be achieved on conventional trickling filters because of denitrification in anoxic zones, which reduces the denitrification capacity required in the additional process (Pearce 2004).

5.3.1.2 Rotating Biological Contactors (RBC)

Rotating biological contactors (RBC) consist of a set of closely spaced disks mounted on a horizontal rotating shaft. The disks are partially immersed in a tank through which wastewater flows continuously. A biofilm grows on the disks. As the disk rotate, the biofilm is exposed to air and oxygen is transferred (Patwardhan 2003).

Like trickling filters, RBCs have been designed for carbon removal but they are able to produce a fully nitrified effluent, either in a combined oxidation and nitrification with an RBC unit or using a tertiary nitrifying RBC (Halling-Sørensen 1993). Recirculation and step-feed in a three stage system were shown to improve carbon and ammonia removal (Ayoub and Saikaly 2004).

Hanhan *et al.* (2005) showed the denitrification potential of RBC for treatment of municipal wastewater.

5.3.1.3 Submerged biofilters

Submerged biofilters also called biological aerated filters (BAF) are filters where a fixed media is packed and continuously submerged in wastewater. The biomass grows on the carrier material of normally about 2 up to 8 mm grain size, depending on the application (Rother and Cornel 2004). Biofiltration serves two purposes (Pujol *et al.* 1994):

1. Biological conversion of dissolved organic matter and nitrogen by the biofilm
2. Physical retention of suspended particles by filtration through the filter bed

Differences in technology between the different competing companies on the market are mainly based on the carrier material (density lower or higher than water), flow direction (up- or down-flow) and the backwash procedure (Rother and Cornel 2004).

The key-points of this process are the control of the biofilm thickness to avoid clogging and the supply of oxygen in aerobic filters (Harremoës and Henze 2002). To control biofilm thickness, backwashing is included as a part of the normal operation of the process. Moreover, advanced pre-treatment (coagulation and primary settling) are recommended for minimizing periodicity of back-washes. In this case, a secondary carbon source as methanol becomes necessary for denitrification.

The same process configurations as the ones used with activated sludge systems can be applied including pre- and post-denitrification. However, most full operation scale operated upflow BAFs in advanced municipal wastewater treatment are secondary nitrification stages after an activated sludge process or tertiary post denitrification filters with external carbon dosage (Rother et al. 2002).

5.3.1.4 Moving-Bed Bioreactors (MBBR)

The MBBR was developed in Norway in the late 1980s and early 1990s. The biomass grows as a biofilm on plastic carriers that move freely in the water volume of the reactor (Ødegaard 2006). Nitrogen removal in MBBR plants may be achieved by several combinations such as pre-denitrification, post-denitrification or a combination of the two, like in a conventional activated sludge system. High nitrification and denitrification rates have been demonstrated, even at low temperatures.

Introduction of the same kind plastic carrier in activated sludge is a useful option to upgrade the process for enhanced nitrogen removal (Christenson and Welander 2004).

5.3.1.5 Hybrid processes

Hybrid processes combine suspended biomass and biofilms in the same reactor. This is an economically attractive solution, when activated sludge processes need to be upgraded for enhanced nitrogen removal. Nitrification takes place mainly in the biofilm whereas BOD removal takes place in the suspended sludge. Since long sludge age for nitrification is kept in the biofilm, the sludge age of the activated sludge can be kept short, making it possible to obtain nitrogen removal in much smaller volumes than for a conventional activated sludge plant (Christensson and Welander 2004). The slow growing organisms that would be washed out without the carrier are maintained in the reactor.

Initially, hybrid processes mainly used immobile carriers (Chudoba and Pannier 1994, Su and Ouyang 1996, Müller 1998). More recently, mobile plastic carriers, similar to those used in MMBR, have been used (Andreottola *et al.* 2000; Christensson and Welander 2004).

5.3.2 Design criteria

Contrary to activated sludge process, for which design criteria for the treatment of municipal wastewater are well defined, the rules for the design of biofilm processes for nitrogen removal may be very different depending on the reference used (Harremoës and Henze 2002). This is due to the complexity of these systems in terms of kinetics, hydrodynamics and mass transfer.

5.3.2.1 Trickling filters

The design of trickling filters should incorporate four critical parameters: hydraulic loading, influent nitrogen, effects of recycling and wastewater temperature (Gullicks and Cleasby 1993). Pearce (2004) especially focused on hydraulic loading.

In the case of combined carbon oxidation-nitrification, Parker and Richards (1986) showed that nitrification could be initiated only if soluble BOD_5 concentrations are less than 20 mg/L in the filter. This is due to competition between heterotrophic bacteria and nitrifiers. Effluent recycling will cause denitrification to occur.

The maximum removal rate, regardless of loading rate, is 3 g $NH_4/m^2 \cdot d$, and availability of oxygen is considered as the limiting factor (Okey and Albertson 1989).

Pearce (2004) recommended a maximum load of 0.15 kg $BOD/m^3 \cdot d$ and a filter depth of 1.5–1.8 m. From results obtained on 9 full scale plants, he obtained the following regression for TN removal from primary effluent:

$$\%TN \text{ removal} = 0.644 + (4.3Bv) - (0.25r/q) - (0.076iv)$$
$$+ (4.96TONv) - (6.74TKNv) - (0.072g) \qquad (5.14)$$

where

Bv = volumetric BOD load (kg/m³ filter volume.d)   Range 0.18–0.02
r/q = recirculation flowrate/feed flowrate   Range 0–0.85
iv = irrigation velocity (m³total flow/m² plan area.day) Range 1.4–4.2
TONv = volumetric oxidized nitrogen loading   Range 0–0.034
    (kg TON/m³ filter volume.day)
TKNv = Total Kjeldahl nitrogen loading   Range 0.03–0.06
    (kg TKN/m³ filter volume.day)
g = nominal media size (mm)

5.3.2.2 Rotating Biological Contactors (RBC)

To get nitrification, flow rates should be as low as 0.03 to 0.08 $m^3/m^2 \cdot d$, which is about one third of the capacity of an RBC when applied only to carbon oxidation (Halling-Sørensen 1993).

The nitrification rate for the RBC fed with domestic wastewater ranges between 1.69 and 2.56 g $N/m^2 \cdot d$.

The following equation can be used to calculate the ammonium removal rate (Halling-Sørensen 1993):

$$Q \cdot (S_{NH,IN} - S_{NH,OUT}) = r_{S,NH4} \cdot A' \qquad (5.15)$$

where

Q = influent wastewater flow-rate (m^3/d)
A' = total disk area (m^3)
$r_{s,NH4}$ = ammonium removal rate (g/$m^2 \cdot$d)

Tertiary nitrification in a trickling filter can give ammonia removal of 80% at a loading rate of 1.5 g NH_4-$N/m^2 \cdot$d (Degrémont 2005).

5.3.2.3 Submerged biofilters

The main advantage of biofiltration compared to conventional activated sludge processes are the high biomass content and the high volumetric reaction rates of the systems, resulting in small footprints (Rother and Cornel 2004). For full treatment of municipal wastewater with C, N and eventually P removal about 40–60 liters of reactor volume/p.e. are necessary, compared to 250–380 l/p.e. for the single stage activated sludge process according to German design standards (ATV-DVWK cited by Rother and Cornel 2004). However, the necessity for backwashes requires equipment, energy and water (15–30% of treated water) and is the cause of a discontinuous operation.

The following elements have to be considered for the design of biofilters for nitrogen removal when compared to the activated sludge process (Rother and Cornel 2004):

- high oxygen concentrations (>5–6 mg/l) for high nitrification rates,
- because of the pretreatment needed to limit clogging, internal carbon sources limited for pre-denitrification,
- relatively higher specific energy consumption due to hydraulic losses in filter bed, backwashes and higher DO levels.

When used for simultaneous carbon removal and nitrification, only partial nitrification can be achieved (<70%) with actual ammonia loads of 0.4–0.6 kgN/$m^3 \cdot$d. (Degrémont 2005).

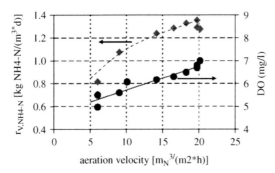

Figure 5.5. Nitrification rate r_V and DO in supernatant as a function of aeration velocity ($v_W = 8.5$ m/h, T = 18–20°C, carrier: $CaCO_3$. 1.6–2.8 mm, flocculated raw wastewater, composite samples) (Rother and Cornel 2004)

If higher performance is needed (80 to 95%), tertiary nitrification must be used. Loads as high as 1.4 kg $N-NH_4/m^3 \cdot$ d have been reported (Pujol *et al.* 1994). The hydraulic retention time, even when very short (10 minutes), does not influence the reactor performance since the highest nitrification rates are obtained at high water velocities (10 $m^3/m^2 \cdot$ h) (Pujol *et al.* 1998). The performances have been shown to be clearly related to the aeration velocity which affects the DO concentration in supernatant of the filter (Figure 5.5). The air needed for nitrification in an up flow biofilter on floating media was estimated to be about 70 N m^3 of air per kg NH_4-N removed (Payraudeau *et al.* 2000).

BAF operated without aeration can be used in pre- or post-denitrification combined with a nitrifying filter. The admissible loads given by Degrémont for their upflow biofiltration process (Biofor) are 1–1.2 kg $NO_3-N/m^3 \cdot$ d for predenitrification (Pujol and Tarallo 2000) and 3.5–5 kg $NO_3-N/m^3 \cdot$ d for post-denitrification with methanol (Degrémont 2005) whereas lower denitrification rates are given by Aesoy *et al.* (1998) in the same process at pilot-scale with ethanol (2.5 kg $NO_3-N/m^3 \cdot$ d) and acetic acid (<2 kg $NO_3-N/m^3 \cdot$ d) as an electron donor for denitrification.

Another alternative to remove nitrogen is to operate an upflow BAF with the aerators placed part way up the floating media to allow for both an aerated and a non aerated zone (Rogalla and Bourbigot 1990; Toettrup *et al.* 1994). More recently, simultaneous nitrification denitrification in an aerated filter was proposed, based on a real time aeration control (Puznava *et al.* 2001). To allow denitrification, the DO concentration in the bulk has to be maintained between 0.5 and 3 mg/l saving up to 50% aeration energy. Ammonia applied loads were as high as 1.8 kg $NH_4-N/m^3 \cdot$ d and the total residual nitrogen was below 20 mg/l.

A new control strategy proposed by Lemoine *et al.* (2006) allows 5% increase in nitrogen removal performance and a reduction of 15–20% in air requirement compared with the previous control.

5.3.2.4 Moving-Bed Bioreactors (MBBR)

In the case of nitrification, three factors primarily determine the nitrification rate: the load of organic matter, the ammonium concentration and the oxygen concentration (Ødegaard 2006). The influence of these parameters is shown in Figure 5.6.

Similar to other aerobic biofilm processes, oxygen concentration is the most important parameter. It can limit nitrification rate even at concentrations above 2–3 mg/l. Nitrification rates as high as 1.2 g NH_4-$N/m^2 \cdot$ d were obtained at low temperature (11°C) while denitrification rates were as high as 3.5 g NO_3-$N/m^2 \cdot$ d (Ødegaard 2006). In tertiary nitrification, the process was shown to be influenced by the temperature but only in oxygen limiting conditions (Salvetti *et al.* 2006).

Pre-denitrification is often limited by the carbon source available and is low. Post-denitrification has a very high rate due to the addition of an easily biodegradable source (Figure 5.7). Ødegaard (2006) recommends a combined pre- and post-denitrification process, which is superior in terms of process control and performance. Depending on the extent of pretreatment applied, the total HRT of the MBBR for nitrogen removal is in the range of 3 to 5 hours.

5.3.2.5 Hybrid processes

The design of hybrid processes includes the choice of the biofilm carrier material, the volume percentage of the biofilm carrier material with respect to

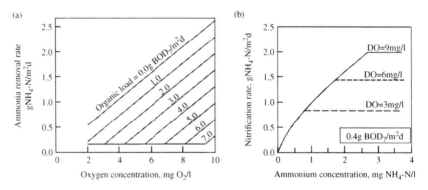

Figure 5.6. Influence of BOD_7, oxygen and ammonium on nitrification rate (Ødegaard 2006)

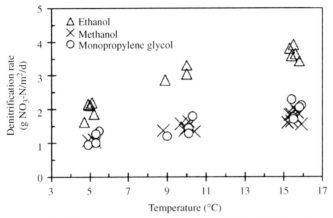

Figure 5.7. Denitrification rate versus temperature, obtained with various external carbon sources (Rusten *et al.* 2006)

the total reactor volume and the specific amount of biofilm biomass per unit of biofilm carrier.

In the case of submerged fixed bed devices, the reactor volume ratio filled with carriers usually ranges between 16% and 28% (Müller 1998). When a stable nitrification is obtained, a simultaneous nitrification-denitrification may be assumed in the deeper layer of the biofilm that can be thick. Thus, the anoxic zone might be smaller than in classical pre-denitrification of activated sludge processes (Müller 1998).

In the case of hybrid MBBRs, the filling fraction of carrier reported is higher, 43% (Christensson and Welander 2004; Falletti and Conte 2007) but lower than the maximum of 70% recommended by Ødegaard (2006) in MBBRs.

5.3.3 Case study: Frederikshavn central wastewater treatment plant (Denmark) (Thogersen and Hansen 2000)

In connection with the reconstruction and capacity extension of the Frederkshavn Central Wastewater Treatment Plant between 1989 and 1995, the existing activated sludge (AS) plant was supplemented with a biological aerated filter (BAF). The parallel operation of the two systems has made a comparison possible.

The Wastewater Treatment Plant has a design capacity of 16,500 m^3/d, or approximately 130,000 PE on average. The raw wastewater flows through a screen to the inlet well of the treatment plant and is pumped to the pre-treatment section, which consists of a combined grit and grease chamber and 3 primary rectangular settlers. The water then flows by gravity into a flow splitting chamber between the AS and the BAF.

The maximum hydraulic load on the pretreatment section as a whole is 1,530 m^3/h.

The backwash sludge from the BAF is returned to the primary settlers. Primary and secondary sludge are anaerobically digested, and the produced methane gas is used for generation of heat and power. In dry weather situations, the filters are washed once every 24 hours, which results in a total sludge flow of approximately 3,000 m^3/d.

Wastewater composition
Approximately 50% of the wastewater comes from the fish processing industry. The remaining 50% is a mixture of other industrial wastewater and domestic wastewater. The average composition of the wastewater after pre-treatment is given in Table 5.3.

Biological treatment
The activated sludge plant:

- Max. hydraulic load from pretreatment: 700 m^3/h
- Nitrification volume: 2,300 m^3
- Denitrification volume: 1,240 m^3
- Volume of clarifiers: 3,000 m^3
- Surface of clarifiers: 750 m^2

The BAF technology is an upflow floating filter with 3 mm polystyrene carrier beads (Biostyr process), that offers the possibility of aeration inside the media to separate aerobic and anoxic zones, as well as countercurrent down flow gravity flush for backwashing.

Table 5.3. Composition of the wastewater after pre-treatment (concentrations in mg/L)

COD	250	COD(f)	130
Total N	36	NH$_4$-N	22
Total P	4	PO$_4$-P	1.5
SS	92	BOD	200

- Max. hydraulic load from pretreatment: 830 m³/h
- Max. hydraulic capacity, wet weather: 3,600 m³/h
- Filter surface: 63 m²
- Nitrification volume (6 filters in recirculation): 794 m³
- Denitrification volume (6 filters in recirculation): 340 m³
- Total filter volume: 1,134 m³
- Max. surface load: 9.6 m³/m². h

The operation of the BAF can be combined in several ways:

1. Nitrification/denitrification in all filters
2. Nitrification/denitrification in 5 filters, post-denitrification in 1 filter. (Normal dry weather operation)
3. Bottom aeration with full nitrification in all filters. (Operation during maximum wet weather load)
4. Controlled oxygen set-point with simultaneous nitrification/denitrification in all filters.

Both processes are fully instrumented for measurement of DO, ammonia and nitrate. The measuring results are used partly for control of the oxygen set-point and partly for adding methanol for denitrification.

Comparison of AS and BAF performances

Table 5.4 gives the average diurnal load and the nitrogen and SS reduction in 1998. The performances are similar for TN loading rates of 0.08 kg TN/m³· d in the activated sludge and 0.36 kg TN/m³· d in the BAF.

Figure 5.8 shows the impact of temperature on nitrification activity in both systems. It appears that the BAF is less affected by winter temperatures than the activated sludge.

For the purpose of optimizing the operation, tests were conducted to increase the hydraulic capacity of the plant during rain and control strategies for the plants were considered for dry weather as well as wet weather situations. The paper discusses in details the experience gained from the plant operation as well as the

Table 5.4. Load and reduction in the BioStyr unit and the activated sludge unit respectively.

	Load TN kg/d	Reduction TN %	Load SS kg/d	Reduction SS %
BAF	404	88	998	94
Activated sludge	290	86	690	93

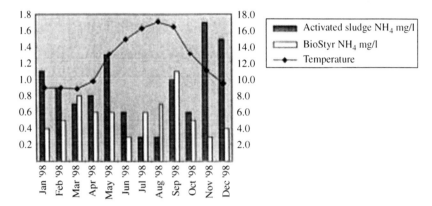

Figure 5.8. Nitrification as a function of temperature, monthly average (Thogersen and Hansen 2000)

considerations made in relation to future control strategies and possibilities of increasing the total hydraulic capacity during rain.

The main conclusions of Thogersen and Hansen (2000) are the following:

- The AS as well as the BAF are operationally reliable and easy to handle and maintain. The AS has a better reduction of phosphorus than the BAF, which, on the other hand, is less sensitive to fluctuations in both hydraulic load and substance load. The BAF also maintains a better nitrification than the activated sludge plant, particularly at low wastewater temperatures.
- Under comparable conditions, the operating costs of the units in Frederikshavn are almost similar in relation to price per m^3 treated wastewater and price per kg removed nitrogen. When including return on and depreciation of the construction costs it will, however, be to the advantage of the BioStyr unit.
- The implementation of an advanced control system in the AS, not presented here but in the paper (Thogersen and Hansen 2000), has contributed to an operational optimization and therefore considerable cost reductions. The implementation of an equivalent system at the BAF is expected to have the same effect and will also improve the effluent quality from the unit.
- In addition, a better and more expedient functioning during rain can be obtained by changing the control strategy for the entire plant, changing the post-denitrification filter at the BAF and returning the sludge liquor from the BAF to the AS.

REFERENCES

Aesoy, A., Odegaard, H., Bach, K., Pujol, R. and Hamon, M. (1998) Denitrification in a packed bed biofilm reactor (BIOFOR) - Experiments with different carbon sources Water Res. **32**, 1463–1470.

Andreottola, G., Foladori, P. and Ragazzi, M. (2000) Upgrading of a small wastewater treatment plant in a cold climate region using a moving bed biofilm reactor (MBBR) system. Water Sci. Technol. **41**(1), 177–185.

Ayoub, G.M. and Saikaly, P. (2004) The combined effect of step-feed and recycling on RBC performance. Water Res. **38**(13), 3009–3016.

Christensson, M. and Welander, T. (2004) Treatment of municipal wastewater in a hybrid process using a new suspended carrier with large surface area. Water Sci. Technol. **49**(11–12), 207–214.

Chudoba, P. and Pannier, M. (1994) Nitrification kinetics in activated sludge with both suspended and attached biomass. Water Sci. Technol. **29**(7), 181–184.

Daigger, G.T., Heinemann, T.A., Land, G. and Watson, R.S. (1994) Practical experience with combined carbon oxidation and nitrification in plastic media trickling filters. Water Sci. Technol. **29**(10–11), 189–196.

Degrémont (2005) Mémento technique de l'eau, Vol 2, 10th edition.

Falletti, L. and Conte, L. (2007) Upgrading of activated sludge wastewater treatment plants with hybrid moving-bed biofilm reactors. Ind. Eng. Chem. Res. **46**(21), 6656–6660.

Gullicks, H.A. and Cleasby, J.L. (1993) Design of trickling filter nitrification towers. J. Water Pollut. Control Fed. **58**(1), 60–67.

Halling-Sørensen, B. (1993) Attached growth reactors. In: The removal of nitrogen compounds from wastewater, Studies in Environmental Science 54, Elsevier, Amsterdam, London, New York, Tokyo.

Hanhan, O., Orhon, D., Krauth, K. and Gunder, B. (2005) Evaluation of denitrification potential of rotating biological contactors for treatment of municipal wastewater. Water Sci. Technol. **51**(11), 131–139.

Harremoës, P. and Henze, M. (2002) Biofilters. In: Wastewater treatment – Biological and chemical processes (eds. U. Förstner, R.J. Murphy and W.H. Rulkens). Springer-Verlag, Berlin, Heidelberg, New-York.

Henze, M. (2002) Wastewater, volumes and composition. In: Wastewater treatment – Biological and chemical processes (eds. U. Förstner, R.J. Murphy and W.H. Rulkens). Springer-Verlag, Berlin, Heidelberg, New-York.

Lemoine, C., Payraudeau, M. and Meinhold, J. (2006) Aeration control for simultaneous nitrification-denitrification in a biological aerated filter using internal model approach. Water Sci. Technol. **54**(8), 129–136.

Mauret, M., Ferrand, F., Boisdon, V., Spérandio, M. and Paul, E. (2001) Process using DO and ORP signals for biological nitrification and denitrification: validation of a food-processing industry wastewater treatment plant on boosting with pure oxygen. Water Sci. Technol. **44**(2–3), 163–170.

Metcalf and Eddy (1991) Wastewater Engineering, Treatment, Disposal, Reuse. 3rd Ed., McGraw-Hill, New-York.

Müller, N. (1998) Implementing biofilm carriers into activated sludge process - 15 years of experience. Water Sci. Technol. **37**(9), 167–174.

Ødegaard, H. (2006) Innovations in wastewater treatment: the moving bed biofilm process. *Water Sci. Technol.* **53**(9), 17–33.

Okey, R.W. and Albertson, O.E. (1989) Evidence for oxygen-limiting conditions during tertiary fixed-film nitrification. *J. Water Pollut. Control Fed.* **61**, 510–519.

Parker, D.S. and Richards, T. (1986) Nitrification in trickling filters. *J. Water Pollut. Control Fed.* **58**(9), 896–902.

Patwardhan, A.W. (2003) Rotating biological contactors: A review. *Ind. Eng. Chem. Res.* **42**(10), 2035–2051.

Payraudeau, M., Paffoni, C. and Gousailles, M. (2000) Tertiary nitrification in an up flow biofilter on floating media: influence of temperature and COD load. *Water Sci. Technol.* **41**(4–5), 21–27.

Pearce, P. (2004) Trickling filters for upgrading low technology wastewater plants for nitrogen removal. *Water Sci. Technol.* **49**(11–12), 47–52.

Pujol, R., Hamon, M., Kandel, X. and Lemmel, H. (1994) Biofilters: flexible, reliable biological reactors. *Water Sci. Technol.* **29**(10–11), 33–38.

Pujol, R., Lemmel, H. and Gousailles, M. (1998) A keypoint of nitrification in an upflow biofiltration reactor. *Water Sci. Technol.* **38**(3), 43–49.

Pujol, R. and Tarallo, S. (2000) Total nitrogen removal in two-step biofiltration. *Water Sci. Technol.* **41**(4–5), 65–68.

Puznava, N., Payraudeau, M. and Thornberg, D. (2001) Simultaneous nitrification and denitrification in biofilters with real time aeration control. *Water Sci. Technol.* **43**(1), 269–276.

Rogalla, F. and Bourbigot, M.M. (1990) New developments in complete nitrogen removal with biological aerated filters. *Water Sci. Technol.* **22**(1/2), 273–280.

Rother, E., Cornel, P., Ante, A., Kleinert, P. and Brambach, R. (2002) Comparison of combined and separated biological aerated filter (BAF) performance for pre-denitrifcation/nitrification of municipal wastewater. *Water Sci. Technol.* **46**(4–5), 149–156.

Rother, E. and Cornel, P. (2004) Optimising design, operation and energy consumption of biological aerated filters (BAF) for nitrogen removal of municipal wastewater. *Water Sci. Technol.* **50**(6), 131–139.

Rusten, B., Eikebrokk, B., Ulgenes, Y. and Lygren, E. (2006) Design and operations of the Kaldnes moving bed biofilm reactors. *Aquacult. Eng.* **34**(3), 322–331.

Salvetti, R., Azzellino, A., Canziani, R. and Bonomo, L. (2006) Effects of temperature on tertiary nitrification in moving-bed biofilm reactors. *Water Res.* **40**(15), 2981–2993.

Su, J.L. and Ouyang, C.F. (1996). Nutrient removal using a combined process with activated sludge and fixed biofilm. *Water Sci. Technol.* **34**(1–2), 477–486.

Thogersen, T. and Hansen, R. (2000) Full scale parallel operation of a biological aerated filter (BAF) and activated sludge (AS) for nitrogen removal. *Water Sci. Technol.* **41**(4–5), 159–168.

Toettrup, H., Rogalla, F., Vidal, A. and Harremoës, P. (1994) The treatment trilogy of floating filters: from pilot to prototype to plant. *Water Sci. Technol.* **29**(10–11), 23–32.

Wareham, D.G., Hall, K.J. and Mavinic, D.S. (1993) Real-time control of aerobic-anoxic sludge digestion ORP. *J. Environ. Eng. ASCE* **119**(1), 120–136.

6

Environmental technologies to remove nitrogen from high-strength wastewaters

B. Wett, N. Jardin and D. Katehis

6.1 INTRODUCTION

Talking about high-strength wastewaters in the context of nitrogen removal, we usually consider ammonia concentrations in the hundreds or even thousands of mg N/L. Typically we see such high nitrogen concentration in industrial effluents, in sludge liquors returning from anaerobic digesters and in landfill leachates. In most cases, these high-strength nitrogen-containing streams derive from processes with elevated temperatures. Both conditions –high concentration and temperature– bear the potential for advantageous and efficient removal technologies, which are hardly applicable for low-strength streams: A good reason to dedicate a separate chapter to this issue.

© 2009 IWA Publishing. *Environmental Technologies to Treat Nitrogen Pollution: Principles and Engineering*, Edited by Francisco J. Cervantes. ISBN: 9781843392224. Published by IWA Publishing, London, UK.

What is the source of concentrated ammonia streams generated from digestion processes? Nitrogen contained in organic matter, specifically in proteins, gets released when organic compounds are degraded. Under aerobic conditions, a substantial portion of the organic feed serves as a substrate for cell growth and the available nitrogen again becomes incorporated into cell proteins. Therefore, no significant change in nitrogen compounds in the bulk liquid is observed. Different from anaerobic digestion processes, where mainly nitrogen-free products are formed (CH_4, CO_2, H_2, H_2S) and only a small biomass yield is achieved. Consequently, most of the bounded nitrogen becomes released as ammonia (Figure 6.1) and is object to consecutive nitrogen conversion routes.

Considering the sludge train of a wastewater treatment plant, nitrogen release can be tracked back: Solids separated by different unit processes are usually mixed before anaerobic stabilisation and digestion parameters (e.g. biodegradability and nitrogen content) represent mean values of the composite sludge. Separate digestion of primary and secondary sludge converts substrates with significantly different values in terms of bounded nitrogen and possible volatile suspended solids (VSS) reduction. Figure 6.2 compares typical initial values of VSS- and nitrogen content of primary and waste activated sludge and the shift of these parameters during digestion (Wett *et al.* 2006b). Obviously, waste activated sludge contains less organic matter with less availability to anaerobic metabolism than primary sludge. Due to the higher initial portion of immobilised nitrogen (ca. 0.05 g N/g TSS comparing to 0.03 g N/g TSS total suspended solids) the release of nitrogen in the digester is still higher for secondary sludge (ca. 0.025 g N/g TSS comparing to 0.015 g N/g TSS).

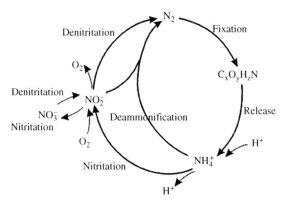

Figure 6.1. Nitrogen cycle showing N-fixation, ammonia release (e.g. in the course of digestion) and metabolic conversion routes back to N_2

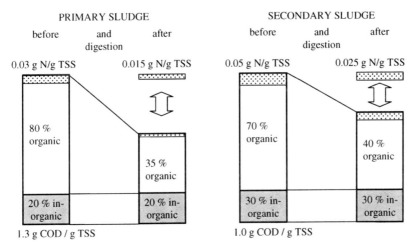

Figure 6.2. Release of nitrogen bounded in primary and secondary sludge (values related to initial total suspended solids TSS in digester feed)

6.2 TYPES OF WASTEWATER

6.2.1 Reject water – sludge liquor from mesophilic digestion

The flow-scheme of a wastewater treatment plant (WWTP, Figure 6.3) displays the interactions of the liquid and the solids train, which are characterised by relatively low hydraulic flows but high nitrogen fluxes. About 40% of the influent nitrogen load is transferred to the thickening units. There, the amount of flow is reduced from about 2–5% of the influent flow down to about 1%. The difference – the thickener overflow representing about 3% – is recycled to the mainstream and shows similar ammonia concentrations as the influent flow. The solids stream (about 1% of the liquid flow) is fed to the digesters and afterwards the high ammonia concentrated bulk liquid is separated from digested solids.

Therefore, 2 return loads can be distinguished:

- Thickener over-flow: Diluted stream at ambient temperature representing only about 1–2% of the plant's total nitrogen load, which is not suitable for separate treatment.
- Reject water from sludge dewatering: High-strength ammonia stream at mesophilic temperature representing 12–25% of the influent load, which is suitable for separate treatment.

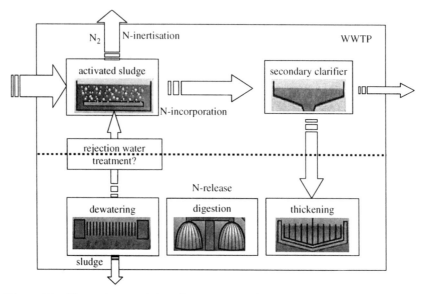

Figure 6.3. Flow-scheme of the nitrogen return load from the solids train to the liquid train of a WWTP

Data Analysis of WWTPs in Germany (Figure 6.4) yields a specific return load of 1.5 g N/PE. i.e. the average N-return load amounts to 15% of the influent load with 2-stage activated sludge plants usually producing ~20% higher internal loads. Ammonia concentrations in reject water typically are in the range of 700 to 2000 mg N/L and the molar ammonia/alkalinity ratio is about 1. The average ratio of chemical oxygen demand (COD) against nitrogen, COD/N, is about 0.5–2.0 with a low COD-biodegradability of only about 25%.

6.2.2 Liquors from (thermophilic) digestion of high solids streams

Wet fermentation processes, as performed in municipal WWTPs, are usually fed by pump-able solids streams (< 10% TSS) resulting in a reject water quality as discussed in section 6.2.1. In case of pre-treatment by sludge disintegration technologies, like thermal hydrolysis, higher solids concentrations can be achieved. This results not only in smaller digester volume requirements, but also in ammonia concentrations higher than 2500 mg/L (Kepp *et al.* 2000; Phothilangka *et al.* 2008). High ammonia level, in combination with high temperature and high pH, causes free ammonia inhibition on methanogenic

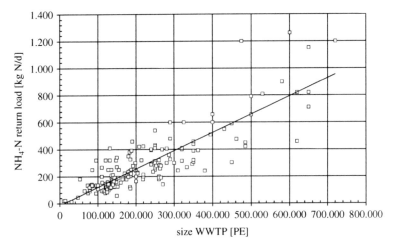

Figure 6.4. Ammonia return load at 204 WWTPs with anaerobic mesophilic digestion (Jardin *et al.* 2005; Person Equivalent (PE) represents a load equivalent of 60 g bio-chemical oxygen demand (BOD) per capita)

organisms. A shift in sludge liquor characteristics and growing concentrations of organic acids are the consequence of those inhibiting impacts.

Similar conditions can be observed in dry fermentation processes, especially of municipal biodegradable organic solids and food wastes. High concentrations of particulate and soluble COD, including a high portion of organic acids, mean an additional challenge to the centrate- or filtrate treatment, respectively.

6.2.3 Landfill leachates

Leachates from landfills or other organic piles represent filtrates from large deposit fermenters. Leachate characteristics are objects of significant timely developments. Depending on the age of the landfill, the initial leachate quality is similar as described in section 6.2.2 (ammonia NH_4-N $\approx$ 3–4000 mg/L; COD $\approx$ 10–20.000 mg/L; COD/BOD $\approx$ 2) and successively gets shifted towards properties shown in section 6.2.1 (NH_4-N $\approx$ 500–1.500 mg/L; COD up to 2–3.000 mg/L; COD/BOD $\approx$ 4). In chapter 8, specific focus is put on nitrogen removal during leachates treatment.

6.2.4 (Agro-) industrial effluents

Agro-industrial effluents, like piggery effluents or fishery discharges, frequently combine high concentrations both in ammonia and organic matter. Therefore, in most cases pre-treatment steps targeting on removal of organics are required before efficient nitrogen conversion processes are applicable.

Anaerobic treatment for carbon degradation – either conventional digestion or Upflow Anaerobic Sludge Blanket (UASB) – and decanting or filtration for solids separation seem the most appropriate technologies (Karakashev *et al.* 2008). Then, in a consecutive step, anaerobic ammonia oxidation becomes feasible. Alternatively, composting of the solid fraction of manure (Imbeah 1998) and liquid treatment in a pond system represents a simpler but less efficient approach.

6.3 REACTOR CONFIGURATIONS AND TREATMENT CONCEPTS

Treatment of the dewatering reject liquor load in the main plant will reduce the plant's overall C/N ratio and alkalinity available for nitrification, triggering the need for additional facilities, and operational complexity in the main plant. Treating this warm, high ammonia side-stream separately of the main plant flow can allow the plant to use these characteristics of the flow to remove the nitrogen load while minimizing electrical power and chemical usage. Beyond that, treating the side-stream separately can allow an increase in the main plant's alkalinity to influent nitrogen ratio and the carbon to influent nitrogen ratio, thereby reducing the need for supplemental alkalinity and carbon facilities in the main plant.

Introduction of side-stream treatment, with the associated reduction in power and chemical consumption, will reduce the carbon footprint of the facility, enhancing the sustainability of the overall wastewater treatment process. Passive strategies that target solely equalization of the nitrogen loadings can also provide for some process enhancements; however, it is difficult to evaluate improvements of the main plant performance and to access the significant operating cost savings that side-stream treatment allows.

Separate side-stream treatment processes for nitrogen removal can be generally grouped into two categories – physical-chemical- and biological treatment (Figure 6.5).

6.3.1 Physical-chemical side-stream treatment options

The major advantage of physical-chemical side-stream treatment options is based on the fact that nutrients are not recycled by an inertisation process to the atmosphere, but recovered in a bio-available form. At high pH-level, ammonia can be transferred into the gas phase and subsequently accumulated in an extremely

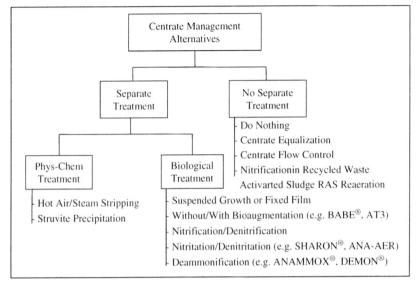

Figure 6.5. Overview of centrate treatment processes (modified from Constantine *et al.* 2005) as described in this chapter

concentrated solution. Another option –struvite precipitation– generates an even more valuable product containing both ammonia and phosphorus. Process details of ammonia stripping and struvite precipitation will be given in chapters 8 and 9, respectively.

6.3.2 Biological treatment options

Apart from the conventional nitrogen removal pathway, side-stream conditions open the possibility of more efficient process schemes (Table 6.1):

- N-removal directly via nitrite (**nitritation/denitritation**)
- Partial nitritation and anaerobic ammonia oxidation (**deammonification**)
- Nitritation/denitritation of centrate combined with augmentation of nitrifiers to the mainstream treatment (**bioaugmentation**)

Figure 6.6 describes the reaction steps performed by the listed N-removal processes. The objective of the control strategy of the **nitritation/denitritation** concept is a complete repression of the growth of nitrite oxidizing organisms

Table 6.1. Advantages and disadvantages of side-stream treatment options

	Advantages	Disadvantages
Nitritation/ Denitritation	Reduces loadings to main plant	Reactor cooling may be required
	Reduced energy (~30–40%) chemical (~40%) consumption conventional alternatives	External carbon is required
Deammonification	Reduces loadings to main plant Minimizes energy (~65–70%) and chemical (90–100%) consumption relative to conventional alternatives	Dependent on retention and growth of slow growing organisms Heating and cooling may be required
Bioaugmentation	Reduces loadings to main plant Uses centrate nitrogen load to fortify main plant BNR process	Reduced sidestream process reliability Tied in with main plant performance

(NOB) in order to decrease the required energy and carbon resources. Further oxidation of nitrite to nitrate has to be repressed by ammonia inhibition of NOBs (Turk and Mavinic 1987), high temperature and limited aerobic Sludge Retention Time (SRT) and Dissolved Oxygen (DO) concentration.

Deammonification is a two-part autotrophic reaction involving two distinct biomass populations. In the first step, aerobic ammonia oxidizing bacteria (AOB) partially nitrify ammonia to produce nitrite. In the second step, anaerobic ammonia oxidizing (anammox) microorganisms autotrophically denitrify these products to nitrogen gas. While high ammonia influent concentration facilitates optimised

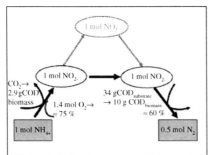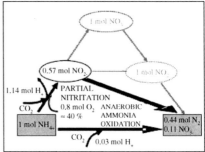

Figure 6.6. Shortcuts in metabolic pathways for nitrogen removal either by nitritation/denitritation (left) or deammonification (right) for savings in energy- and carbon resources

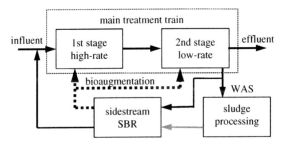

Figure 6.7. Flow scheme of bioaugmentation for 2-stage treatment plants

metabolic routing, accumulation of nitrite concentrations endangers process stability due to toxic impact on anammox organisms (Strous *et al.* 1999a).

The biochemical reactions underlying the **bioaugmentation** concept are identical to the ones of the nitritation/denitritation concept. The only difference is a return activated sludge or waste activated sludge (WAS) recycle strategy (Katehis *et al.* 2002; Salem *et al.* 2004; Parker and Wanner 2007) to transfer acclimated nitrifier populations between mainstream and side-stream treatment for enhanced nitrification capacity at limited SRT (Figure 6.7).

Stoichiometry

Stoichiometry of nitritation and denitritation can be described by following the equations, which show a 2-molar alkalinity requirement for oxidising ammonia to nitrite and a 1-molar alkalinity recovery from its reduction:

Nitritation: $2NH_4^+ + 3O_2 \rightarrow 2NO_2^- + 4H^+ + 2H_2O + \text{Energy}$
Denitritation: $2NO_2^- + 2H^+ + 1.5CH_2O \rightarrow 2N_2 + 1.5CO_2 + 2.5H_2O + \text{Energy}$

The existence of anaerobic ammonia oxidation in nature was unknown for a long time and only postulated by Broda (1977). Finally, Strous *et al.* (1999b) managed to identify the missing lithotroph as a new planctomycete, which catalyses the ANAMMOX process according to the following equation:

$$NH_4^+ + 1.32NO_2^- + 0.066HCO_3^- + 0.13H^+ \rightarrow$$
$$0.26NO_3^- + 1.02N_2 + 0.066CH_2O_{0.5}N_{0.15} + 2.03H_2O$$

Stoichiometric coefficients of this reaction have been derived in closer detail on base of an elemental balancing approach (Takacs *et al.* 2007).

These two reaction steps – aerobic and anaerobic ammonia oxidation – can be conducted in two individual units providing different SRT and conditions where nitrite produced in the aerobic reactor and residual or by-passed ammonia are fed to the anaerobic reactor (Van Dongen *et al.* 2001). In an alternative approach, both process steps are operated in a single-sludge system (see sections 6.4 and 6.5).

It is important to note that the alkalinity requirement for nitrogen removal is independent on the selection of the metabolic route. i.e. the available buffer capacity of the liquid allows certain nitrogen removal efficiency with the ion balance summing up to zero irrespective of metabolism (Figure 6.8). Process selection has a significant influence on the demand of resources like energy and carbon, but not on caustic. Usually, sludge liquors provide just enough alkalinity for almost complete removal efficiency while landfill leachates mostly show a surplus of alkalinity from other sources than ammonia release.

Kinetics

Understanding and controlling process rates is important for sizing equipment. However, determining process rates is a complex task. For example, the major variables that control the nitritation process rate include substrate concentrations (oxygen, ammonia and inorganic carbon) as well as process conditions (pH and temperature) as shown in the following equation:

Process rate for oxidation of ammonia (applied in Figure 6.9).

$$r = r_{max} * Ar^{T-20} \frac{S_O}{S_O + k_O} * \frac{k_{NH3}}{k_{NH3} + S_{NH3}} * \frac{\exp(S_{HCO3} - k_{HCO3})}{1 + \exp(S_{HCO3} - k_{HCO3})}$$

Figure 6.9 illustrates the sensitivity of autotrophic growth rates on individual operation parameters. Considering the overlapping impacts of e.g. DO, ammonia and bicarbonate the rate-difference between AOB- and NOB-growth is substantial and increases with temperature while uninhibited growth rates remain very close over the whole temperature range. The application of side-stream conditions and appropriate control performance can take advantage of such interactions for a reliable repression of the NOB population.

The complexity of the equation system involved makes a unique identification of kinetic parameters difficult, which results in high parameter uncertainty, specifically in high-strength N-removal processes or nitrite modeling, respectively,

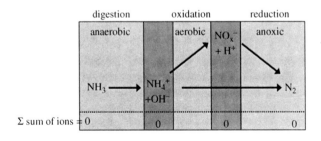

Figure 6.8. Alkalinity balance of nitrogen removal processes

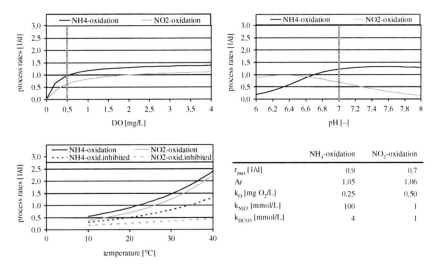

Figure 6.9. Variability of autotrophic growth rates: Individual impacts of DO and pH level (top) and superposition of all impacts (assuming DO and pH operation range indicated, ammonia at 100 mg/L and inorganic carbon at 7 mmol/L in the reactor and given kinetic parameters)

found in literature. Facing this enormous variability in kinetic model parameters in this field, Sin *et al.* (2008) suggest not varying core parameters like yield, growth and decay rates, but parameters reflecting operational conditions like inhibition and affinity coefficients.

Additional to the analysis of monitored data, also hardly measurable variables – process states – can be visualised. The composition of the biomass is of particular interest for a detailed investigation of the process behaviour. Due to the significant difference in SRT, different biomass characteristics can be seen: At a SRT of 10 days and due to carbon dosage, methanol utilising microorganisms make up about 56% of the biomass solids (Figure 6.10). In the deammonification system at a SRT of about 48 days, inert solids and endogenous products are the main fractions and autotrophs dominate the active biomass (Wett *et al.* 2008).

6.4 DESIGN CRITERIA

The optimised design of side-stream treatment systems requires efficient use of the reactor and equipment and a robust process control philosophy. Flexibility in the implementation of process configurations is key, as the needs of the WWTP hosting the side-stream reactor may change over time.

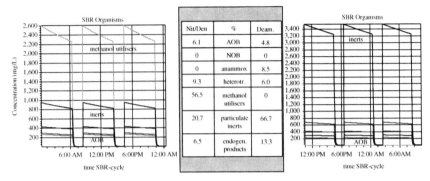

Figure 6.10. Comparison of simulated biomass composition in % of particulate COD during 3 SBR cycles at nitritation/denitritation mode at Alexandria Sanitation Authority (left; dominance of methylotrophs; SRT = 10 days; NH4-load = 0.51 g N/L.d) and deammonification mode (right; dominance of inert particulate COD; SRT = 48 days; NH4-load = 0.49 g N/L.d)

The key design considerations for side-stream treatment are:

1. Maximum volumetric loading that can be achieved in the system based on process rates,
2. Minimum SRT to reliably establish the required autotrophic organisms,
3. Maximum oxygen uptake rate (OUR), energy demand and chemical needs.

6.4.1 Design considerations for SBRs (suspended growth)

Sequencing batch reactor (SBR) systems are known for their flexibility allowing a sequence of process steps within a single tank volume (Artan *et al.* 2001; Hopkins *et al.* 2001). Side-stream treatment of high-strength digestion liquors represents a special process engineering field with proven SBR applicability. A survey of 50 existing side-stream treatment plants in Central Europe (Jardin *et al.* 2005) suggests that 16 of 42 biological systems use SBRs. Such a SBR system can provide environments for basically all 3 different biological side-stream process options discussed above. Table 6.2 gives key design parameters for these process options.

The differences between all 3 process options does not so much concern the reactor configuration itself, but the mode of operation. Repression of NOB as a common process goal is addressed by the following control features:

- High reactor temperature (>25°C, preferable >30°C).
- Low DO (e.g. DO level of 0.5 mg/L compromises sufficient process rates and NOB repression).

- Limited aeration time: Depending on the current load, aeration should be restricted by means of time- or pH-control, otherwise extended aeration allows complete oxidation to nitrate.

Table 6.2. Process design parameters for SBR side-stream treatment at temperatures higher than 30°C

sidestream option	volumetric loading [kg NH$_4$-N/m^3]	SRT [d]	HRT* [d]	sludge product.**(*) [kg TSS/kg N]	energy [kWh/kgN]	carbon*** [kgCOD/kgN]
nitritation/ denitritation	0.6	5–10	1.0–2.5	0.3–0.7	1.8–2.5	2.5–3.0
deammonification	0.6	20–30	1.0–2.5	0.1–0.3	0.8–1.3	0
bioaugmentation	0.6	2–5	1.0–2.5	–	2.5–3.0	0–2.0

depending on *ammonia and **TSS concentration in the sludge liquor and type of ***carbon source

It is notable (Table 6.2) that despite the huge range of target SRT of 3, 8 and 25 days for bioaugmentation, nitritation/denitritation and deammonification, respectively, the design calculations result in similar SBR volumes due to significant differences in sludge production.

Deammonification requires additional control tasks. Aerobic and anaerobic process steps need to be balanced in a way that accumulation of toxic nitrite concentrations is prevented (see case studies in section 6.5). Operational settings and appropriate decant equipment have to ensure reliable sludge retention in order to achieve target SRT.

In case of bioaugmentation, WAS is recycled to the side-stream system in order to enrich AOBs and to decrease temperature effects and activity losses when seeding back to the main plant. On the other hand, lay-out of the flow scheme needs to maintain advantageous side-stream conditions. The WAS stream recycled to the SBR for AOB enrichment shows a fourfold impact on the population dynamics in the reactor:

- Solids mass flow reduces SRT substantially.
- Recycled WAS represents a reasonable carbon source for denitritation.
- Blending WAS and SBR sludge usually means a reduction in the resulting temperature.
- Introduction of WAS biomass leads to intrusion of NOBs in the system.

Three out of the 4 mentioned impacts can limit or disturb the SBR performance and therefore, the WAS recycle rate is highly relevant for the design.

A typical control objective is to prevent a drop in reactor temperature below 25°C and to limit nitrate formation (NO_3/NO_x ratio) below 50% of oxidised ammonia. Depending on concentration and temperature of each individual stream, the WAS recycle rate is expected to be in the range of about 50% of the reject water flow (1:3 blend ratio). At high ambient temperatures, the controlled WAS recycling functions as an efficient reactor cooling measure.

6.4.2 Design of chemostats (suspended growth)

Chemostats are flow-through systems without sludge retention; i.e. the SRT equals the hydraulic retention time. The basic principle of design and operation of a chemostat for side-stream nitritation/denitritation (SHARON) relies on control of retention time in order to provide just sufficient time for AOBs to establish a stable population but still a limited period to out-compete NOBs (Van Kempen *et al.* 2001). This strategy means making optimum use of ammonia inhibition on NOBs without acclimation effects. Depending on reactor temperature, the typical retention time for SHARON-systems is 2 to 3 days.

The design of the heating- or cooling equipment, respectively, needs to take into account that additional heat is generated from biological processes. Assuming a perfectly insulated reactor, a temperature increase of 10°C would be experienced in case of an ammonia concentration of about 1000 mg N/L and a COD/N ratio of about 1 in the centrate. The exothermic nature of performed heterotrophic metabolism (about 60% of produced heat) shows more significant impact than the autotrophic one (about 40%).

As a specific feature of chemostats, these systems are not sensitive to a break-through of solids from dewatering units entering the reactor. Due to their flow-through characteristics, the mixed liquor solids (MLSS) concentration in the reactor develops according the influent concentration without any consequences for the reactor design. On the other hand, the lack of sludge retention and corresponding minimum SRT means no chance for adaptation of organisms to peak loads of any toxic compound.

6.4.3 Design of moving bed systems (attached growth)

Some of the first full scale implementations of the deammonification process were realised using fixed film systems. This was mainly due to the assumption that, in order to establish a significant mass of anammox organisms, it is of crucial importance to differentiate hydraulic retention time from the sludge age of the biomass, which is most easily achieved in fixed film systems. By using rotating contactors, deammonification was successfully established to treat landfill leachate in full scale (Helmer *et al.* 1999).

Besides rotating contactors, also moving bed systems have been used for the deammonification process in order to reduce the nitrogen load of return flow streams of municipal wastewater treatment plants. The initial idea in using a moving bed system for nitrogen elimination of high-strength wastewater was to establish the process in two steps (Rosenwinkel *et al.* 2005). The first moving bed reactor was dedicated for nitritation only and two consecutive reactors were designed for the anammox reaction. The challenging task was to restrict the first process step to the oxidation of ammonium only by using the classical principles of competitive low DO and elevated NH_3-concentrations in order to inhibit the growth of NOBs in the system. Although this strategy works well, it was not possible to suppress the growth of anammox organisms in the biofilm with the final result that a differentiation between NO_2-production in one reactor and the anammox reaction in another biofilm compartment was not achievable (case study in section 6.5.2).

Today, biofilm moving bed systems with the process aim of deammonification are designed most preferable as a two reactor in series system, in which in both reactors the concurrent reactions of NH_4-oxidation and anammox reaction take place. The following design guidelines will allow a stable and reliable operation of the process:

- Specific density of the carrier material slightly below the density of water, e.g. 0.98 kg l^{-1} for completely mixed systems.
- Slow speed vertically mounted mixers in order to avoid high shear stresses.
- Intermittent aeration to inhibit the growth of NOBs.
- Filling degree with carriers 40 to 60%.
- Specific nitrogen design load up to 0.75 kg N m^{-3} d^{-1} (60% filling degree).

6.5 CASE STUDIES

6.5.1 Case Study: DEMON® deammonification process

A control system for a single sludge SBR system has been developed in order to adjust 3 impacts – ammonia inhibition, nitrite toxicity and inorganic carbon limitation. The controller is based on three main mechanisms – time-, pH- and DO-control – listed according to their order of hierarchy:

- Time control defines operation cycles of e.g. 8 hours each, involving a fill/react phase, a settling period and a decanting period. During the react period of about 6 hours of the SBR cycle, both deammonification processes – partial nitritation and anaerobic ammonia oxidation – are operated.

- These two successive processes conversely impact pH. The partial nitritation reaction depresses the pH and the anaerobic ammonia oxidation reaction elevates the pH. The actual duration of aeration intervals are ruled by the pH-signal, which characterises the current state of reactions (ph-control).

- The set-point of DO control is specified at a low range, close to 0.3 mg/L in order to prevent rapid nitrite accumulation and to maintain a continuous repression of the second oxidation step of nitrite to nitrate.

Described deammonification process has been implemented at the WWTP Strass, Austria, in an SBR tank (Figure 6.11) with a maximum volume of 500 m^3 and at loading rates up to 340 kg NH$_4$-N/d. The aeration system is activated only within a very tight pH-bandwidth of 0.01. Due to oxygen input, nitritation runs at a higher rate than anaerobic ammonia oxidation and H$^+$ production drives the pH-value to the lower set-point and aeration stops. While DO is depleted, all the

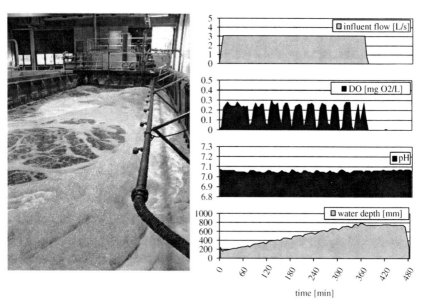

Figure 6.11. Profiles of process variables (flow rate, DO, pH and water level) of 1 SBR cycle displaying the control of intermittent aeration by a tight interval of pH-set-points (right) and photo of the full-scale DEMON-system at WWTP Strass (left)

nitrite that has been accumulated during the aeration interval is used for oxidising ammonia. Alkalinity recovers from this biochemical process and the continuous rejection water feed flow until the pH-value reaches the upper set-point and aeration is switched on again.

Probably the most impressive impact of the DEMON-system concerns the energy balance of the whole plant. From 1997 to 2004 a SBR-strategy for nitritation/denitritation was operated and primary sludge served as a carbon source. In 2004, the DEMON®-process for deammonification without any requirements of carbon was implemented (Wett 2006a). Since then, all solids have been fed to the digesters and, on the other hand, the specific oxygen demand for side-stream treatment was decreased from about 2.9 kWh/kg N to 1.1 kWh/kg N. The total benefit, in terms of savings in aeration energy and additional methane, sums up to about 12% of the plant-wide energy balance. The resulting net electrical energy production of the Strass plant amounts to 108% of the demand on site and the excess energy is fed to the grid (Figure 6.12).

6.5.2 Case Study: Biofilm deammonification (Hattingen)

In 2000, a full-scale plant for the treatment of process water, using a fixed film system, was built by the Ruhrverband corporation at the WWTP of Hattingen, Germany. After an intensive research phase focusing on nitritation and denitritation, deammonification was established on the WWTP in 2003.

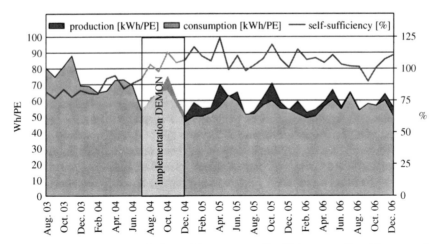

Figure 6.12. Percentage of plant-wide energy self-sufficiency as the difference of demand and production of electrical energy after implementation of deammonification at the Strass WWTP, Austria

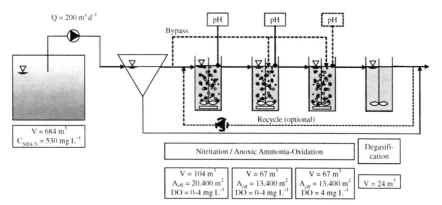

Figure 6.13. Flow sheet of the full-scale deammonification plant at Hattingen WWTP

During the last years, attention has been concentrated on process stabilization and optimization. The flow scheme of the full-scale plant is shown in Figure 6.13.

The three biological reactors, the first holding 104 m³ and originally designated for nitritation only, the second and third holding 67 m³ each and dedicated to the anammox process, may be either aerated or, using agitators, operated under anoxic conditions. A 24-m³-degassing basin is available to prevent the dispersal of oxygen in the recycle stream (optional). The plant is charged by external feed pumps from the sludge-liquor storage tank with a capacity of 684 m³ and, via external by-pass pipelines, the influent can be allocated to the three reactors.

As carriers needed to establish a biofilm, Kaldnes[R] Moving Bed material was used. It consists of polyethylene rings with a stable internal cross and with a density slightly lower than that of water, which, therefore, can be suspended in the wastewater to allow a perfect circulation. The plastic elements provide a specific surface area of no less than 750 m²/m³. With the biofilm mainly forming at the 'sheltered' inside, a value of 500 m²/m³ bulk volume can be assumed. To ensure adequate mixing of the Kaldnes[R] carrier elements also during deammonification under low oxygen concentrations, all three reactors are provided with both bubble aeration systems and agitators. Furthermore, each reactor is equipped with a pH-measuring and a dosing system for acid and caustic soda addition to control the pH-value, as well as with an oxygen probe to monitor the DO concentration in the biological processes.

The plant is fed from the front side, with a bypass of ~30% directed to the third reactor. As there is no need for denitritation, neither a recycle stream nor an external carbon source is required. The sludge liquor treatment plant has been designed to treat a partial stream of the daily sludge liquor influent, corresponding to an average wastewater flow rate of 200 m³/d and a nitrogen load of 120 kg/d.

After several attempts to establish a stable nitritation in the first reactor and to provide adequate substrate (nitrite and ammonium) for the anammox designated reactors 2 and 3, finally in all reactors a concurrent reaction of nitritation and anammox has been established. The overall performance of the system now operated mainly in parallel is shown in Figure 6.14, demonstrating that a stable nitrogen elimination of up to 85% can be achieved even at temperatures of 20°C.

6.5.3 Case Study: Bioaugmentation using the AT-3 Process

The process at New York City's 26th Ward WWTP intentionally uses free ammonia and free nitrous acid toxicity in the side-stream system to force nitritation and prevent the biological conversion of the reject water ammonia to nitrate. The side-stream reactor is operated aerobically at low DO and pH levels that enhance the preferential production of nitrite (Anthonisen 1973; Katehis 2003; Wett 2003). Nitrifying biomass produced in the side-stream reactor is used

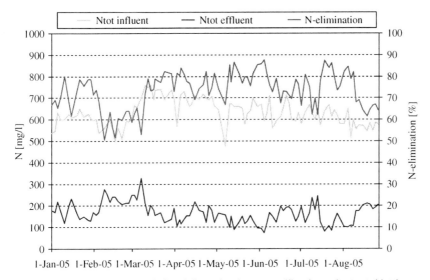

Figure 6.14. Performance of the full-scale deammonification plant at Hattingen WWTP

to seed the main plant resulting in nitrogen removals higher than 50% while operating the main plant at very low SRTs (1–3 days).

Utilisation of a plug flow reactor with the ability to introduce supplemental alkalinity in multiple locations along the length of the reactor and flexibility to modify the aerated fraction of the reactor in response to seasonal temperature changes and influent loading conditions is necessary to allow successful implementation of this process configuration. An internal recycle from the end to the head of the reactor is used to recycle nitrite rich flows, which in conjunction with operating pH set-points, allow for control of free ammonia levels within the bioaugmentation reactor.

The temperature in the reactor is a function of the temperatures and flows of the WAS and dewatering centrate introduced into the reactor and range for $17°C$ in the winter to $30°C$ in the summer. Minimal changes in temperature occur across the reactor length under typical full scale conditions. The operating pH of the reactor is maintained at above neutral levels (7.5–7.8) in the front portion of the reactor, where ammonia concentrations are high, to maintain elevated free unionized ammonia levels (NH_3). The pH is monitored at multiple points along the plug flow reactor and is allowed to drop to below neutral levels in the latter portion of the reactor, where nitrite levels are elevated, thereby resulting in elevated concentrations of free nitrous acid (HNO_2). The blend of recycled WAS and the biomass enriched under side-stream conditions with the combination of selective mechanisms applied results in the oxidation of 50–80% of the ammonia to nitrite, the balance being converted to nitrate.

Care is exercised to avoid a significant pH depression in the back end of the plug flow reactor as this leads to conversion of the residual carbonate alkalinity to CO_2 and thus loss of the CO_2 through stripping (Wett 2003), posing challenges to cost effective pH control within the reactor, using caustic soda (which is used at 26th Ward) or lime. However, the presence of the heterotrophic biomass, which continues to respire endogenously, helps provide a low level of CO_2 to the system, countering the stripping effect, and thus allow for enhanced stability of the process.

The treated centrate stream from the side-stream system, consisting primarily of ammonia and nitrite oxidising autotrophs, would then be used to seed a portion of the first stage aeration tanks (Constantine et al. 2005). Typical data from the NYC side-stream reactor at the 26th Ward Plant that has been operating since 1994 is shown in Figure 6.15.

A more detailed review of this process, including operating results, is provided in Katehis (2002). Molecular probing of the activated sludge from this type of process configuration has shown a reduction in diversity of the ammonia oxidising species in the activated sludge of the main plant reactor, when the

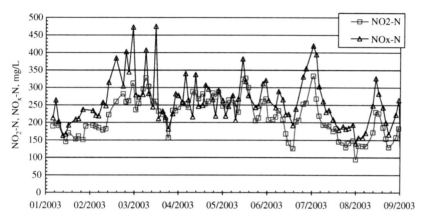

Figure 6.15. Preferential Nitritation in the Side-stream Reactor at the 26th Ward WPCP (Adapted from Fillos *et al.* 2005)

bioaugmentation process is operational. The Bioaugmentation processes using dewatering liquor can also be extended to the specialised heterotrophic biomass, such as denitrifying methanol degrading bacteria by integrating denitrification into the bioaugmentation process.

REFERENCES

Andrews, J.F. (1993) Modeling and simulation of wastewater treatment processes. *Water Sci. Technol.* **28**(11/12), 141–150.

Artan, N., Wilderer, P., Orhon, D., Morgenroth, E. and Özgür, N. (2001): The mechanism and design of sequencing batch reactor systems for nutrient removal - the state of the art. *Water Sci. Technol.* **43**/3, 53–60.

Bowden, G., Wett, B., Takacs, I., Musabyimana, M., Murthy, S. and Deur, A. (2007): Evaluation of the Single Sludge Deammonification Process for the Treatment of Plant Recycle Streams Containing High Ammonia Concentration. *Proc. WEFTEC*, San Diego.

Broda, E. (1977). Two kinds of lithotrophs missing in nature. *Z. Allg. Mikrobiologie*, **17**/6, 491–493.

Constantine, T., Murthy, S., Bailey, W., Benson, L., Sadick, T. and Daigger, G. (2005) Alternatives for treating high nitrogen liquor from advanced anaerobic digestion at the Blue Plains AWTP. *Proc. WEFTEC*, Washington.

Van Dongen, L.G.J.M., Jetten, M.S.M. and van Loosdrecht, M.C.M. (2001). The combined Sharon/Anammox process. *STOWA report, IWA Publishing, ISBN 1 84339 0000.*

EnviroSim 2007: BioWin[TM].

Fillos, J., Ramalingam, K. and Carrio, L.A. (2005) Integrating Separate Centrate Treatment with Conventional Plant Operations For Nitrogen Removal. *Proc. WEFTEC*, Washington.

Hellinga, C., Schellen, A.A.J.C., Mulder, J.W., van Loosdrecht, M.C.M. and Heijnen, J.J. (1998). The SHARON process: An innovative method for nitrogen removal from ammonium-rich wastewater. *Water Sci. Technol.* **43**(11), 135–142.

Helmer, C., Tromm, C., Hippen, A., Rosenwinkel, K.-H., Seyfried, C.F. and Kunst, S. (1999): Einstufige biologische Stickstoffelimination durch Nitritation und anaerobe Ammoiumoxidation im Biofilm. *gwf* 140, Nr. 9, 622–632.

Hopkins, L.N., Lant, P.A. and Newell, R.B. (2001): Using the flexibility index to compare batch and continuous activated sludge process. *Water Sci. Technol.* **43**/3, 35–43.

Imbeah, M. (1998): Composting piggery waste – a review. *Biores. Technol.* **63**, 197–203.

Jardin, N., Arnold, E., Beier, M., Grömping, M., Kolisch, G., Kühn, V., Otte-Witte, R., Rolfs, T., Schmidt, F. and Wett, B. (2005): Return load from sludge treatment – process strategies (in German). *Proc. DWA 4. Klärschlammtage*, Würzburg.

Jones, R.M., Dold, P., Takács, I., Chapman, K., Wett, B., Murthy, S. and O'Shaughnessy, M. (2007): Simulation for operation and control of reject water treatment processes. *Proc. WEFTEC*, San Diego.

Karakashev, D., Schmidt, J.E. and Angelidaki, I. (2008): Innovative process scheme for removal of organic matter, phosphorus and nitrogen from pig manure. *Water Res.* **42**, 4083–4090.

Katehis, D., Stinson, B., Anderson, J. Gopalakrishnan, K., Carrio, L. and Pawar A. (2002): Enhancement of Nitrogen Removal Thru Innovative Integration of Centrate Treatment. Proc. *WEFTEC*, Chicago.

Van Kempen, R., Mulder, J.W., Uijterlinde, C.A. and Loosdrecht, M.C.M. (2001): Overview: full scale experience of the SHARON[R] process for treatment of rejection water of digested sludge dewatering. *Water Sci. Technol.* **44**/1, 145–152.

Kepp, U., Machenbach, I., Weisz, N. and Solheim, O.E. (2000): Enhanced stabilisation of sewage sludge through thermal hydrolysis – three years of practical experience with full scale plants. *Water Sci. Technol.* **42**/9, 89–96.

Parker, D. and Wanner, J. (2007): Review of methods for improving nitrification through bioaugmentation. *Water Practice* **1**/5, 1–16.

Phothilangka, P., Schoen, M., Huber, M., Luchetta, P., Winkler, T. and Wett, B. (2008): Prediction of thermal hydrolysis pretreatment on anaerobic digestion of waste activated sludge. *Water Sci. Technol.* **58**/7, 1467–1473.

Rosenwinkel, K.-H., Cornelius, A. and Thöle, D. (2005) Full scale application of the deammonification process for the treatment of sludge water, *Proc. IWA Specialised Conference on Nutrient Management in Wastewater Treatmemt Processes and Recycle Streams, Krakow*, 483–491, ISBN 83-921140-1-9.

Salem, S., Berends, D.H., van der Roest, H.F., van der Kuij, R.J. and van Loosdrecht, M.C. (2004): Full-scale application of the BABE technology. *Water Sci. Technol.* **50**/7, 87–96.

Sin, G., Kaelin, D., Kampschreur, M.J., Takács, I., Wett, B., Gernaey, K.V., Rieger, L., Siegrist, H. and van Loosdrecht, M.C.M. (2008): Modelling nitrite in wastewater treatment systems: A discussion of different modelling concepts. *Proc. 1st IWA/WERF Wastewater Treatment Modelling Seminar*, Quebec.

Strous, M., Kuenen, J.G. and Jetten, M.S.M. (1999a): Key physiology of anaerobic ammonium oxidation. *Appl. Environm. Microbiol.* **65**/7, 3248–3250.

Strous, M., Fuerst, J.A., Kramer, E.H.M., Logemann, S., Muyzer, G., van de Pas-Schoonen, K.T., Webb, R., Kuenen, J.G. and Jetten, M.S.M. (1999b). Missing lithotroph identified as new planctomycete. *Nature* **400**, 446–449.

Takács, I., Vanrolleghem, P.A., Wett, B. and Murthy, S. (2007). Elemental balancing-based methodology to establish reaction stoichiometry in environmental modeling. *Water Sci. Technol.* **56**/9, 37–41.

Turk, O. and Mavinic, D.S. (1987): Benefits of using selective inhibition to remove nitrogen from highly nitrogenous wastes. *Envirn. Tech. Letters,* 8, 419–426.

Wett, B. and Rauch, W. (2003). The role of inorganic carbon limitation in biological nitrogen removal of extremely ammonia concentrated wastewater. *Water Res.* **37**/5, 1100–1110.

Wett, B. (2006a): Solved scaling problems for implementing deammonification of rejection water. *Water Sci. Technol.* **53**/12, 121–128.

Wett, B., Eladawy, A. and Ogurek, M. (2006b). Description of nitrogen incorporation and release in ADM1. *Water Sci. Technol.* **54**/4, 67–76.

Wett, B., Murthy, S., O'Shaughnessy, M., Sizemore, J., Takács, I., Musabyimana, M., Sanjines, P., Katehis, D. and Daigger, G.T. (2008): Design considerations for SBRs used in digester liquor sidestream treatment. *Proc. WEFTEC*, Chicago.

7

Environmental technologies to remove recalcitrant N-pollutants from wastewaters

F.J. Cervantes

7.1 INTRODUCTION

Several industrial sectors require different N-substituted aromatic compounds, such as nitroaromatics, azo dyes and aromatic amines, for the intensive production of dyes, explosives, pesticides and pharmaceutical products. These chemicals are intentionally designed to remain unaffected under conventional product service conditions and it is this property, linked with their toxicity to microorganisms, which makes their biodegradation difficult. The release of these recalcitrant N-pollutants into the environment may create serious health and environmental problems. Certainly, many N-substituted aromatic pollutants have been shown to be toxic for different aquatic species and to have mutagenic or carcinogenic activity. Moreover, due to their hydrophobic nature, many recalcitrant N-pollutants have the risk to bio-accumulate in the food chain.

Due to the recalcitrance and toxicity of many different N-substituted aromatic contaminants, several conventional wastewater treatment systems have shown poor performance or have even collapsed when applied for the treatment of wastewaters containing these pollutants. In most cases, therefore, a single treatment operation unit does not warrant proper removal of recalcitrant N-pollutants from wastewaters.

Technologies currently available to remove recalcitrant N-pollutants from wastewaters include physical processes, such as ultra-filtration and adsorption; chemical treatment systems, such as advanced oxidation processes (AOP), ion exchange and electro-kinetic coagulation; and biological treatment systems based on anaerobic or aerobic biodegradation, bio-augmentation or phytoremedition.

The environmental technologies available to remove recalcitrant N-pollutants from wastewaters will be discussed in this chapter. The chapter will emphasize the application of these technologies for the treatment of the main industrial effluents containing recalcitrant N-pollutants (originated from the production of textiles, explosives and pharmaceuticals).

7.2 TEXTILE WASTEWATER

7.2.1 Introduction

The textile industry is a major economical sector around the world. For instance, the European industry is composed of more than 110 thousands companies (principally based in Italy, Germany, the United Kingdom, France and Spain) with an average turn-over of nearly €200 billion a year, making it the world's leading exporter of textiles and the third largest exporter of clothing (Dos Santos *et al.* 2006). During the last years, Asia has been the largest textiles importer and exporter, and the leading exporter of clothing, while U.S.A. and Canada together are the largest clothing importers (WTO 2004). The worldwide distribution of imports and exports of textiles and clothing is shown in Figure 7.1. In Table 7.1, the values of textiles and clothing exports and imports, in 1995 and 2003, are given for the European Union, China and the U.S.A., showing the differences in how the textile market has developed.

Unfortunately, linked to the great economical benefits given by the textile sector to several countries, severe environmental problems have been created due to discharge or inadequate disposal of textile wastes. The most important N-contaminants released by the textile industry are dyes, which enter the environment mainly via discharge of wastewaters. To judge the relative share of the different dye classes in the wastewater of textile-processing industries, dye

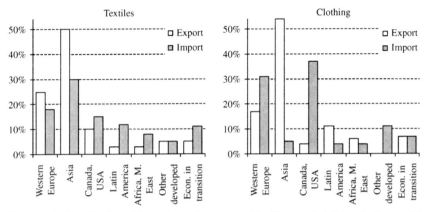

Figure 7.1. Distribution of imports and exports of textiles and clothing in the world, divided by region, in 2003 (WTO 2004)

consumption data should be considered together with the degree of fixation of the different dye classes. Recent statistics on the global production and use of dyes and on the relative distribution among the different dye classes are not readily available. Table 7.2 shows the most recent data reported on the total sales of dyes in Western Europe and worldwide, which could reasonably be assumed as the production and consumption rates (Van der Zee 2002). Table 7.3 summarizes the estimated degree of fixation for different dye/fibre combinations, which combined with the data presented on Table 2, indicate that approximately 75% of the dyes discharged by Western-Europe textile-processing industries

Table 7.1. Exports and imports of textiles and clothing, in billion dollars, rounded off at 0.5 billion. E.U. (15) = Austria, Belgium, Denmark, Finland, France, Germany, Greece, Ireland, Italy, Luxembourg, Netherlands, Portugal, Spain, Sweden and the United Kingdom (WTO 2004)

		Textiles		Clothing	
		Export	Import	Export	Import
World	1995	112	–	125	–
	2003	137	–	185	–
E.U. (15)	1995	22	17	15	40.5
	2003	26	20	19	60.5
China	1995	14	11	24	1
	2003	27	14	52	1.5
U.S.A.	1995	7.5	10.5	6.5	41.5
	2003	11	18.5	5.5	71.5

Table 7.2. Total sales of dyes in 1991 (adapted from Van der Zee 2002)*

Dye Class	Western Europe (1000 tonnes)	World (1000 tonnes)
Acid (and Mordant)	24	100
Azoic	2	48
Basic	8	44
Direct	9	64
Disperse	22	157
Reactive	13	114
Sulphur	3	101
Vat	4	40
Sum	85	668
Relative share (%)	13	

*With the exclusion of solvent and pigment dyes.

belong to the classes of reactive (~36%), acid (~25%) and direct (~15%), all of which are dye classes primarily composed by azo dyes. In fact, azo dyes, which are characterised by the presence of one to several $-N = N-$ links in their structure, constitute the largest (~70% in weight) class of dyes applied in textile processing (Van der Zee 2002).

7.2.2 Environmental problems associated with the discharge of textile wastewaters

Most textile wastewaters are highly coloured because they are typically discharged with a dye contain in the range 10–200 mg/L and many dyes are visible in water at concentrations as low as 1 mg/L. Thus, discharge of these effluents in open water bodies not only represents an aesthetic problem, but also may limit

Table 7.3. Estimated degree of fixation for different dye/fibre combinations (Van der Zee 2002)

Dye Class	Fibre	Degree of fixation (%)	Loss to effluent (%)
Acid	Polyamide	80–95	5–20
Basic	Acrylic	95–100	0–5
Direct	Cellulose	70–95	0–10
Disperse	Polyester	90–100	2–10
Metal-complex	Wool	90–98	10–50
Reactive	Cellulose	50–90	10–40
Sulphur	Cellulose	60–90	5–20
Vat	Cellulose	80–95	–

photosynthesis in aquatic plants. As dyes are designed to be chemically and photolytically stable, they are highly persistent in natural environments. The release of dyes may therefore present eco-toxic hazard and introduces the potential danger of bioaccumulation that may eventually affect man by transport through the food chain.

7.2.2.1 Bioaccumulation

Intensive research has documented the bioaccumulation tendency of dyestuffs in fish. The bio-concentration factors (BCF's) of 75 dyes from different application classes were determined and related to the partition coefficient n-octanol/water (K_{OW}) of each different compound. Water-soluble dyes with low K_{OW} (e.g. ionic dyes like acid, reactive and basic dyes) did not bio-accumulate (generally log BCF < 0.5). For these water-soluble dyes, log P (log K_{OW}) showed a liner relationship with log BCF; thus, it was expected that dyestuffs with higher K_{OW} would bio-accumulate. Nonetheless, water-insoluble organic pigments with extremely high partition coefficients did not bio-accumulate probably due their extremely low water and fat solubility and also the BCF values for disperse dyes, i.e. scarcely soluble compounds with a moderately lipophilic nature, were much lower than expected. In all cases, log BCF < 2, which indicates that none of the dyes tested showed any substantial bio-accumulation (Van der Zee 2002).

7.2.2.2 Toxicity of dyestuffs

The toxicological effects of several distinct dyes have been intensively investigated during the last decades. These toxicity studies (with respect to their mortality, genotoxicity, mutagenicity and carcinogenicity effects) diverge from tests with aquatic organisms (fish, algae, bacteria, etc.) to tests with mammals. Furthermore, research has also documented the inhibitory effects of dyes on the activity of both aerobic and anaerobic bacteria in wastewater treatment consortia.

The acute toxicity of dyestuffs is generally low. Algal growth (photosynthesis), tested with respectively 56 and 46 commercially dyestuffs, was generally not inhibited at dye concentrations below 1 mg/L. The most acutely toxic dyes for algae are –cationic–basic dyes. Fish mortality tests showed that 2% out of 3000 commercial dyestuffs tested had LC_{50} (lethal concentration causing 50% of mortality) values below 1 mg/L. The most acutely toxic dyes for fish are basic dyes, especially those with a tri-phenyl-methane structure. Fish also seem to be relatively sensitive to many acid dyes. Mortality tests with rats evidenced that only 1% out of 4461 commercial dyestuffs tested had LC_{50} values below 250 mg/kg body weight. Therefore, acute sensitisation reactions by humans to dyestuffs often occur. Especially, some disperse dyes have been found to cause allergic reactions, i.e. eczema or contact dermatitis (Van der Zee 2002).

Chronic effects of dyes have been studied for several decades with special emphasis on food colorants, usually azo compounds. Furthermore, the effects of occupational exposure to dyestuffs of human workers in dye manufacturing and dye utilising industries have received attention. Azo dyes in purified form are hardly ever directly mutagenic or carcinogenic, except for some azo dyes with free amino groups in their structure. However, several aromatic amines readily produced through reductive azo cleavage are known as mutagens and carcinogens (see below).

7.2.3 Characterisation of textile wastewaters

The production of textiles involves many different steps and most of these processes generate highly contaminated liquid streams. The amount and composition of these wastewaters depend on many different factors, including the processed fabric and the type of process. Type of machinery, chemicals applied and other characteristics of the processes also determine the amount and composition of the generated wastewater. Most studies dealing with the treatment of textile wastewaters have been reported on the whole mixed effluents generated from this industrial sector. Nevertheless, considering water and chemicals savings, the processes should be regarded separately, although this does not mean that all of the streams originated from textile processing have to be treated independently. Combining selected streams can lead to a better treatable wastewater. For instance, mixing of acid and alkaline discharges could be a suitable strategy to obtain a neutralized effluent. Eventually, a stream could be separated from the rest to facilitate the recovery of water or chemicals, or to prevent dilution of a compound difficult to remove.

Figure 7.2 shows the sequence of textile processes as they generally take place in a factory. Some processes hardly generate wastewater, such as yarn manufacture, weaving (some machines use water), and singeing (just some lightly polluted cooling water). The amount of wastewater produced in a process like sizing is small, but very concentrated. On the other hand, processes like scouring, bleaching and dyeing generate large amounts of wastewater, varying much in composition.

Although effluent characteristics differ greatly even within the same process, some general values can be given. However, the individual wastewater characteristics of all textile processing steps are not easily found. The ones generating smaller amounts of wastewater or less polluted effluents are hardly investigated. Table 7.4 gives an overview of the large fluctuation of important

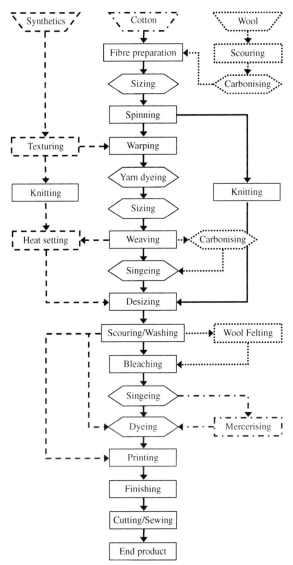

Figure 7.2. General flowchart for processes taking place in textile manufacturing. Square boxes are processes that always take place in that order, and 'diamond' shaped boxes are processes that can occur at different places in the chain. The line style indicates if the process is only used for a certain fabric type: (—) for all fibres, (- - -) for synthetic fibres, (- · -) for cotton and (· · ·) for wool (Adapted from Bisschops and Spanjers 2003).

Table 7.4. Average and peak values in mg/L for selected parameters of most reported textile processes (adapted from Dos Santos *et al.* 2006)

Parameter	Desizing	Scouring	Bleaching	Dyeing
COD	3580–5900	3200–40000	250–6000	550–8000
COD peak	11000	90000	13500	40000
BOD	200–5200	300–8000	80–400	11–2000
BOD peak		60000	2800	27000
TS	7600–42000	1100–30000	900–14000	200–2000
TS peak		65000		14000
SS	400–800	200–20000	35–900	25–200
SS peak	6000	40000	25000	
TDS	1600–6900		40–5000	
TDS peak			20000[1]	
Lipids	190–750	100–9000		
Lipids peak		108000		
pH	6–9	7–14	6–13	3.5–12

[1]Denim stone wash involving pumice stone. COD, Chemical Oxygen Demand; BOD, Biochemical Oxygen Demand; TS, Total Solids; TDS, Total Dissolved Solids; SS, Suspended Solids.

parameters from textile wastewaters. Mixed textile wastewater generally contains high levels of COD and colour, and usually has a high pH. Because its composition depends on the types of processes and fabrics handled in the factory, it is difficult to present general values. For example, COD concentrations varying from approximately 150 mg/L to over 10000 mg/L and pH values from 4.5 to 13, are reported (Dos Santos *et al.* 2006).

7.2.4 Technologies for the treatment of textile wastewaters

The different technologies available to treat textile effluents are mainly focused on the removal of COD and colour. As this chapter deals with recalcitrant N-pollutants and considering that azo dyes are by far the most frequently applied dyestuffs in textile processing, strategies to remove these colorants from textile effluents will be discussed here. As COD and colour originate from almost all textile processes and due to the different characteristics of every stream, the options for achieving COD and colour removal should ideally be discussed for each separate wastewater stream. However, to keep the information presented in this chapter concise, the methods for COD and colour removal will be described for textile wastewater in general.

Before deciding on a certain type of treatment, it is important to know the nature of the COD load with respect to the different fractions that contribute to the total COD, such as soluble, colloidal and suspended, and if they are biodegradable. For mixed textile effluents, generally all COD fractions are present. Therefore, these characteristics will help to determine which technique is suitable for the wastewater treatment. However, currently there is no single and economically attractive treatment that can remove both COD and colour from textile wastewaters. Therefore, the wastewater treatment plant (WWTP) will consist of different processes, which are following described.

7.2.4.1 Biological wastewater treatment systems

The biodegradation of azo dyes has been reported by a wide variety of microorganisms including fungi, yeasts and bacteria both in pure and mixed populations. However, successful applications (e.g. in full-scale) have only been reported for wastewater treatment systems combining bacterial populations under anaerobic and aerobic conditions. Several research papers and full-scale applications have suggested that combined anaerobic-aerobic biological treatment systems are the most suitable technologies available for the treatment of textile effluents, based on COD and colour removal efficiencies and on operational costs. Table 7.5 summarizes the performance of different wastewater treatment systems applied on textile effluents based on anaerobic-aerobic treatment in separate reactors. Table 7.6 show the application of sequencing batch reactors (SBR) for the same industrial effluents based on temporal separation of the anaerobic and the aerobic phase. Meanwhile, Table 7.7 lists other systems, either hybrids with aerated zones or micro-aerobic systems based on the principle of limited oxygen diffusion in microbial biofilms. Following, the main operational parameters affecting the performance of the anaerobic and aerobic steps in combined anaerobic-aerobic treatment systems applied for textile wastewaters will be discussed.

Anaerobic step Azo dyes generally prevail unaffected in conventional aerobic wastewater treatment systems. In contrast, reductive cleavage of azo bonds readily occurs under anaerobic conditions, resulting in the formation of colourless aromatic amines (Field et al. 1995). Considering that a major fraction (~70%) of dyestuffs discharged in textile wastewaters are azo compounds, it is reasonable to assume that an important removal of colour may occur in anaerobic wastewater treatment systems. In fact, the efficiency of decolourisation achieved in the anaerobic steps of the studies reported in Tables 7.5–7.7 was usually higher than 70% and in several cases almost 100%, whereas further decolourisation was very limited in the subsequent aerobic stage.

Table 7.5. Sequential anaerobic-aerobic reactor systems treating azo dye-containing wastewater (Van der Zee and Villaverde 2005)

Anaerobic		Aerobic		Wastewater characteristics				Color removal		Aromatic amines			References
Type[a]	HRT (h)	Type[b]	HRT (h)	ww[c]	Dye[d]	Conc. (mg/l)	Substrates[e]	Anaerobic	Aerobic[f]	Recovery anaerobic[g]	Removal aerobic[h]	Detect. method[j]	
3[j]	36	1[j]	36	S[k]	AO10, ma ABk1, da DR2, da DR28, da	5–100 10–100 25–200 25–200	glucose	90–100% 100% 95–100% 80–100%	+	25–50%	max. 100%	2	(Rajaguru et al. 2000)
1	24	1	19	S	h-RR141, da, mct	450	starch and acetate	64%	+11%	+	+	1	(O'Neill et al. 2000b)
1	24	1	19	S	h-RR141, da, mct	150–750	starch and acetate	38–59%	+7–18%	n.e.	n.e.		(O'Neill et al. 2000a)
1	24–48	1	NM	S	h-RR141, da, mct	1500	starch and acetate	~78%	–7%	n.e.	n.e.		(O'Neill et al. 1999a)
5	34–84	1	NM	S	h-RR141, da, mct	1500	starch and acetate	max. 62%	–7%	n.e.	n.e.		(O'Neill et al. 1999a)
2	1–24	4	NM	S	AO7, ma >AO8, ma AO10, ma AR14, ma	5–40 5–40 5–40 5–40	ME, peptone, YE, trout, chow	20–90%[l]	0	+	n.e.	n.m.	(Seshadri et al. 1994)
2	31	4	3.1	S	AO10, ma AR14, ma AR18, ma	10 10 10	ME, peptone, YE, trout chow	~62% ~90% ~90%	+	<1%	+	1-MS	(FitzGerald and Bishop 1995)
4	15	2 (2)	7.5	S	h-RV5, ma, vs	650–1300	acetate and YE	90–95%	–	~69–83%	~100%[m]	1	(Sosath and Libra 1997)
4	31	2 (2)	7.5	S	h-RBk5, ma, vs	600	acetate and YE	~70%	+	n.e.	–[n]	4	(Sosath et al. 1997; Wiesmann et al. 2002)

Anaerobic		Aerobic		Wastewater characteristics				Color removal		Aromatic amines			References
Type[a]	HRT (h)	HRT (h)	Type[b]	ww[c]	Dye[d]	Conc. (mg/l)	Substrates[e]	Anaerobic	Aerobic[f]	Recovery anaerobic[g]	Removal aerobic[h]	Detect. method[i]	
4	15	7.5	2	S	(h-)RBk5, ma, vs	530	acetate and YE	~100%[o]	-35%[o]	~100%[p]	+[p]	1-MS, 4	(Libra et al. 2004)
1	8–20	23	1[q]	S	AY17, ma (BB3, ox) (BR2, az)	40 / 40 / 40	glucose	20% (72%) (78%)	+0–13%	n.e.	n.e.		(An et al. 1996)
1	4–10	6.5	1	R/S	ww from a dye-manufacturing factory, mixed with simulated municipal wastewater			70–80%	+10–20%	+[r]	+[r]		(An et al. 1996)
4	7–8	4.5–5	2	R	textile dye wastewater with PVA and LAS as main COD			60–85%		n.e.	n.e.		(Zaoyan et al. 1992)
1	6–10	6	1	R	textile dye wastewater with PVA and LAS as main COD			90–95%	+ (max. 96%)	n.e.	n.e.		(Jianrong et al. 1994)
3	6	7.7 + 8.6	3 (2)	S	AY17, ma BR22, ma	25 / 200	starch and glucose	0% / >99%	+	n.e.	n.e.		(Basibuyuk and Forster 1997)
1	25	10	1	S	MY10, ma	100–200	ethanol	~100%	0	~100%[s]	~100%[s]	1	(Tan et al. 2000)
3[t]	var.	12–24	3[t]	S	DisB79, ma	25–150	glucose and acetate or none	max. 100%	n.m.	~40%[u]	65%[v]	1, 2, 3	(Cruz and Buitrón 2001)
3	w	w	3	S	h-RR198, ma, vs + mct	5000	starch, wax and acetate	97%	+2–3%	+	+ (~100%)	1	(Sarsour et al. 2001)
1[x]	24–48	w	1[p]	R	highly colored textile wastewater			70–90%	+ (~100%)	n.e.	n.e.		(Kuai et al. 1998)
1	15–16.5	55–60	1	S	DBk38, ta	100–3200	glucose	80–100%	+	85–95%	~50%	2	(Sponza and Işik 2005)
1	86.4	432	1	S	DBk38, ta	3200	glucose	81%	+13%	74%	81%	2[y]	(Işik and Sponza 2004c)
1	3–30	10–30	1	S	RBk5, da, vs	100	glucose	82–98%	n.m.	n.e.	n.e.		(Sponza and Işik 2002a)

(continued)

Table 7.5. *Continued*

Anaerobic		Aerobic		Wastewater characteristics				Color removal		Aromatic amines			References
Type[a]	HRT (h)	Type[b]	HRT (h)	ww[c]	Dye[d]	Conc. (mg/l)	Substrates[e]	Anaerobic	Aerobic[f]	Recovery anaerobic[g]	Removal aerobic[h]	Detect. method[i]	
1	3–30	1	10–108	S	RBk5, da, vs	100	glucose	82–98%	–	n.e.	n.e.	3	(Sponza and Işik 2002b)
1	2.6–26	1	10–102	S	DR28, da	100–4000	glucose	97–100%	n.m.	40–95%	80–100%	2	(Işik and Sponza 2003)
1	3–30	1	10–108	S	RBk5, da, vs	100	glucose	87–98%	–10–20%	n.e.	n.e.		(Işik and Sponza 2004b)
1	2.5–19	1	9–67	S	DR28, da	100	glucose	92–97%	–1–5%	n.e.	n.e.		
1	30	1	108	R	cotton mill wastewater (CMW)			46–55%	–10%– +25%	+	35–90%	1,2	(Işik and Sponza 2004a)
					CMW + mixture of azo dyes (250–500 mg/l) and glucose			60–75%	+1–15%	+	40–80%		
3	12–72	1	10	S	RR195, ma, vs+mct	50–400	molasses	60–100%[c]	+max. 15%	n.e.	n.e.		(Kapdan et al. 2003)
3[z]	12–72	1	10	R	textile wastewater with added glucose and nutrients			60–85%	+10%	n.e.	n.e.		(Kapdan and Alparslan 2005)
1	24	5	21.5 (24)	S	AO7, ma	60–300	glucose +peptone	60–97%	+	n.e.	n.e.		(Ong et al. 2005)
(6) 1	(15–18) 7	1	15–18	R	bleaching, scouring, dyeing (+ desizing) wastewater containing 10–150 mg/l dyes			(50–70%) 80–95%	–5 – +5%	n.e.	n.e.		(Frijters et al. 2005)
3	26–90	(6)	(480)	R	1. reactive dyebath waste and ww with starch & PVA 2. split flows from yarn processing			89–94% 81–92%	+1–2% +1–7%	n.e.	n.e.		(Minke and Rott 2002)

a Anaerobic reactor types: 1. Upflow Anaerobic Sludge Bed; 2. Anaerobic Fluidized Bed; 3. Anaerobic Filter; 4. Anaerobic Rotating Disc; 5. Inclined Tubular Digester; 6. pre-acidification tank.

b Aerobic reactor types: 1. Aerobic Tank; 2. Aerobic Rotating Disc; 3. Aerobic Filter; 4. Swisher; 5. Sequential Batch Reactor; 6. Aerobic biodegradability tests (BOD_{20}).

c Wastewater type (ww): S, synthetic wastewater; R, real wastewater.

d Dyes: First abbreviation refers to Colour Index generic names: A, acid; B, basic; D, direct; Dis, disperse; M, mordant; R, reactive; B, Blue; Bk, black; O, orange; R, red; V, violet; Y, yellow. Second abbreviation refers to amount of azo linkages: ma, monoazo; da, disazo; ta, triazo; (ox, oxazine; az, azine). Third abbreviation refers to reactive groups (reactive dyes only): vs, vinylsulfone; mct, monochlorotriazine. The prefix 'h-' means hydrolyzed (reactive dyes only). Libra et al. (2004) investigated both hydrolyzed, partially hydrolyzed and non-hydrolyzed Reactive Black 5.

e Substrates, abbreviations: YE, yeast extract; PVA, polyvinylalcohol; LAS, linear alkyl benzene sulfonate; ME, meat extract.

f Color removal aerobic: positive values express the additional color removal as percentage of the influent color, negative values express development of color (autoxidation) as percentage of influent color. 'n.m.' not mentioned.

g Anaerobic aromatic amine recovery: '+' indicates non-quantified sign of recovery: 'n.e.' not evaluated.

h Aerobic aromatic amine removal: '+' indicates non-quantified sign of removal: percentages express removal of recovered aromatic amines; 'n.e.' not evaluated.

i (Main) detection method aromatic amines: 1. HPLC; 1-MS. LC-MS; 2. diazotization based colorimetric method; 3. UV spectophotometry; 4. DOC measurements.

j Both anaerobic and aerobic reactor inoculated with a mixture of four pseudomonads isolated from dyeing effluent-contaminated soils.

k Nitrogen-free medium.

l Depending on dye, dye concentration and HRT. All dyes >80% decolorization at high HRT.

m Complete removal of the metabolites from anaerobic treatment, probably mostly due to autoxidation.

n Presumably no removal of dye metabolites: hardly any DOC removal and only slight decrease of toxicity.

o Data refer to fully hydrolyzed RBk5, less color removal for partially hydrolyzed RBk5.

p Fully hydrolyzed RBk5 was completely converted in the anaerobic phase, to p-aminobenzene-2-hydroxyleethylsulfonic acid (2 mol p-ABHES per mol RBk5) and 1,2,7-triamino-8-hydroxynaphthalene-3,6-disulfonic acid (1 mol TAHNDS per mol RBk5). In the aerobic phase, p-ABHES was mineralized while TAHNDS autoxidized to 1,2-ketimino-7-amino-8-hydroxynaphthalene-3,6-disulfonic acid. Partially hydrolyzed RBk5 was not completely converted in the anaerobic phase. p-ABHES and TAHNDS were detected, but in relatively small amounts. There was no removal of p-ABHES in the aerobic phase.

q Semi-continuous system.

r Increased BOD_5/COD-ratio after anaerobic treatment may point at formation of biodegradable dye metabolites.

s Almost complete recovery of one of the dye metabolites, sulfanilic acid; partial anaerobic degradation of the other, 5-aminosalicylate. In the aerobic reactor complete mineralization of 5-aminosalicylate; after bioaugmentation also complete mineralization of sulfanilic acid.

t Discontinuously fed reactors.

u Percentage expresses HPLC-recovery of 2-bromo-4,6-dinitroaniline (BDNA). Additional thin layer chromatography measurements indicate anaerobic transformation BDNA.

v Percentage based on total amine measurements (diazotization method).

w HRT total system 96 h.

x Sludge bed amended with granular activated carbon.

y Additional support of aerobic AA removal from HPLC, GC-MS and nitrate analyses.

z Inoculated with a facultative anaerobic consortium (PDW, Alcaligenes faecolis and Comamonas acidourans).

Table 7.6. Anaerobic-aerobic Sequenced Batch Reactor (SBR) systems treating azo dye-containing wastewater (Van der Zee and Villaverde 2005)

Cycle				Wastewater characteristics			Color removal		Aromatic amines			References
Anaerobic (h)	Aerobic (h)	Total time (h)	ww[a]	Dye[b]	Conc. (mg/l)	Substrates[c]	Anaerobic	Aerobic[d]	Recovery anaerobic[e]	Removal aerobic[f]	Detect. meth.[g]	
13	8	24	S	h-RV5, ma, vs	60–100	starch	30–90%[h]	+/0	+	+[i]	1	(Lourenço et al. 2000)
9–12	8–12	24	S	h-RV5, ma, vs	60–100	starch	20–90%[j]	n.m.	+	+[i]	1	(Lourenço et al. 2001)
				h-RBk5, da, vs	30		65%[k]		n.e.	n.e.		
9–13	8–12	24	S	h-RV5, ma, vs	100	starch	max. 90%	n.m.	+	+[i]	1	(Lourenço et al. 2003)
10.5	10	24	S	h-RV5, ma, vs	100	starch	90–99%	n.m.	+	n.e.	1	(Albuquerque et al. 2005)
10.5–17	3.5–10	24	S	AO7, ma	25	starch	5–55%		n.e.			
10.5	10	24	S	AO7, ma	25	starch + lactate	max. 95%[l]		n.e.			
0–12	8–12	24	R	wool dyeing effluent with azo and anthraquinone dyes			+	+	n.e.	+	3	(Cabral Gonçalves et al. 2005)
18	5	24	S	RBk5, ma, vs	20–100	glucose and acetate	58–63%	+	+	–	1	(Luangdilok and Panswad 2000; Panswad and Luangdilok 2000)
				(RB19, aq, vs)	20–100		(64–32%)					
				(RB5, aq, mct)	20–100		(66–41%)					
				(RB198, ox, hh)	20–100		(—)[m]					

Cycle			Wastewater characteristics				Color removal		Aromatic amines			References
Anaerobic (h)	Aerobic (h)	Total time (h)	ww[a]	Dye[b]	Conc. (mg/l)	Substrates[c]	Anaerobic	Aerobic[d]	Recovery anaerobic[e]	Removal aerobic[f]	Detect. meth.[g]	
18	5	24	S	RBk5, da, vs	10	NB + acetate or glucose	68–72%	+2–8%	+	n.e.	3	(Panswad et al. 2001a)
							63–68%	+8–11%				
0 – 8	3–11	12	S	RBk5, da, vs	10–80	NB + acetate or NB + glucose	30–61%	+2–17%	+	n.e.	3	(Panswad et al. 2001b)
18.5	0.5	24	S	h-RBk5, da, vs	533	starch, PVA, CMC	86–96%	+	+	+/0	3	(Shaw et al. 2002)

[a] Wastewater type (ww): S. synthetic wastewater; R. real wastewater.

[b] Dyes: First abbreviation refers to Colour Index generic names: A, Acid; R, reactive; B, Blue; Bk, black; O, Orange; V, violet. Second abbreviation refers to amount of azo linkages: ma, monoazo; da, disazo; (aq, anthraquinone; ox, oxazine). Third abbreviation refers to reactive groups (reactive dyes only): vs, vinylsulfone; mct, monochlorotriazine; hh, halogenohetrocyclic. The prefix 'h-' means hydrolyzed (reactive dyes only).

[c] Substrates, abbreviations: NB, nutrient broth; PVA, polyvinylalcohol; CMC, carboxymethylcellulose.

[d] Color removal aerobic: positive values express the additional color removal as percentage of the influent color, negative values express development of color (autoxidation) as percentage of influent color. 'n.m.' not mentioned.

[e] Anaerobic aromatic amine recovery: '+' indicates non-quantified sign of recovery; 'n.e.' not evaluated.

[f] Aerobic aromatic amine removal: '+' indicates non-quantified sign of removal; 'p' non-quantified sign of partial removal; percentages express removal of recovered aromatic amines; 'n.e.' not evaluated.

[g] (Main) detection method aromatic amines: 1. HPLC; 3. UV spectophotometry.

[h] ~90% color removal at a sludge concentration of 2.0 gVSS l⁻¹ and SRT = 15 days, ~30% color removal at a sludge concentration of 1.2 gVSS l⁻¹ and SRT = 10 days

[i] no degradation of RV5's constituent naphthalene-based amine; (bio)transformation but no mineralization of its benzene-based amine.

[j] ~90% color removal at a sludge concentration = 2.0 gVSS l⁻¹, SRT = 15 days and feed dye concentration = 60 mg l⁻¹, ~20% color removal at a sludge concentration = 1.2 gVSS l⁻¹, SRT = 10 days and feed dye concentration = 100 mg l⁻¹.

[k] no effect of changing the SRT.

[l] highest color removal achieved with addition of anthraquinone-2,6-disulfonate.

[m] could not be quantified.

Table 7.7. Integrated anaerobic-aerobic reactor systems treating azo dye-containing wastewater (Van der Zee and Villaverde 2005)

| System | | Wastewater characteristics | | | | Color removal | | Aromatic amines | | | References |
Reactor type[a]	Total time (h)	ww[b]	Dye[c]	Conc. (mg/l)	Substrates[d]	Anaerobic	Aerobic	Recovery anaerobic[e]	Removal aerobic[f]	Detect. meth.[g]	
EGSB with oxygenation of recycled effluent	36–43	S	MY10, ma	59–65	ethanol	~100%		+[i]	+[i]	1	(Tan et al. 1999; Tan 2001)
	26–34		4-PAP, ma	50		<100%[h]		+[i]	+[i]	1	
UASB with aerated upper part	1–100	S	DY26, da	300	ethanol	40–70%	+10–20%	+[j]	+[j]	3	(Kalyuzhnyi and Sklyar 2000)
RAD	0.16–3	S	AO7, ma	0–22	ME, peptone, YE, trout chow	18–97%[k]		+	+	1,5	(Harmer and Bishop 1992)
RAD	2	S	AO8, ma AO10, ma AR14, ma	n.m.	ME, peptone, YE, trout chow	20–90%[k]	max. 60% max. 60%	n.e.	n.e.		(Jiang and Bishop 1994)
Baffled reactor with anaerobic and aerobic compartments	48+18	S	h-RBk5, da, vs	500	starch, PVA, CMC	84–88%		+	+[j]	3	(Gottlieb et al. 2003)

[a] Reactor types: EGSB, Expanded Granular Sludge Bed; UASB, Upflow Anaerobic Sludge Blanket; RAD, Rotating Annular Drum.

[b] Wastewater type (ww): S. synthetic wastewater.

[c] Dyes: 4-PAP is 4-phenylazophenol. For the other dyes, the first abbreviation refers to Colour Index generic names: A, acid; D, direct; M, mordant; R, reactive; Bk, black; O, orange; R, red; Y, yellow. The second abbreviation refers to the amount of azo linkages: ma, monoazo; da, disazo. The third abbreviation refers to the reactive groups (reactive dyes only): vs, vinylsulfone. The prefix 'h-' means hydrolyzed (reactive dyes only).

[d] Substrates, abbreviations: YE, yeast extract; PVA, polyvinylalcohol; ME, meat extract; CMC, carboxymethylcellulose.

[e] Anaerobic aromatic amine recovery: '+' indicates non-quantified sign of recovery; 'n.e.' not evaluated.

[f] Aerobic aromatic amine removal: '+' indicates non-quantified sign of removal; percentages express removal of recovered aromatic amines; 'n.e.' not evaluated.

[g] (Main) detection method aromatic amines: 1. HPLC; 3. UV spectophotometry; 5. GC-MS.

[h] residual color due to autoxidation of 4-aminophenol (one of 4-PAP's constituent aromatic amines).

[i] aromatic amines from MY10: almost complete recovery of sulfanilic acid, partial anaerobic degradation of 5-aminosalicylate; aromatic amines from 4-PAP: complete mineralization of aniline, autoxidation of 4-aminophenol.

[j] one of the dye's aromatic amine (5-aminosalycilate) was partially degraded in the anaerobic part and underwent autoxidation in the aerobic part.

[k] at high oxygen/low COD flux, dye removal probably (partly) due to aerobic degradation.

[l] decrease of toxicity after addition of adapted biomass may indicate biological degradation of aromatic amines.

Analysing the reactor studies listed in tables 7.5–7.7 it can be concluded that long hydraulic retention times (HRT) are required in anaerobic wastewater treatment systems to achieve efficient decolourization of textile wastewaters. Two main factors have been identified as the main causes for the long HRT demanded during the reductive decolourisation of azo dyes: (i) the slow transfer of reducing equivalents from an external electron donor to the azo dyes, which act as terminal electron acceptors in the decolourising process, and (ii) the toxicity effects imposed by several recalcitrant azo dyes on anaerobic consortia. Several strategies, however, have been proposed to overcome these limitations.

For instance, the application of redox mediating compounds to shuttle electron equivalents from the electron-donating primary substrate to the electron-accepting dyes has been demonstrated, in several studies, as a suitable method to accelerate the reductive decolourisation of azo dyes. Certainly, continuous dosing of soluble quinoid redox mediators at catalytic concentrations has been demonstrated to strongly increase the azo dye reduction efficiencies of lab-scale anaerobic bioreactors (Cervantes et al. 2001; Van der Zee et al. 2001; Dos Santos et al. 2003; Dos Santos et al. 2004a; Dos Santos et al. 2005) and of a lab-scale SBR (Albuquerque et al. 2005). Nevertheless, continuous addition of redox mediators should be supplied in anaerobic bioreactors in order to achieve increased conversion rates, which raise the costs of treatment. An approach to disregard the continuous supply of redox mediators is to create a niche for their immobilization in anaerobic bioreactors. One of the alternatives available considers the application of activated carbon as a natural source of quinoid redox mediators and its potential for reducing azo dyes have been explored in anaerobic bioreactors promoting a greater decolourisation efficiency in the bioreactor configurations explored (Van der Zee et al. 2003; Mezohegyi et al. 2007). Another immobilizing approach reported is by inserting quinoid redox mediators within different materials through polymerization procedures (Guo et al. 2007). Quinoid redox mediators entrapped within the synthesized polymer were shown to accelerate the reductive decolourisation of several azo dyes. However, the immobilizing methods explored so far are not ready to be implemented at full-scale as redox mediators are gradually lost during the decolourising processes due to washout or disruption of the immobilizing materials.

Another tactic to promote the efficient decolourisation of textile effluents is the application of reactor configurations with a high biomass concentration. Indeed, by increasing the biomass concentration from 1.2 to 2 g VSS/L and the solids retention time from 10 to 15 days in dye-treating SBR resulted in a considerable increase, from ∼30% to ∼90%, of the colour removal efficiency (Lourenço et al. 2000). Systems with a higher biomass retention capacity, such as UASB and EGSB reactors, might therefore perform a better colour removal efficiency than treatment systems with a lower biomass retention capacity, such as SBR systems. The superior performance of high-rate anaerobic reactors, such as UASB and EGSB systems, for the reductive decolourisation of azo dyes has previously been suggested due to a number of operational aspects (Razo-Flores et al. 1997). Firstly, reductive azo cleavage produces colourless aromatic amines, which are generally reported to be less inhibitory for anaerobic consortia compared to their precursor azo compounds (Donlon et al. 1995). Thus, by creating a proper niche for a high biomass concentration promotes a faster reductive decolourisation of azo dyes preventing their accumulation, then minimising inhibitory effects by the latter. Furthermore, the application of redox mediators in reductive decolouring processes has recently been demonstrated as a proper strategy to overcome the toxicological effects of azo dyes on different syntrophic groups in methanogenic sludge (Cervantes et al. 2008). Secondly, the good hydraulic and gas mixing conditions prevailing in high rate anaerobic treatment systems minimises biological dead-space and prevents localised high concentrations of toxic azo dyes. Thirdly, the protection of methanogenic microorganisms created inside the granules where aromatic amines predominate due to the reducing conditions established, in contrast to the toxic compounds prevailing in the bulk of the reactor (azo compounds).

A prerequisite to promote the reductive decolourisation of textile wastewaters in anaerobic wastewater treatment systems is the presence of a suitable primary electron donor. Several research studies have evidenced that both the amount and the type of substrate available in decolourising processes are important parameters, which determine the efficiency of decolourisation. In theory, the required amount of electron donating primary substrate is low, four reducing equivalents per azo linkage, i.e. 32 mg COD per mmol monoazo dye. However, the rather slow process of azo dye reduction may benefit kinetically by a higher primary substrate concentration. For the studies listed in Tables 7.5–7.7, the primary substrate (biodegradable COD) was usually in large excess of the stoichiometric amount required for azo dye reduction. Only in one of the studies (Cruz and Buitrón 2001) has the reactor been operated for a while under clear sub-stoichiometric conditions, when the reactor influent contained no other electron donor than the dye's dispersing agent. Consequently, the reactor's anaerobic colour removal efficiency dropped. Sub-stoichiometric conditions might also explain the low colour removal efficiency before raising the level of primary substrate in another reactor system (Sosath and Libra 1997). Other studies report sub-optimal primary substrate levels in large excess of the stoichiometric level. For instance, O'Neill et al. (2000a) reported improved colour removal efficiencies when doubling the primary substrate concentration, even though the original primary substrate concentration was already 60–300 times in excess of the stoichiometric amount.

The reactor studies report a wide variety of primary electron donating compounds, from simple substrates like acetate, ethanol and glucose to more complex ones, including relevant constituents of textile-processing wastewaters like starch, polyvinylalchol (PVA) and carboxymethylcellulose (CMC). In all cases, azo dye decolorization occurred, which suggests that the process is relatively non-specific with respect to its electron donor. Nevertheless, some primary substrates promote a faster and greater extent of decolourisation of textile wastewaters. For instance, several research studies suggest glucose, lactate, formate and hydrogen as the most appropriate primary substrates to sustain decolourising processes; whereas methanol and complex compounds like PVA and CMC are not suitable to achieve high decolourisation efficiencies (Van der Zee and Villaverde 2005).

Textile wastewaters generally contain alternative electron acceptors, besides azo dyes, which may significantly affect the reductive decolourisation of these industrial effluents, although the literature reports heterogeneous results depending on the type and concentration of the electron acceptors available. Nitrate, a common constituent of textile wastewater, was shown to compete with azo dyes for the reducing equivalents available in wastewater treatment systems decreasing the decolourising capacity ((Lourenço et al. 2000; Panswad and Luangdilok 2000). Those results are agreement with previously published data from batch experiments to azo dye decolorization in the presence of nitrate (Carliell et al. 1995; Carliell et al. 1998) and nitrite (Wuhrmann et al. 1980). However, the inhibitory effects of nitrate on decolourising processes is not universal as evidenced in a recent report in which nitrate (up to 310 mg NO_3^-/L) did not significantly affect the reductive decolourisation of Reactive Red 2 in EGSB reactors (Dos Santos et al. 2008).

Textile-processing wastewaters usually contain moderate to high sulphate concentrations, which may have different effects on the reduction of azo dyes. Firstly, sulphate may compete with the dyes, as an electron acceptor, for the electrons available, depending on the capacity of the inocula to carry out sulphate-reduction and on the concentration of sulphate. Secondly, electron equivalents may be generated through the reduction of sulphate via anaerobic substrate oxidations; thus, reduced cofactors involved in sulphate-reduction may also play a role on azo dye reductions. Finally, sulphide generation, via sulphate-reduction, may also contribute to the reduction of azo dyes (Cervantes et al. 2007). The last observation is supported by Alburquerque et al. (2005), who observed a sharp decline of the anaerobic azo dye decolorization efficiency in a SBR when selectively inhibiting its sulphate reducing activity by the addition of molybdate. Moreover, biogenic sulphide was lately shown to play a major role on the reductive decolourisation of azo dyes during decolourising batch processes mediated by riboflavin (Cervantes *et al.* 2007).

Aerobic step The reductive decolourisation of azo dyes generally produces aromatic amines, which are seldom further metabolised in anaerobic wastewater treatment systems. Scarce information is currently available documenting the anaerobic mineralization of aromatic amines, previously generated from the reduction of azo dyes, which contain hydroxyl and carboxyl groups in their structure (Donlon *et al.* 1997; Razo-Flores *et al.* 1997). Therefore, to achieve efficient removal of azo dyes from textile wastewaters, combined anaerobic-aerobic wastewater treatment systems have been proposed (Field *et al.* 1995).

In contrast to the poor biodegradation of aromatic amines, which generally occur in anaerobic bioreactors, several of the studies listed in Tables 7.5–7.7 indicate the partial or complete removal of many aromatic amines in the aerobic stage. However, some aromatic amines may not be removed. Especially cleavage products from the reactive azo dyes Reactive Black 5 and Reactive Violet 5 were often reported not to be removed aerobically (Sosath *et al.* 1997; Lourenço *et al.* 2000; Luangdilok and Paswad 2000; Panswad and Luangdilok 2000; Lourenço *et al.* 2001; Shaw *et al.* 2002; Lourenço *et al.* 2003; Libra *et al.* 2004). Also a relatively large fraction (~50%) of the aromatic amines from the benzidine-based dye Direct Black 38 resisted removal in the aerobic stage (Sponza and Işik 2005).

Enrichment of specific bacteria capable of degrading aromatic amines seems a suitable approach to enhance the removal of these contaminants in aerobic bioreactors. For instance, the degradation of an aromatic amine, sulphanilic acid, could only be achieved after bio-augmentation with a proper bacterial culture (Tan *et al.* 2000). Moreover, the observation that addition of sludge from a textile wastewater treatment facility decreased the toxicity of the effluent from an azo dye treating baffled reactor (Gottlieb *et al.* 2003) suggests the involvement of specific bacteria.

The main limitations identified during the biodegradation of aromatic amines in aerobic wastewater treatment systems are: 1) the uncertainty of the fate of aromatic amines during the aerobic processes, and 2) the prevalence of recalcitrant aromatic amines, which remain unaffected even at long HRT. For the first point, it is not clear whether the removal of aromatic amines is due to biodegradation, adsorption or chemical reactions in many cases and following these phenomena is a difficult task. Surprisingly, although auto-oxidation of aromatic amines during aerobic treatment has been suggested in some studies (Sosath and Libra 1997; Kalyuzhnyi and Sklyar 2000; Cruz and Buitrón 2001; Tan 2001; Işik and Sponza 2004b; Libra et al. 2004), a minor decrease of the colour was much more often observed (Jianrong et al. 1994; An et al. 1996; Basibuyuk and Forster 1997; Sosath et al. 1997; Kuai et al. 1998; Luangdilok and Paswad 2000; O'Neill et al. 2000b; Panswad and Luangdilok 2000; Rajaguru et al. 2000; Panswad et al. 2001a; Sarsour et al. 2001; Shaw et al. 2002; Kapdan et al. 2003; Işik and Sponza 2004c; Sponza and Işik 2004; Kapdan and Alparslan 2005; Ong et al. 2005). Since many of the azo dyes treated in these studies yield aromatic amines that are expected to auto-oxidise, the latter observation suggests, in several cases, removal of these compounds or their auto-oxidation products from the water phase. However, in several cases it will not be possible to achieve complete removal of dye residues, nor of colour, from the wastewater. The often-observed further colour removal in the aerobic phase of the bioreactor studies, in contrast to the expectation that recalcitrant aromatic amines will tend to auto-oxidise forming coloured products, suggests that many aromatic amines can in fact be removed from the water phase. Further research should be focused on the competition between biodegradation and auto-oxidation of aromatic amines in aerobic bioreactors (Van der Zee and Villaverde 2005).

Furthermore, the nature and potential hazardousness of recalcitrant dye residues in the final effluent of anaerobic-aerobic biological treatment systems needs to be clarified. The reactor studies that evaluated toxicity all showed promising results, i.e. far-reaching decrease or even complete loss of toxicity in the aerobic phase. However, the amount of data is limited and most data concern direct toxicity, usually bioluminescence suppression or respiration inhibition, in contrast to genotoxicity or mutagenicity. Only one paper also evaluated geno-toxicity during anaerobic-aerobic treatment of an azo dye. The dye, hydrolyzed Reactive Black 5, was found weakly genotoxic, both before and after reduction, but no genotoxicity could be detected in the effluent of an anaerobic-aerobic baffled reactor that treated the dye (Gottlieb et al. 2003). The results of this study also showed severe genotoxicity of (1-amino-2-naphthol from) reduced Acid Orange 7, in contrast to no genotoxicity of 1-amino-2-naphthol-6-sulfonate, which is in agreement with the lower level or absence of genotoxicity/mutagenicity/carcinogenicity of sulfonated aromatic amines in comparison to some of their non-sulphonated analogues (Chung and Cerniglia 1992; Jung et al. 1992). This is something highly relevant for the toxicity of textile-processing wastewaters, since most water-soluble dyestuffs used for textile dyeing contain one or more sulphonate groups. Furthermore, it should be noted that over the last decade, azo dyes that could breakdown to carcinogenic aromatic amines have been largely phased out in Europe (Pinheiro et al. 2004). Although it may thus be expected that the fraction of recalcitrant azo dye residues in the final effluent of anaerobic-aerobic biological treatment systems would not pose a hazard, careful evaluation of its (geno)toxicity still remains recommendable.

Combining compound analysis with (geno)toxicity measurements will deliver insight in the size, composition, and potential harm of the recalcitrant fraction, thereby providing the information needed to judge whether discharge can be allowed or whether tertiary treatment (e.g. advanced oxidation, adsorption, coagulation) will have to be applied.

A realistic scenario indicates that combined anaerobic-aerobic wastewater treatment systems are not able to completely remove both COD and colour from textile effluents. The last observation is pertinent considering the data shown in Tables 7.5–7.7 and the fact that there are several constituents of textile wastewaters, besides azo dyes, which may persist biological degradation. Therefore, bioreactors should be combined with physical-chemical treatment processes in order to achieve legislation standards.

7.2.4.2 Non-Biological wastewater treatment systems

Physical-chemical methods

The coagulation-flocculation method is one of the most widely used processes in textile wastewater treatment plants in many countries. It can be used either as a pre-treatment, post-treatment or even as a main treatment system (Gähr et al. 1994; Marmagne and Coste 1996). By these methods, chemicals such as ferric salts or aluminium polychloride are dosed to form flocs with the dyes, and then separated by filtration or sedimentation. Another strategy considers the addition of polyelectrolyte during the flocculation phase to improve the settleability of the flocs (Vandevivere et al. 1998). Marmagne and Coste (1996) reported that coagulation-flocculation methods were successfully applied for colour removal of sulphur and disperse dyes, whereas acid, direct, reactive and vat dyes presented very low colour removal by this method. The main disadvantages of physical-chemical methods are: 1) low colour removal efficiency with some dyes, 2) production of high volumes of polluted sludge, which then need to be treated as well, and 3) demand of large quantities of chemicals (Robinson et al. 2001).

Chemical Methods

Chemical oxidation typically involves the use of an oxidising agent such as ozone (O_3), hydrogen peroxide (H_2O_2), permanganate (MnO_4) etc., to change the chemical composition of a compound or a group of compounds, e.g. dyes (Metcalf and Eddy 2003). Among these oxidants, ozone is the most widely used because of its high reactivity with many dyes, usually providing good colour removal efficiencies (Alaton et al. 2002). The oxidation with ozone can be designed in such a way that only —N$=$N— bond scission occurs, and biodegradable compounds remain non-oxidised, in a process called selective oxidation (Boncz 2002). However, disperse dyes and those insoluble in water represent a drawback for the process, as well as the high cost of ozone (Hassan and Hawkyard 2002). The relatively low efficiency of conventional chemical oxidation techniques to achieve both colour and COD removal have been overcome by the development of the so-called AOP. In this process, oxidizing agents such as O_3 and H_2O_2 are used with catalysts (Fe, Mn and TiO_2), either in the presence or absence of an irradiation source (Alaton et al. 2002). Consequently, an improvement in the generation and use of the free hydroxyl radical ($HO^{\bullet}$) is obtained, representing a rate increase of one to several orders of magnitude, compared with normal oxidants in the absence of a catalyst (Ince and Tezcanli 1999). Table 7.8 gives an indication of the oxidative power of the hydroxyl radical based on the electrochemical oxidation potential (EOP) capacity, compared with other oxidants.

At present, many different combinations of these AOP have been investigated for colour removal, all of which are capable of producing $HO^{\bullet}$ radicals. The first example is a reaction called the **Fenton reaction**, in which hydrogen peroxide is added in an acid solution (pH 2–3) containing Fe^{+2} ions:

$$Fe^{+2} + H_2O_2 \rightarrow Fe^{3+} + HO^{\bullet} + HO^{-} \tag{7.1}$$

In comparison to ozonation, this method is relatively cheap and also presents high COD removal and decolourisation efficiencies (Van der Zee 2002). The main process drawbacks are the high sludge generation due to the flocculation of reagents and dye molecules (Robinson et al. 2001), as well as the need for decreasing the bulk pH to acidic conditions.

Hassan and Hawkyard (2002) reported that a pre-ozonation of coloured wastewaters prior to Fenton reaction not only considerably accelerated the overall colour removal rates, but also decreased the sludge generation.

Table 7.8. Oxidation capacity of different oxidants in terms of electrochemical oxidation potential (EOP) (Adapted from Metcalf and Eddy 2003)

Oxidizing agent	EOP (V)
Fluorine	3.06
Hydroxyl radical	2.80
Oxygen (atomic)	2.42
Ozone	2.08
Hydrogen peroxide	1.78
Hypochlorite	1.49
Chlorine	1.36
Chlorine dioxide	1.27
Oxygen (molecular)	1.23

In the **H_2O_2/UV** process, $HO^\bullet$ radicals are formed when water-containing H_2O_2 is exposed to UV light, normally in the range of 200–280 nm (Metcalf and Eddy 2003). The H_2O_2 photolysis follows the reaction:

$$H_2O_2 + UV \ (\text{or } h\nu, \lambda \approx 200\text{–}280\,\text{nm}) \rightarrow HO^\bullet + HO^\bullet \qquad (7.2)$$

This process is the most widely used AOP technology for the treatment of hazardous and refractory pollutants present in wastewaters, mainly because no sludge is formed and a high COD removal is achieved in a short HRT (Safarzadeh et al. 1997). Additionally, it has successfully been applied for colour removal. For instance, more than 95% decolourisation was achieved during the treatment of wastewaters containing reactive, basic, acid and direct dyes at pH 5, whereas disperse and vat dyes were only partially decolourised (Yang et al. 1998). A comparative study between ozone and H_2O_2/UV was carried out in treating a concentrated reactive dyebath from a textile factory. The H_2O_2/UV system presented decolourisation rates close to those rates obtained with ozone but with a lower cost (Alaton et al. 2002). In some cases, however, the H_2O_2/UV process presents low COD and colour removal efficiency due to inefficient use of UV light (mainly for highly coloured wastewaters) (Moraes et al. 2000), or because of the low molar extinction coefficient of H_2O_2 (specific oxidation capacity), requiring high dosages of the latter.

The addition of catalysts, such as TiO_2, to UV-based methods has also been shown to distinctly enhance colour removal (So et al. 2002; Grzechulska and Morawski 2002). Thus, different combinations such as Ozone/TiO_2, Ozone/TiO_2/H_2O_2 and TiO_2/H_2O_2, have been investigated, but the performance of these systems is enormously influenced by the type of dye, dye concentration and pH (Galindo et al. 2000). Recently, the utilization of solar technologies instead of UV-based methods has been attracting attention (Wang 2000).

Physical Methods

Filtration methods, such as ultrafiltration, nanofiltration and reverse osmosis have been used for water reuse and for the recovery of chemicals. In the textile industry, these filtration methods can be used for both filtering and recycling not only for pigment-rich streams, but also for mercerising and bleaching wastewaters. The specific temperature and chemical composition of the wastewater determine the type and porosity of the filter to be applied (Porter 1997). The main drawbacks of membrane technologies are the high investment costs, the potential membrane fouling and the production of a concentrated streams which need to be treated (Robinson et al. 2001). The recovery of concentrates from membranes can attenuate the treatment costs. For instance, caustic recovery, i.e.

recovery of the sodium hydroxide used in the mercerising step, or the recovery of sizing agents such as PVA (Porter 1997). The water reuse from dyebath effluents has successfully been achieved by using reverse osmosis. However, a coagulation and micro-filtration pre-treatment was necessary to avoid membrane fouling (Vandevivere *et al.* 1998).

Adsorption methods to achieve colour removal are based on the high affinity of many dyes for adsorbent materials. Decolourisation by adsorption is influenced by some physical-chemical factors such as dye-adsorbent interactions, adsorbent surface area, particle size, temperature, pH and contact time (Mattioli *et al.* 2002). The main criteria for the selection of an adsorbent should be based on characteristics such as high affinity and capacity for target compounds and the possibility of adsorbent regeneration (Karcher *et al.* 2001). Activated carbon (AC) is the most common adsorbent and can be very effective to remove many dyes (Walker and Weatherley 1997). However, its efficiency is directly dependent upon the type of carbon material used and the wastewater characteristics, i.e. types of dyes present in the stream. Additionally, AC is relatively expensive and has to be regenerated offsite with losses of about 10% in the thermal regeneration process. In order to decrease the adsorbent losses during regeneration, new adsorbent materials have been tested for their ability for on-site regeneration. Karcher *et al.* (2001) studied alternative materials such as zeolites, polymeric resins, ion exchangers and granulated ferric hydroxide. It was found that zeolites and micro-porous resins were unsuitable due to their low sorption capacity. Although the ion exchangers provided good sorption capacity, regeneration was sometimes difficult. A number of low-cost adsorbent materials such as peat, bentonite clay, fly ash, and agriculture wastes, have been investigated for colour removal (Ramakrishna and Viraraghavan, 1997; Elizalde-González and Hernández-Montoya, 2008). However, the efficiency of these materials varied with the dye class. For instance, fly ash presented high sorption affinity for acid dyes, whereas peat and bentonite presented high affinity for basic dyes. Table 7.9 summarises the advantages and drawbacks of some non-biological decolourisation processes applied to textile wastewaters.

7.2.5 Case study: Biological treatment of textile wastewater

7.2.5.1 Description

The textile factory described in this case study is located in The Netherlands and produces fabrics and technical textiles for clothing. A wide range of processes

Table 7.9. Advantages and drawbacks of some non-biological decolourisation processes applied to textile wastewaters (after Robinson *et al.* 2001)

Physical/ chemical methods	Method description	Advantages	Disadvantages
Fenton reagents	Oxidation reaction using mainly H_2O_2-Fe(II)	Effective decolourisation of both soluble and insoluble dyes	Sludge generation
Ozonation	Oxidation reaction using ozone gas	Application in gaseous state: no alteration of volume	Short half-life (20 min)
Photochemical	Oxidation reaction using mainly H_2O_2-UV	No sludge production	Formation of by-products
NaOCl	Oxidation reaction using Cl^+ to attack the amino group	Initiation and acceleration of azo-bond cleavage	Release of aromatic amines
Electrochemical destruction	Oxidation reaction using electricity	Breakdown compounds are non-hazardous	High cost of electricity
Activated carbon	Dye removal by adsorption	Good removal of a wide variety of dyes	Very expen-sive
Membrane filtration	Physical separation	Removal of all dye types	Concentrated sludge production
Ion exchange	Ion exchange resin	Regeneration: no adsorbent loss	Not effective for all dyes
Electrokinetic coagulation	Addition of ferrous sulphate and ferric chloride	Economically feasible	High sludge production

take place in the factory, including weaving, desizing and scouring, bleaching, mercerizing, dyeing and finishing (Dos Santos *et al.* 2006).

7.2.5.2 WWTP set-up and performance

Part of the factory's wastewater, consisting of a mixture of effluents from dyeing, scouring, bleaching and desizing processes, is treated in an anaerobic-aerobic treatment plant to reduce its COD load and colour. In the plant, approximately 360 m^3 of combined textile wastewater (with a COD concentration average of 4500 mg/L) is treated per day. The dyes used in the factory are mainly disperse, vat and reactive dyes, of which 70% are insoluble. Additionally, the wastewater

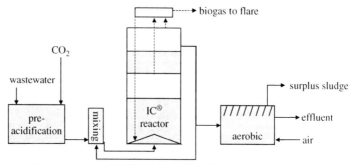

Figure 7.3. Scheme of the WWTP located in The Netherlands

contains concentrations of sulphate varying from 200 to 1000 mg/L. The first part of the treatment plant is an equalisation/pre-acidification tank, from which waste-water is pumped into a mixing tank for pH and temperature control (Figure 7.3).

After the necessary adjustments, the wastewater is transferred to a mesophilic anaerobic Internal Circulation (IC^R) reactor (see Figure 7.4), resulting in a COD removal efficiency of 35–55% and a decolourisation of 80–95%. COD removal was affected by the presence of sulphate; at high sulphate concentrations less COD was removed. Post-treatment of the anaerobic effluent takes place in an aerobic basin equipped with plate settlers, after which a total COD reduction of

Figure 7.4. Photograph of the anaerobic reactor at the textile factory located in The Netherlands

80%–90% is achieved. During the aerobic treatment step a large part of the anaerobically non-degradable dyes and dye degradation products are either degraded or adsorbed to the sludge. Finally, the treated effluent is discharged into the sewer, which takes the wastewater to the municipal WWTP. The remaining factory wastewater that is not treated in the previously described facilities is aerated and neutralised in a separate basin.

Many dyes and their degradation products are toxic, as well as other chemicals used in the textile industry, and therefore the toxicity of the wastewater before and after biological treatment was investigated. Using the Microtox[R] method, it was shown that both the pre-acidified wastewater and the anaerobic effluent were toxic, but that after aerobic treatment the toxicity had disappeared. When the anaerobic reactor was bypassed, the aerobic effluent was found to be toxic, indicating a main contribution of anaerobic treatment in detoxifying the textile effluent. Dos Santos *et al.* (2006) report several strategies to achieve water reuse and products recovery in this WWTP.

7.3 EXPLOSIVE WASTE TREATMENT

7.3.1 Introduction

Several explosive compounds were developed during the first decades of the 20th century. Testing and use of these explosives has extensively contaminated soil, sediments, and water with toxic explosive residues at several government installations. During the Cold War, the production and application of explosives greatly increased causing direct and indirect contamination of the environment (Chen *et al.* 2007). Explosives are classified as environmental hazards and priority pollutants by the U. S. Environmental Protection Agency (USEPA) due to their high toxicity and potential mutagenicity (Druker 1994; Noyes 1996; U.S. Army Environmental Center 2000).

Large amounts of highly explosive wastes (HEW) need to be treated. The United States, Europe, and the former Soviet Union have produced more than 360,000 tons of HEW per year. It is estimated that 2,4,6-trinitrotoluene (TNT) alone is produced in amounts close to 1×10^6 kg a year. Several research studies have documented that the toxicity of these energetic chemicals can be accumulated in bodies of aquatic organisms and terrestrial species, such as earthworms and mammals (Chen *et al.* 2007). Under the Intermediate-Range Nuclear Forces Treaty (INF) and Strategic Arms Reduction Treaties, the U.S. Department of Defence demilitarisation program destroys a significant amount of weapons, which generates large amounts of high explosive contaminated wastewaters that need to be treated (Heilmann *et al.* 1994; Maloney *et al.* 1999; Zoh and

Stenstrom 2002). In this section, the characteristics of explosive wastewaters and the available technologies to treat them will be described.

7.3.2 Characterisation of wastewaters containing explosives

Energetic materials include "explosive wastes" generated during the production and handling of explosive chemicals, which are mainly composed of propellants, explosives, and pyrotechnics (PEP). These materials are susceptible to initiation, or self-sustained energy release, when they are present in sufficient quantities and exposed to stimuli (*e.g.* heat, shock, and friction). Each reacts differently; all will burn, but explosives and propellants can detonate under certain conditions (Chen *et al.* 2007). Figure 7.3 outlines the various categories of energetic materials.

Military installations discharge wastewaters containing different types of explosives due to improper handling during manufacturing, loading, assembly, and packing (LAP) of munitions. Further release of explosives from these sites has been associated to washout or deactivation/demilitarisation operations. The LAP generates wastewaters from cleanup operations. Deactivation is accomplished by washing out or steaming out the explosives from bombs and shells. Water streams contaminated with explosives are usually subdivided into two categories based on the colour of the wastewater:

- Red water, which exclusively originates from the manufacture of TNT.
- Pink water, which includes any water stream associated with LAP operations or with the deactivation of munitions involving contact with finished TNT.

Table 7.10 lists the most common highly explosive compounds produced for military purposes. Explosives and contaminants associated with their production and handling enter natural sub-surfaces from several types of sources:

- Production facilities (*e.g.* wastewater lagoons and filtration pits).
- Solid waste destruction facilities (*e.g.* burn pits and incineration wastes).
- Packing and warehouse facilities.
- Dispersed, unexploded ordnance (*e.g.* from firing ranges)

The wastewater from manufacturing and packing operations has posed the greatest threat to groundwater. In the United States, royal demolition explosives (RDX) are currently the most important military highly explosive compounds

Table 7.10. Common highly explosive compounds (Chen *et al.* 2007)

TNT (2,4,6-Trinitrotoluene)	Picrates
RDX (Cyclo-1,3,5-trimethylene-2,4,6-trinitroamine)	TNB (Trinitrobenzenes)
Tetryl (*N*-Methyl-*N*, 2,4,6-tretanitrobenzeneamine)	DNB (Dinitrobenzenes)
2,4-DNT (2,4-Dinitrotoluene)	Nitro-glycerine
2,6-DNT (2,6-Dinitrotoluene)	Nitro-cellulose
HMX (1,3,5,7-Tetranitro-1,3,5,7-tetraazocyclooctane)	AP (Ammonium perchlorate)
Nitro-aromatics	

since TNT is no longer produced, although it still is a widely used military explosive. Energetic materials are classified into propellants, explosives, and PEP as shown in Figure 7.3. Characterisation and environmental impact of PEP are summarised in the following sections.

7.3.2.1 Explosives

Explosives are commonly classified as primary and secondary based on their susceptibility to initiation. Primary explosives, such as lead azide and lead styphenate are highly susceptible to initiation; as they are used to ignite secondary explosives, they are referred to as initiating explosives (Chen *et al.* 2007).

Secondary explosives include TNT, cyclo-1,3,5-trimethylene-2,4,6-trinitro-amine, commonly known as royal demolition explosives (RDX or cyclonite), cyclo-1,3,5,7-tetramethylene-2,4,6,8-tetranitroamine, commonly known as high melting explosives (HMX), and tetryl. Secondary explosives are much more prevalent at military installations than primary explosives (Chen *et al.* 2007). The general chemical and physical properties of TNT, RDX, and HMX are summarised in Table 7.11.

The nitro-aromatic contaminants found in water streams from army ammunition plants are highly toxic and suspected carcinogens. Therefore, these wastewaters, if released untreated, would cause serious environmental problems in surface and sub-surface water ecosystems, thereby affecting public health. For example, HMX has adverse effects on the central nervous system in mammals and has also been classified as a class D carcinogen by the USEPA. Thus, army facilities are prohibited from discharging wastewaters into the environment prior to meeting a discharge limit of 1 mg/L of total nitro-aromatic compounds. The USEPA has ambient criteria of 0.06 mg/L for TNT and 0.3 mg/L for RDX/HMX, and drinking water criteria of 0.049 mg/L for TNT and 0.035 mg/L for RDX/HMX (Heilmann *et al.* 1994; Maloney *et al.* 1999; Zoh and Stenstrom 2002, U.S. Department of Defence 2003).

Table 7.11. Properties of important explosives wastes (Chen *et al.* 2007)

Munitions	TNT	RDX	HMX
Chemical formula	$C_7H_5N_3O_6$	$C_3H_6N_6O_6$	$C_4H_8N_8O_8$
Molecular weight	227.15	222.26	296.16
Physical properties	Light yellow or buff crystalline solid; soluble in alcohol, ether and hot water, detonates at around 240°C	Colourless polycrystalline material; soluble in acetone, insoluble in water, alcohol, carbon tetrachloride, and carbon disulphide; slightly soluble in methanol and ether	Colourless polycrystalline material; similar to replaced RDX, but has a higher density and much higher melting point; highly soluble in dimethylsulphoxide
Usage	Most common military explosive due to its easy of manufacture and its suitability for melt loading, either as the pure explosive or as binary mixtures with RDX or HMX	Used extensively as the base charge in detonators. Its most common uses are as an ingredient in TNT-based binary explosives and as the primary ingredient in plastic-bonded explosives. Mixtures are used as the explosives fill in almost all types of munitions	Because of its high density, HMX has replaced RDX in explosive applications for which energy and volume are important. It is used in TNT-based binary explosives, as the main ingredient in high-performance plastic bonded explosives, and in high-performance solid propellants
Hazard	Flammable, dangerous fire risk, moderate explosive risk. Toxic by ingestion, inhalation, and skin absorption	Highly explosive, easily initiated by mercury fulminate. Toxic by inhalation and skin contact. 1.5-fold more powerful than TNT	Highly explosive; toxic by inhalation and skin contact

7.3.2.2 Propellants

Propellants include both rocket and gun propellants. Most rocket propellants are either based on: (a rubber binder, ammonium perchlorate (AP) oxidiser and a powdered aluminium fuel; or (b) a nitrate ester, usually nitro-glycerine (NG), nitro-cellulose (NC), HMX, AP, or polymer-bound low NC. If a binder is used, it is normally an isocyanate-cured polyester or polyether. Some propellants contain combustion modifiers, such as lead oxide. Gun propellants are usually single base (NC), double base (NC and NG), or triple base (NC, NG and nitro-guanidine). Some of the newer, lower vulnerability gun propellants contain binders and crystalline explosives and are similar to PBX (Chen *et al.* 2007).

7.3.2.3 Pyrotechnics

Pyrotechnics include illuminating flares, signalling flares, coloured and white smoke generators, tracers, incendiary delays, fuses, and photo-flash compounds. Pyrotechnics are composed of an inorganic oxidiser and metal powder in a binder. Illuminating flares contain sodium nitrate, magnesium, and a binder. Signalling flares contain barium strontium, or other metal nitrates (Chen *et al.* 2007).

7.3.3 Treatment technologies for wastewaters containing explosives

The selection of treatment technologies for wastewaters containing explosives must consider a site-by-site basis (Hampton and Sisk 1997). For instance, the toxicity of explosives limits the applicability of biological treatment processes when concentrations are high, or the treatment process may produce recalcitrant reaction by-products (Schmelling *et al.* 1996). Conversely, energy-intensive chemical treatments, such as incineration, may be too expensive at low concentrations, or may cause other environmental problems, such as NOx emissions (see chapter 11). In the following sections, the application of different treatment processes for the removal of explosives from wastewaters will be discussed.

7.3.3.1 Adsorption

Several commonly used remediation technologies for the removal of explosives from wastewaters are based on separations rather than destructive processes. Separation techniques include resin adsorption, surfactant complexing, liquid-liquid extraction, ultra-filtration and reverse osmosis, which concentrate the explosives compounds from water into another medium for further treatment or for land disposal, but do not modify them to non-hazardous materials (Rodgers and Bunce 2001).

Granular activated carbon (GAC) is a non-specific adsorbent that can treat contaminated water streams at explosives manufacturing plants, although it is unsuited to treating red water (Wujcik *et al.* 1992; Edwards and Dennis 1999). During adsorption processes, water is pumped through columns packed with carbon, to which contaminants adsorb. GAC is more suitable as a polishing technique since the carbon bed has high affinity for organic compounds, but finite capacity. The performance of a GAC bed depends on the type of carbon, pore size, hydraulic loading and operating temperature. Oil and grease must be absent, as they foul the bed, and removal of suspended solids is important to avoid excessive pressure drop (Rodgers and Bunce 2001).

Adsorption of contaminants in GAC columns usually consists of two beds in series, with the first being taken out of service when break-through occurs. The spent GAC is classified as a K045 hazardous waste and must be further treated (Levsen et al. 1993). Options include disposal in a secure hazardous landfill, incineration, or regeneration by partial (5–15%) oxidation of the GAC, during which explosives are desorbed and burned. An alternative regeneration technology involves bioremediation of the spent GAC column off-line by inoculation with explosive-degrading microorganisms and nutrients, followed by aerobic transformation at 55°C (NDCEE 1995).

Furthermore, GAC has also been used as the supporting medium in fluidised-bed bioreactors, in which a microbial biofilm on the GAC creates anaerobic conditions suitable to achieve the reductive transformation of nitroaromatic explosives to aromatic amines (Maloney et al. 1998).

7.3.3.2 Advanced Oxidation Processes (AOPs)

Several AOPs have been applied for the removal of explosives from contaminated waters. For instance, UV oxidation (UVO) with ozone has been used in batch and continuous modes to treat both process waters and groundwater contaminated with nitroaromatic explosives, with destruction efficiencies of up to 99.9% for TNT and TNB (Edward and Dennis 1999b). Although UVO has been proved at several test sites, it lags far behind GAC in implementation (Edward and Dennis 1999a). One criticism of UV+ozone technology is that despite removing explosives from contaminated waters, it does not necessarily ameliorate toxicity. Mcphee et al. (1993) claimed that TNB was one of the degradation products of TNT, which presumably maintained the toxicity of the solution during a 96 h rainbow trout test, and required much higher UV doses to remove it.

Beltran et al. (1998a) found that (non-photolytic) ozonation of nitrobenzene and 2,6-DNT proceeded best below 30°C and at pH 7–9, but the reaction was inefficient because nitro groups deactivate aromatic compounds towards electrophilic attack. Ozonation was almost completely inhibited by hydroxyl radical scavengers, indicating that attack of OH$^•$ on the substrate is the principal reaction pathway. Nitro-phenols appeared as main ozonation products from nitro-benzene and di-nitro-benzaldehydes from DNT. Mcphee et al. (1993) suggested that UV/O$_3$ was superior to UV/H$_2$O$_2$ for removing TNT from contaminated groundwater;

since both techniques generate reactive OH$^•$ radicals, the reason may be better light absorption by ozone.

Fenton's reagent has also been applied for the removal of nitroaromatic explosives from wastewaters. For example, Li *et al.* (1998) used Fenton's reaction, alone or in combination with UV, for remediating contaminated waters containing nitroaromatic explosives. UV promoted a faster oxidation and higher extent of mineralisation of the explosives, and was more effective at pH 6 than at pH 3 (possibly because Fe^{3+} precipitates above pH~5). Furthermore, the reaction rate depended on the aromatic ring: 2-nitrotoluene > 4-nitrotoluene > 2,4-DNT > 2,6-DNT > TNT, consistent with electrophilic attack by the hydroxyl radical.

Spanngord *et al.* (2000a) compared the (non-photolytic) rates of oxidation of 2-amino-2,4-DNT and 4-amino-2,6-DNT (both biological reduction products of TNT) by ozone and Fenton's reagent in buffers and in natural water. The reaction with ozone was faster and more efficient, even though the rate constants for reaction with OH$^•$ were much larger than those for O_3; when both oxidants were present, H_2O_2 inhibited the oxidation by ozone. The proposed mechanism of oxidation by ozone is an initial 1,3-dipolar cyclo-addition between O_3 and the amino-dinitrotoluene, based on the evidence of a 2:1 stoichiometry (O_3: substrate) and the formation of pyruvate and glyoxylate as primary products. The nitrogen atoms of the substrates were eventually converted to NO_2^- and NO_3^- (Spanngord *et al.* 2000b).

Hydrothermal technologies, which include wet air oxidation and supercritical water oxidation, are AOPs that involve mineralisation of organic contaminants by oxygen or H_2O_2 at high temperatures and pressures (Rodgers and Bunce 2001). This treatment concept was applied in a pilot plant, with the capacity to treat 0.4 GPM of contaminated water, for the oxidation of DNT and TNT at the Radford army ammunition plant (Loeffler 1998). DNT concentrations were decreased from 128,000 to 5 ppb (99.996% removal) obtaining molecular nitrogen, water and carbon dioxide as reaction products. Problems with this technique include energy intensity and NOx emissions; previous problems with corrosion of the pressure vessels were avoided by using a dual shell reactor (Loeffler 1998). A pilot scale system operating at 275°C achieved > 99.9% destruction of TNT, with no build-up of intermediates and no residual toxicity (Hawthorne *et al.* 2000).

7.3.3.3 Chemical reduction

Nitroaromatic compounds are catalytically hydrogenated (reduced) to amines under acidic conditions. Hydrogenation alone is not a complete remediation approach since anilines are toxic and must be treated further, for example by oxidative polymerisation, which promotes their precipitation, or by biological transformation. Noble metal catalysts, such as palladium, are too costly for water treatment (Kralik *et al.* 1998); moreover, in certain instances the catalyst dissolves in the presence of nitroaromatic explosives (Neri *et al.* 1997).

Rajashekharam *et al.* (1998) hydrogenated 2,4-DNT on 10% Ni/Zeolite Y, using a trickle-bed reactor at 45–55°C, and developed a model to predict hydrogenation rates at different particle sizes and gas/liquid velocities. Malyala and Chaudhari (1999) found that the reaction tended to zero kinetic order in 2,4-DNT at high organic and H_2 concentration, indicating that the reaction was not mass transfer limited; adsorbed 2,4-DNT was proposed to react with dissociatively adsorbed hydrogen in the rate-limiting step.

Hintze and Wagner (1992) reduced nitroaromatic explosives to amines in the cathode compartment of an electrochemical cell, but did not discuss the disposal of the amine which, as previously mentioned, is by itself toxic. Rodgers *et al.* (1999) reduced TNT and the DNT isomers at modest potentials at a variety of cathodes, with high efficiencies and essentially quantitative material balance, and suggested subsequent electrochemical oxidation to polymerize the amines to insoluble by-products.

7.3.3.4 Biological treatment processes

Several research articles have reported the capacity of different microorganisms to degrade nitroaromatic explosives under both aerobic and anaerobic conditions. Rodgers and Bunce (2001) report an extensive description of microorganisms capable to degrade nitroaromatic explosives and the main parameters affecting the efficiency of biodegradation. Nevertheless, scarce information is available documenting the removal of these contaminants from wastewaters in continuous treatment processes. The biological treatment processes tested include anaerobic fluidised bed reactors (Maloney *et al.* 2002) and anaerobic membrane bioreactors (Zoh and Stenstrom 2002).

7.3.4 Case study: Biological treatment of explosive process wastewater

7.3.4.1 Description

A pilot scale study was conducted to evaluate the capacity of an anaerobic fluidised bed reactor (AFBR) containing GAC to treat *pink water* originated at McAlester Army Ammunition Plant (MCAAP, OK). Different hydraulic conditions were established in the GAC-AFBR and the removal efficiency of TNT and RDX, the main nitro-aromatic components of *pink water*, was monitored. This treatment technology was proposed as a pre-treatment process for *pink water* prior to its discharge to an aerobic wastewater treatment system (Maloney *et al.* 2002).

7.3.4.2 Pilot plant set-up and performance

The GAC-AFBR was a 20 in. (51 cm) diameter column with an overall height of 15 ft (4.9 m) and a bed of GAC occupying 11 ft (3.4 m). Water was recycled through the column continuously at 114 l/min to keep the GAC fluidised, and *pink water* for treatment was pumped into the recycling line (Figure 7.5). Nutrients and co-substrate (electron donor) were also fed into the recycling line. The treatment system was operated with a pH set point of 6.8–7 and a temperature of 32°C.

Table 7.12 summarises the performance of the GAC-AFBR under different experimental conditions regarding TNT biodegradation. Several hydraulic conditions were tested to determine the proper mass loading rate, which the

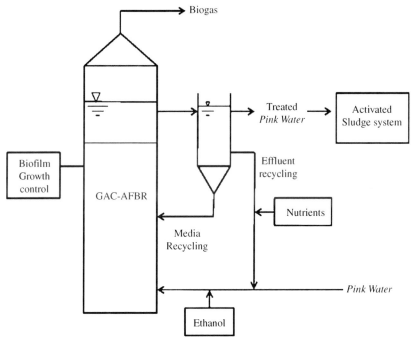

Figure 7.5. GAC-AFBR applied for the treatment of *pink water* (Maloney *et al.* 2002)

Table 7.12. Performance of a GAC-AFBR during the treatment of *pink water* under different experimental conditions (Maloney *et al.* 2002)

Period (days)	Flow (gpm)	Temperature (°C)	HRT (min)	TNT$_{in}$ (mg/L)	TNT$_{out}$ (mg/L)	TNT LR (kg/m^3–d)
1 (1–44)	0.5	41.4	375	31.7	<0.03	42
2 (45–87)	1.0	32.2	188	43.4	<0.03	26.3
3 (87–127)	1.5	32.2	125	·29.2	<0.03	27.8
4 (127–177)	1.0	19.4	188	56.2	2.8	24.4
5 (249–289)	1.0	32.2	188	15.1	0.6	8.6
6 (290–353)	1.0	32.2	188	3.5	<0.03	283

gpm, gallons per minute; HRT, hydraulic residence time; TNT$_{in}$, influent concentration of TNT; TNT$_{out}$, effluent concentration of TNT; TNT LR, TNT loading rate.

reactor could handle while meeting the discharge limitations. Operational period 4 was the only phase in which the discharge limit for TNT (1 mg/L) was exceeded. During this period, the effluent was recycled back through the system for further treatment. The main causes identified for the removal efficiency diminishment were a temperature drop to 19.4°C and a sharp increase in TNT concentration from 29.2 to 56.2 mg/L. Moreover, during this operational period the buffer provided by the GAC became exhausted.

Based on the tests performed, the primary design parameter, TNT loading rate, was set at approximately 0.33 kg/m^3-d. A 50% safety factor was applied to this value decreasing the design applied loading rate to 0.22 kg/m^3-d. For design purposes a system sized at this loading rate was then used to develop a cost comparison between the GAC-AFBR and the existing GAC adsorption treatment.

7.3.4.3 Cost comparison

A cost estimate was developed to compare the existing treatment process, adsorption on GAC, with biological treatment using the GAC-AFBR. The total estimated operating costs for the GAC-AFBR are ~US$19,000 per year. No estimate has been made for manpower costs, because the operation of the existing plant requires approximately the same level of effort as the GAC-AFBR, and the sampling and analysis efforts would be the same for either system. The capital cost for the GAC-AFBR was estimated to be US$195,000. The amortised capital cost (6%, 20 years) for the GAC-AFBR is US$17,000 per year. Thus, the total yearly cost for the GAC-AFBR is approximately US$36,000 per year, which represents approximately half of the current cost of the GAC adsorption treatment system (Maloney *et al.* 2002).

7.4 WASTEWATERS CONTAINING PHARMACEUTICALS

7.4.1 Introduction

The pharmaceutical industry is a major component of the chemical sector. Pharmaceuticals constitute a large and diverse group of compounds applied as human and veterinary medicines, as well as nutraceuticals, which have extensively been used in significant quantities throughout the world. There are 4,500 pharmaceuticals currently available, including experimental drugs in development, which represent about 70% of the total. Drugs are becoming an increasingly complex component of health care. These are defined as substances responsible for physiological or pharmaceutical action and used in the diagnosis, cure, mitigation, treatment, or prevention of disease or as non-food articles intended to affect the structure or any function of the body of man or other animals (Khetan and Collins 2007). They can cure some diseases (e.g., antibiotics), control symptoms (e.g., analgesics), replace or supplement required drugs (e.g., insulin and vitamins), and control the body's self-regulating systems (e.g., high blood pressure and thyroid drugs). Drugs can also serve as complements to medical procedures (e.g., anticoagulants during heart valve replacement surgery), disease prevention agents (e.g., lipid-lowering drugs that lessen the risk of coronary artery obstruction), and new treatments for emerging diseases (e.g., HIV). Thousands of tons of drugs are also used by people yearly to treat illnesses, to prevent unwanted pregnancy, or to face the stresses of modern life (Khetan and Collins 2007).

Despite the benefits brought by pharmaceuticals to cover a wide range of applications in public health, contamination of surface and ground-waters by these compounds has become a priority concern as some of them are beginning to be associated with adverse developmental effects in aquatic organisms at environmentally relevant concentrations, which are usually thought to be harmless. Furthermore, scarce information is available to document the impacts of human exposure to low-dose mixtures of pharmaceuticals or of low-dose pharmaceuticals mixed with other low-dose synthetic pollutants, which increases the uncertainty of their effects on public health (Khetan and Collins 2007).

Improper disposal, leaking storage tanks and spills have caused contamination of soils, wastewaters, aquifers and sediments by chemicals originated from the pharmaceutical sector. Furthermore, significant fractions of the original substances, which are used for the different purposes previously described, often are excreted in un-metabolised form or as active metabolites via urine or feces ending in the sewage. In fact, patient excretion following therapy is widely considered as the main pathway to the environment.

Recent studies have reported the potential of municipal WWTP to remove pharmaceuticals. It has been obvious from these studies that several pharmaceuticals prevail unaffected in conventional wastewater treatment facilities. Thus, advanced wastewater treatment systems are required in order to efficiently remove these contaminants from wastewaters. Several pharmaceuticals are nitrogenous compounds, which are persistent in the environment and may generate serious public health problems. Miège *et al.* (2009) developed a comprehensive database documenting the most frequently detected pharmaceuticals, their main concentrations and removal efficiency observed in different treatment technologies applied for municipal wastewater (Figure 7.6). Furthermore, the authors underlined the main operational parameters (SRT and HRT) affecting the removal efficiency of these contaminants. Processes with nitrogen treatment (*i.e.*, primary treatments with physic-chemical treatment, membrane bioreactors, submerged biofilters and some activated sludge systems), which are characterised by high HRT (> 12 h) and high SRT (> 10 d), are more efficient in removing pharmaceuticals than processes without nitrogen treatment (*i.e.*, primary treatment without physic-chemical treatment, fixed biomass reactors, waste stabilisation ponds and some activated sludge systems), which are characterised by low HRT and SRT and which only provide suspended particles and dissolved carbon removal. Certainly, high retention times provide more time for low rate reactions, such as biodegradation and sorption mechanisms, to occur (Miège *et al.* 2009).

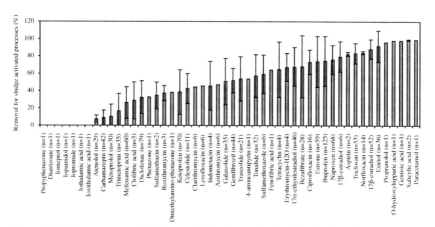

Figure 7.6. Mean removal efficiency (%) and standard deviation for pharmaceutical products in WWTPs with activated sludge processes (Miège *et al.* 2009)

Table 7.13. Performance of different treatment systems applied for high-strength pharmaceutical wastewaters

Treatment system	HRT (h)	COD (mg/L)	TN (mg/L)	Main pharmaceutical products	Treatment system performance	Reference
CSTR[1]	36–50	1010–1290	1016–1062	NR	COD removal, 99%; TN removal, 87–99%	Gupta and Sharma (1996)
TPAT[2]	NR	2800–3650	180–270	NR	COD removal, 48.2% for M-M system; 47.9%, for T-M system	Lapara et al. (2001)
SBR[3]	8–24 (An), 4–12 (Ae)	28400–72200	280–605	Phenols and o-NA	COD removal, 95–97%; TN removal 33%	Buitrón et al. (2003)
UASR[4]	96	6200–7800	314–410	Tylosin and Avilamycin	COD removal, 70–75%; Tylosin removal, 75–95%	Chelliapan et al. (2006)
SAnAeS[5]		9736–19862	NR	Ampicillin and Aureomycin	COD removal, 97.4–98.2%. Ampicillin removal 16–42% (An) and <10% (Ae); Aureomycin removal 25–31% (An) and <10% (Ae)	Zhou et al. (2006)
TPAD-MBR[6]	5, 12, 55	5789–58792	560–980	N,N-DMM and N-N-DME	COD removal, 99.3–99.9%	Chen et al. (2008)

[1]Performance reported for the treatment of pharmaceutical wastewater mixed with urea plant wastewater. [2]Two phase aerobic treatment (TPAT) conducted under mesophilic (M) and thermophilic (T) treatment. [3]SBR operated under sequential anaerobic (An) and aerobic (Ae) conditions. [4]Anaerobic step applied as a pre-treatment in an upflow anaerobic stage reactor (UASR). [5]Sequential anaerobic-aerobic treatment. [6]A CSTR combined with an upflow anaerobic sludge blanket-anaerobic filter (UASBAF) and a subsequent membrane bioreactor.

TN, total nitrogen; NR, not reported; N,N-DMM, N,N-dimethyl-methanamine, N,N-DME, N,N-dimethyl-ethanamine; o-NA, o-nitroaniline.

On the other hand, scarce information is reported in the literature documenting the application of treatment technologies to remove pharmaceuticals from industrial discharges. In the following sections, the main biological technologies proposed to remove pharmaceuticals from industrial wastewaters will be described. Review articles, such as the reported by Ikehata *et al.* (2006) have extensively described the application of physical-chemical treatment processes, such as advanced oxidation processes, for the removal of pharmaceuticals from wastewaters.

7.4.2 Biological treatment processes

The biological technologies explored so far for the treatment of high-strength pharmaceutical wastewaters include completely stirred tank reactors (CSTR), under both anaerobic and aerobic conditions, fluidised bed bioreactors, membrane bioreactors and sequential batch reactors (SBR). Table 7.13 reports the performance of different treatment configurations during the treatment of high-strength pharmaceutical wastewaters. Treatment systems with the largest HRT achieved the highest removal of COD. Nevertheless, the efficiency reported for the removal of particular pharmaceuticals varies depending on the type of contaminant and on the operational conditions prevailing.

Based on the information available in the literature (Table 7.13), the main design criteria to consider for the appropriate treatment of high-strength pharmaceutical wastewaters in biological treatment systems can be underlined:

- HRT: large enough to achieve high removal efficiencies for COD and nitrogen.
- SRT: sufficient to enrich specialised microorganisms to biodegrade recalcitrant pharmaceuticals.
- Loading rate: especially important for highly toxic/inhibitory compounds.

Further studies are essential to clarify the fate of pharmaceuticals in WWTP since only dissolved concentrations of pharmaceuticals are usually reported. This should not be such a problem for hydrophilic molecules, but cannot be overlooked for more hydrophobic compounds.

ACKNOWLEDGEMENTS

The author wants to thank the financial support of the Council of Science and Technology of Mexico through the Grant SEP-CONACYT 55045.

REFERENCES

Alaton, I.A., Balcioglu, I.A. and Bahnemann, D.W. (2002) Advanced oxidation of a reactive dyebath effluent: Comparison of O3, H2O2/UV-C and TiO2/UV-A processes. *Wat. Res.* **36**, 1143–1154.

Albuquerque, M.G.E., Lopes, A.T., Serralheiro, M.L., Novais, J.M. and Pinheiro, H.M. (2005) Biological sulphate reduction and redox mediator effects on azo dye decolourisation in anaerobic-aerobic sequencing batch reactors. *Enzyme Microb. Technol.* **36**, 790–799

An, H., Qian, Y., Gu, X.S. and Tang, W.Z. (1996) Biological treatment of dye wastewaters using an anaerobic-oxic system. *Chemosphere* **33**(12), 2533–2542.

Basibuyuk, M. and Forster, C.F. (1997) The use of sequential anaerobic/aerobic processes for the biotreatment of a simulated dyeing wastewater. *Environ. Technol.* **18**(8), 843–848.

Beltran, F.J., Encinar, J.M. and Alonso, M.A. (1998a) Nitroaromatic hydrocarbon ozonation in water: 1. Single ozonation. *Ind. Engng. Chem. Res.* **37**, 25–31.

Bisschops, I.A.E. and Spanjers, H. (2003) Literature review on textile wastewater characterisation. *Environ. Technol.* **24**, 1399–1411.

Boncz, M.A. (2002) *Selective oxidation of organic compounds in waste water by ozone-based oxidation processes.* Ph.D. Thesis, Wageningen University, Wageningen, The Netherlands.

Buitrón, G., Melgoza, R.M., Jiménez, L. (2003) Pharmaceutical wastewater treatment using an anaerobic/aerobic sequencing batch biofilter. *J. Environ. Sci. Heal. A.* **A38**(10), 2077–2088.

Cabral Gonçalves, I., Penha, S., Matos, M., Santos, A.R., Franco, F. and Pinheiro, H.M. (2005) Evaluation of an integrated anaerobic/aerobic SBR system for the treatment of wool dyeing effluents. *Biodegradation* **16**(1), 81–89.

Carliell, C.M., Barclay, S.J., Naidoo, N., Buckley, C.A., Mulholland, D.A. and Senior, E. (1995) Microbial decolourisation of a reactive azo dye under anaerobic conditions. *Water SA* **21**(1), 61–69.

Carliell, C.M., Barclay, S.J., Shaw, C., Wheatley, A.D. and Buckley, C.A. (1998) The effect of salts used in textile dyeing on microbial decolourisation of a reactive azo dye. *Environ. Technol.* **19**(11), 1133–1137.

Cervantes, F.J., Enríquez, J.E., Galindo-Petatán, E., Arvayo, H., Razo-Flores, E. and Field, J.A. (2007) Biogenic sulphide plays a major role on the riboflavin-mediated decolourisation of azo dyes under sulphate reducing conditions. *Chemosphere* **68**, 1082–1089.

Cervantes F.J., López-Vizcarra M.I., Siqueiros E. and Razo-Flores E. (2008) Riboflavin prevents inhibitory effects during the reductive decolorization of Reactive Orange 14 by methanogenic sludge. *J. Chem. Technol. Biotechnol.* **83**, 1703–1709.

Cervantes, F.J., Van der Zee, F.P., Lettinga, G. and Field, J.A. (2001) Enhanced decolourisation of Acid Orange 7 in a continuous UASB reactor with quinones as redox mediators. *Water Sci. and Technol.* **44**(4), 123–128.

Chelliapan, S., Wilby, T., Sallis, P.J. (2006) Performance of an up-flow anaerobic stage reactor (UASR) in the treatment of pharmaceutical wastewater containing macrolide antibiotics. *Wat. Res.* **40**, 507–516.

Chen, J.P., Zou, S., Pehkonen, S.O., Hung, Y.-T. and Wang, L.K. (2007) Explosive waste treatment. In *Hazardous industrial waste treatment*. Wang, L.K., Hung, Y.-T., Lo, H.H. and Yapijakis, C. (Eds.).Taylor & Francis, London. pp. 429–440.

Chen, Z., Ren, N., Wang, A., Zhang, Z.P., Shi, Y. (2008) A novel application of TPAD-MBR system to the pilot treatment of chemical synthesis-based pharmaceutical wastewater. *Wat. Res.* **42**, 3385–3392.

Chung, K.T. and Cerniglia, C.E. (1992) Mutagenicity of azo dyes: Structure-activity relationships. *Mutat. Res.* **277**(3), 201–220.

Cruz, A. and Buitrón, G. (2001) Biodegradation of Disperse Blue 79 using sequenced anaerobic/aerobic biofilters. *Water Sci. Technol.* **44**(4), 159–166.

Donlon, B., Razo-Flores, E., Field, J. and Lettinga, G. (1995) Toxicity of N-substituted aromatics to acetoclastic methanogenic activity in granular sludge. *Appl. Environ. Microbiol.* **61**, 3889–3893.

Donlon, B.A., Razo-Flores, E., Luijten, M., Swarts, H., Lettinga, G. and Field, J.A. (1997) Detoxification and partial mineralization of the azo dye mordant orange 1 in a continuous upflow anaerobic sludge-blanket reactor. *Appl. Microbiol. Biotechnol.* **47** (1), 83–90.

Dos Santos, A.B., Bisschops, I.A.E. and Cervantes, F.J. (2006) Closing process water cycles and product recovery in textile industry: perspective for biological treatment. In *Advanced Biological Treatment Processes for Industrial Wastewaters: Principles & Applications*. Cervantes, F.J., Pavlostathis, S.G. & van Haandel A.C. (Eds.). IWA Publishing, London. pp. 298–320.

Dos Santos A.B., Braúna C.H., Mota S. and Cervantes F.J. (2008) Effect of nitrate on the reduction of Reactive Red 2 by mesophilic granular sludge. *Water Sci. Technol.* **57**(7), 1067–1071.

Dos Santos, A.B., Bisschops, I.A.E., Cervantes, F.J. and van Lier, J.B. (2004a) Effect of different redox mediators during thermophilic azo dye reduction by anaerobic granular sludge and comparative study between mesophilic (30°C) and thermophilic (55°C) treatments for decolourisation of textile wastewaters. *Chemosphere* **55**(9), 1149–1157.

Dos Santos, A.B., Cervantes, F.J., Yaya-Beas, R.E. and van Lier, J.B. (2003) Effect of redox mediator, AQDS, on the decolourisation of a reactive azo dye containing triazine group in a thermophilic anaerobic EGSB reactor. *Enzyme Microb. Technol.* **33**(7 SUP), 942–951.

Dos Santos, A.B., Traverse, J., Cervantes, F.J. and van Lier, J.B. (2005) Enhancing the electron transfer capacity and subsequent color removal in bioreactors by applying thermophilic anaerobic treatment and redox mediators. *Biotechnol. Bioeng.* **89**(1), 42–52.

Drucker, M.P. (1994) Pollution control and waste minimization in military facilities. In *Professional Engineering for Pollution Control and Waste Minimization*. Wise, D.L. and Trantolo D. (Eds.) Marcel Dekker, New York.

Edward, E. and Dennis, T. (1999a) Federal Remediation Technologies Roundtable: Common Treatment Technologies for Explosives in Soil Sediment and Sludge. Http://www.plaii.com/matrix/section2/2_10_3.html.

Edward, E. and Dennis, T. (1999b) Federal Remediation Technologies Roundtable: Ultraviolet Oxidation. Http://www.plaii.com/matrix/section4/4_56.html.

Elizalde-González, M.P. and Hernández-Montoya, V. (2008) Fruit seeds as adsorbents and precursors of carbon for the removal of anthraquinone dyes. *Int. J. Chem. Eng.* 1, 243–253.

Field, J.A., Stams, A.J.M., Kato, M. and Schraa, G. (1995) Enhanced biodegradation of aromatic pollutants in cocultures of anaerobic and aerobic bacterial consortia. *Antonie van Leeuwenhoek* 67, 47–77.

FitzGerald, S.W. and Bishop, P.L. (1995) Two stage anaerobic/aerobic treatment of sulfonated azo dyes. *J. Environ. Sci. Health., Part A* 30(6), 1251–1276.

Frijters, C.T.M.J., Vos, R.H., Scheffer, G. and Mulder, R. (2005) Decolorizing and detoxifying wastewater of a Dutch textile factory in a full scale combined anaerobic / aerobic system. Submitted for publication.

Gähr, F., Hermanutz, F. and Oppermann, W. (1994) Ozonation - an important technique to comply with new German laws for textile wastewater treatment. *Water Sci. Technol.* 30, 255–263.

Galindo, C., Jacques, P. and Kalt, A. (2000) Photodegradation of the aminoazobenzene acid orange 52 by three advanced oxidation processes: UV/H2O2, UV/TiO2 and VIS/TiO2; Comparative mechanistic and kinetic investigations. *J. Photochem. Photobiol., A* 130, 35–47.

Gottlieb, A., Shaw, C., Smith, A., Wheatley, A. and Forsythe, S. (2003) The toxicity of textile reactive azo dyes after hydrolysis and decolourisation. *J. Biotechnol.* 101(1), 49–56.

Grzechulska, J. and Morawski, A.W. (2002) Photocatalytic decomposition of azo-dye acid black 1 in water over modified titanium dioxide. *Appl. Catal. B-Environ.* 36, 45–51.

Guo, J., Zhoua, J., Wanga, D., Tiana, C., Wanga, P., Uddina, M. S. and Yua, H. (2007) Biocalalyst effects of immobilized anthraquinone on the anaerobic reduction of azo dyes by the salt-tolerant bacteria. *Water Res.* 41, 426–432.

Gupta, S.K. and Sharma, R. (1996) Biological oxidation of high strength nitrogenous wastewater. *Wat. Res.* 30(3), 593–600.

Hampton, M.L. and Sisk, W.E. (1997) Emerging Technologies in Hazardous Waste Management. American Chemical Society, Washington, DC. Vol. 9, pp. 252–257.

Harmer, C. and Bishop, P. (1992) Transformation of azo dye AO-7 by wastewater biofilms. *Water Sci. Technol.* 26(3/4), 627–636.

Hassan, M.M. and Hawkyard, C.J. (2002) Ozonation of aqueous dyes and dyehouse effluent in a bubble-column reactor. *J. Environ. Sci. Heal. A.* A37(8), 1563–1579.

Hawthorne, S.B., Lagadec, A.J., Kalderis, D., Lilke, A.V. and Miller, D.J. (2000) Pilot-scale destruction of TNT, RDX, and HMX on contamined soils using subcritical water. *Environ. Sci. Technol.* 34(15), 3224–3228.

Heilmann, H.M.; Stenstrom, M.K.; Hesselmann, R.P.X.; Wiesmann, U. (1994) Kinetics of the aqueous alkaline homogenous hydrolysis of high explosive 1,3,5,7-tetraaza-tetranitrocyclooctane (HMX). *Water Sci. Technol.* 30, 53–61.

Hintze, J.L. and Wagner, P.J. (1992) *TNT Wastewater Feasibility Study: Phase 1 Laboratory Study.* Gencorp Aerojet Propulsion division US Army Missile Command/production base modernization activity, contract number DAAH01-91-C-0738, PP. 1–730.

Ikehata, K., Naghashkar, N.J., Ei-Din, M.G. (2006) Degradation of aqueous pharmaceuticals by ozonation and advanced oxidation processes: A review. *Ozone: Sci. Eng.* 28, 353–414.

Ince, N.H. and Tezcanli, G. (1999) Treatability of textile dye-bath effluents by advanced oxidation: preparation for reuse. *Water Sci. Technol.* **40**, 183–190.

Işik, M. and Sponza, D.T. (2003) Aromatic amine degradation in a UASB/CSTR sequential system treating Congo Red dye. *J. Environ. Sci. Health., Part A* **38**(10), 2301–2315.

Işik, M. and Sponza, D.T. (2004a) Anaerobic/aerobic sequential treatment of a cotton textile mill wastewater. *J. Chem. Technol. Biotechnol.* **79**(11), 1268–1274.

Işik, M. and Sponza, D.T. (2004b) Decolorization of azo dyes under batch anaerobic and sequential anaerobic/aerobic conditions. *J. Environ. Sci. Health., Part A* **39**(4), 1107–1127.

Işik, M. and Sponza, D.T. (2004c) Monitoring of toxicity and intermediates of C.I. Direct Black 38 azo dye through decolorization in an anaerobic/aerobic sequential reactor system. *J. Hazard. Mater.* **114**(1–3), 29–39.

Jiang, H. and Bishop, P.L. (1994) Aerobic biodegradation of azo dyes in biofilms. *Water Sci. Technol.* **29**(10–11), 525–530.

Jianrong, Z., Yanru, Y., Huren, A. and Yi, Q. (1994) A study of dyewaste treatment using anaerobic-aerobic process. In *Proceedings of the Seventh International Symposium on Anaerobic Digestion*, Cape Town, South Africa, 360–363.

Jung, R., Steinle, D. and Anliker, R. (1992) A compilation of genotoxicity and carcinogenicity data on aromatic aminosulphonic acids. *Food Chem. Toxicol.* **30**(7), 635–660.

Kalyuzhnyi, S. and Sklyar, V. (2000) Biomineralisation of azo dyes and their breakdown products in anaerobic-aerobic hybrid and UASB reactors. *Water Sci. Technol.* **41**(12), 23–30.

Kapdan, I.K. and Alparslan, S. (2005) Application of anaerobic-aerobic sequential treatment system to real textile wastewater for color and COD removal. *Enzyme Microb. Technol.* **36**(2–3), 273–279.

Kapdan, I.K., Tekol, M. and Sengul, F. (2003) Decolorization of simulated textile wastewater in an anaerobic-aerobic sequential treatment system. *Process Biochem.* **38**(7), 1031–1037.

Karcher, S., Kornmuller, A. and Jekel, M. (2001) Cucurbituril for water treatment. Part I: Solubility of cucurbituril and sorption of reactive dyes. *Wat. Res.* **35**, 3309–3316.

Khetan, S.K. and Collins, T.J. (2007) Human pharmaceuticals in the aquatic environment: A challenge to Green Chemistry. *Chem. Rev.* **107**, 2319–2364.

Kralik, M., Fisera, R., Zecca, R., and D'Arrchivio A.A. (1998) Modeling of the deactivation of polymer-supported palladium catalysts in the hydrogenation of 4-nitrotoluene. *Coll. Czechoslovak Chem. Commun.* **63**(7), 1074–1088.

Kuai, L., De Vreese, I., Vandevivere, P. and Verstraete, W. (1998) GAC-amended UASB reactor for the stable treatment of toxic textile wastewater. *Environ. Technol.* **19**(11), 1111–1117.

Lapara, T.M., Nakatsu, C.H., Pantea, L.M. and Alleman, J.E. (2001) Aerobic biological treatment of a pharmaceutical wastewater: Effect of temperature on cod removal and bacterial community development. *Wat. Res.* **35**(18), 4417–4425.

Levsen, K., Mussmann, E., Berger-Preiss, E., Preiss, A., Volmer, D. and Wunsch, G. (1993) Analysis of nitroaromatics and nitramines in ammunition wastewater and in aqueous samples from former ammunition plants and other military sites. *Acta Hydrochim. Hydrobiolog.* **21**, 153–156.

Li, Z.M., Shea, P.J. and Comfort S.D. (1998) Nitrotoluene detruction by UV-catalyzed Fenton oxidation. *Chemosphere* **36**(8), 1849–1865.

Libra, J.A., Borchert, M., Vigelahn, L. and Storm, T. (2004) Two stage biological treatment of a diazo reactive textile dye and the fate of the dye metabolites. *Chemosphere* **56**(2), 167–180.

Loeffler, A. (1998) Evaluation of the NITREM process. *National Defense Industrial Association, 24th Environmental Symposium & Exhibition, Tampa Fl.* Paper No. 3, pp. 1–4, www.ndia.ord/events/past/html.

Lourenço, N.D., Novais, J.M. and Pinheiro, H.M. (2000) Reactive textile dye colour removal in a sequencing batch reactor. *Water Sci. Technol.* **42**(5–6), 321–328.

Lourenço, N.D., Novais, J.M. and Pinheiro, H.M. (2001) Effect of some operational parameters on textile dye biodegradation in a sequential batch reactor. *J. Biotechnol.* **89**(2–3), 163–174.

Lourenço, N.D., Novais, J.M. and Pinheiro, H.M. (2003) Analysis of secondary metabolite fate during anaerobic-aerobic azo dye biodegradation in a sequential batch reactor. *Environ. Technol.* **24**(6), 679–686.

Luangdilok, W. and Paswad, T. (2000) Effect of chemical structures of reactive dyes on color removal by an anaerobic-aerobic process. *Water Sci. Technol.* **42**(3–4), 377–382.

Maloney, S.W., Adrian, N.R., Hickey, R.F. and Heine, R.L. (2002) Anaerobic treatment of pinkwater in a fluidized bed reactor containing GAC. *J. Hazard. Mater.* **92**, 77–88.

Maloney, S.W., Engbert, E.G., Suidan, M.T. and Hickey, R.F. (1998) Anaerobic fluidized-bed treatment of propellant wastewater. *Water Environ. Res.* **70**(1), 52–59.

Maloney, S.W., Meenakshisundaram, D., Mehta, M., Pehkonen, S.O. (1999) *Electrochemical reduction of nitro-aromatic compounds, product studies and mathematical modeling*, technical report; U.S. Army, Corps of Engineers, CERL:Champaign, IL; Report Number 99/85,ADANumber 371059, 01.

Malyala, R.V. and Chaudhari, R.V. (1999) Hydrogenation of 2,4-DNT using a supported Ni catalyst: reaction kinetics and semibatch slurry reactor modeling. *Ind. Engng. Chem. Res.* **38**(3), 906–915.

Marmagne, O. and Coste, C. (1996) Color removal from textile plant effluent. *Am. Dyest. Rep.* **84**, 15–21.

Mattioli, D., Malpei, F., Bortone, G. and Rozzi, A. (2002) Water minimization and reuse in textile industry. In *Water Recycling and resource recovery in industry: Analysis, technologies and implementation.* IWA Publishing, Cornwall, UK. p. 677.

Mcphee, W., Wagg, L., Martin, P. (1993) Advanced oxidation processes of the destruction of ordenance and propellant compounds using Rayox. *Solarchem Environ. Systems*, Ontario, Canada 249–265.

Metcalf and Eddy (2003) Wastewater engineering: treatment and reuse. McGraw-Hill, New York, USA.

Mezohegyi, G., Kolodkin, A., Castro, U. I., Bengoa, C., Stuber, F., Font, J. and Fabregat A. (2007) Effective anaerobic decolorization of azo dye Acid Orange 7 in continuous upflow packed-bed reactor using biological activated carbon system. *Ind. Eng. Chem. Res.* **46**, 6788–6792.

Miège, C., Choubert, J.M., Ribeiro, L., Eusèbe, M., Coquery, M. (2009) Fate of pharmaceuticals and personal care products in wastewater treatment plants – Conception of a database and first results. *Environ. Poll.* doi:10.1016/j.envpol.2008.11.045.

Minke, R. and Rott, U. (2002) Untersuchungen zur innerbetrieblichen anaeroben Vorbehandlung stark farbiger Abwässer der Textilveredelungsindustrie. *Wasser Abwasser* **143**(4), 320–328.

Moraes, S.G., Freire, R.S. and Duran, N. (2000) Degradation and toxicity reduction of textile effluent by combined photocatalytic and ozonation processes. *Chemosphere* **40**, 369–373.

NDCEE (1995) *NDCEE Pink Water Treatment Options: National Defense Center for Environmental Excellence Pink Water Treatment Options.* Technical Report. Contract No. DAAA21-93-C-0046 pp. 1–42.

Neri, G., Musolino, M.G., Bonaccorsi, M.G., Donato, L. and Mercadante L. (1997) Catalytic hydrogenation of 4-(hydroxyamino)-2-nitrotoluene and 2,4-nitroamine isomers. *Ind. Engng. Chem. Res.* **36**, 3619–3624.

Noyes, R. (1996) Chemical Weapons Destruction and Explosive Waste/Unexploted Ordnance Remediation. Noyes Publications, New Jersey.

O'Neill, C., Hawkes, F.R., Esteves, S.R.R., Hawkes, D.L. and Wilcox, S.J. (1999a) Anaerobic and aerobic treatment of a simulated textile effluent. *J. Chem. Technol. Biotechnol.* **74**(10), 993–999.

O'Neill, C., Hawkes, F.R., Hawkes, D.W., Esteves, S. and Wilcox, S.J. (2000a) Anaerobic aerobic biotreatment of simulated textile effluent containing varied ratios of starch and azo dye. *Wat. Res.* **34**(8), 2355–2361.

O'Neill, C., Lopez, A., Esteves, S., Hawkes, F.R., Hawkes, D.L. and Wilcox, S. (2000b) Azo-dye degradation in an anaerobic-aerobic treatment system operating on simulated textile effluent. *Appl. Microbiol. Biotechnol.* **53**(2), 249–254.

Ong, S.A., Toorisaka, E., Hirata, M. and Hano, T. (2005) Decolorization of azo dye (Orange II) in a sequential UASB-SBR system. *Sep. Purif. Technol.* **40**, 2907–2914.

Panswad, T. and Luangdilok, W. (2000) Decolorization of reactive dyes with different molecular structures under different environmental conditions. *Water Res.* **34**(17), 4177–4184.

Panswad, T., Iamsamer, K. and Anotai, J. (2001a) Decolorisation of azo-reactive dye by polyphosphate and glycogen-accumulating organisms in an anaerobic-aerobic sequencing batch reactor. *Bioresour. Technol.* **76** 151–159.

Panswad, T., Techovanich, A. and Anotai, J. (2001b) Comparison of dye wastewater treatment by normal and anoxic+anaerobic/aerobic SBR activated sludge processes. *Water Sci. Technol.* **43**(2), 355–362.

Pinheiro, H.M., Touraud, E. and Thomas, O. (2004) Aromatic amines from azo dye reduction: status review with emphasis on direct UV spectrophotometric detection in textile industry wastewaters. *Dyes Pigments* **61**(2), 121–139.

Porter, J.J. (1997) Filtration and recovery of dyes from textile wastewater, Treatment of wastewaters from textile processing. Schriftenreihe Biologische Abwasserreinigung, Berlin, Germany.

Rajaguru, P., Kalaiselvi, K., Palanivel, M. and Subburam, V. (2000) Biodegradation of azo dyes in a sequential anaerobic-aerobic system. *Appl. Microbiol. Biotechnol.* **54**, 268–273.

Rajashekharam, M.V., Jaganathan, R. and Chaudhari, R.V. (1998) A trickle bed reactor for the hydrogenation of 2,4-DNT: Experimental hydrogenation. *Chem. Engng. Sci.* **53**(4), 787–805.

Ramakrishna, K.R. and Viraraghavan, T. (1997) Dye removal using low cost adsorbents. *Water Sci. Technol.* **36**, 189–196.

Razo Flores, E., Luijten, M., Donlon, B.A., Lettinga, G. and Field, J.A. (1997) Complete biodegradation of the azo dye azodisalicylate under anaerobic conditions. *Environ. Sci. Technol.* **31**(7), 2098–2103.

Robinson, T., McMullan, G., Marchant, R. and Nigam, P. (2001) Remediation of dyes in textile effluent: a critical review on current treatment technologies with a proposed alternative. *Bioresour. Technol.* **77**(3), 247–255.

Rodgers, J.D. and Bunce, N.J. (2001) Treatment methods for the remediation of nitroaromatic eplosives. *Wat. Res.* **35**, 2101–2111.

Rodgers, J.D., Jedral, W. and Bunce, N.J. (1999) Electrochemical treatment of wastewater: a study of the reduction of DNT and oxidation of chlorinated phenols. *Proceedings AWMA 92nd Annual Meeting.* St, Louis, Missouri, Abs. No. 875.

Safarzadeh, A.A., Bolton, J.R. and Cater, S.R. (1997) Ferrioxalate-mediated photodegradation of organic pollutants in contaminated water. *Water Res.* **31**, 787–798.

Sarsour, J., Janitza, J. and Gähr, F. (2001) Biological degradation of dye-containing wastewater (Biologische Abbau farbstoffhaltiger Abwässer). *Wasser, Luft und Boden* (6), 44–46.

Schmelling, D.C., Gray, K.A. and Kamat, P.V. (1996) Role of the reduction in the photocatalytic degradation of TNT. *Environ. Sci. Technol.* **30**, 2547–2555.

Seshadri, S., Bishop, P.L. and Agha, A.M. (1994) Anaerobic/aerobic treatment of selected azo dyes in wastewater. *Waste Manage.* **14**(2), 127–137.

Shaw, C.B., Carliell, C.M. and Wheatley, A.D. (2002) Anaerobic/aerobic treatment of coloured textile effluents using sequencing batch reactors. *Water Res.* **36**(8), 1993–2001.

So, C.M., Cheng, M.Y., Yu, J.C. and Wong, P.K. (2002) Degradation of azo dye Procion Red MX-5B by photocatalytic oxidation. *Chemosphere* **46**, 905–912.

Sosath, F. and Libra, J.A. (1997) Purification of wastewaters containing azo dyes (Biologische Behandlung von synthetischen Abwässern mit Azofarbstoffen). *Acta Hydroch. Hydrob.* **25**(5), 259–264.

Sosath, F., Libra, J. and Wiesmann, U. (1997) Combined biological and chemical treatment of textile dye-house wastewater using rotating disc reactors. In *Treatment of wastewaters from textile processing.* Technische Universität Berlin, Berlin, Germany. pp. 229–243.

Spanngord, R.J., Yao, D. and Mill, T. (2000a) Kinetics of aminodinitrotoluene oxidation with ozone and hydroxyl radical. *Environ. Sci. Technol.* **34**(3), 450–454.

Spanngord, R.J., Yao, D. and Mill, T. (2000b) Oxidation of aminodinitrotoluenes with ozone: products and path-ways. *Environ. Sci. Technol.* **34**(3), 497–504.

Sponza, D.T. and Işik, M. (2002a) Decolorization and azo dye degradation by anaerobic/aerobic sequential process. *Enzyme Microb. Technol.* **31**(1–2), 102–110.

Sponza, D.T. and Işik, M. (2002b) Ultimate azo dye degradation in anaerobic/aerobic sequential processes. *Water Sci. Technol.* **45**(12), 271–278.

Sponza, D.T. and Işik, M. (2004) Decolorization and inhibition kinetic of Direct Black 38 azo dye with granulated anaerobic sludge. *Enzyme Microb. Technol.* **34**(2), 147–158.

Sponza, D.T. and Işik, M. (2005) Reactor performances and fate of aromatic amines through decolorization of Direct Black 38 dye under anaerobic/aerobic sequentials. *Process Biochem.* **40**(1), 35–44.

Tan, N.C.G. (2001) *Integrated and sequential anaerobic/aerobic biodegradation of azo dyes*. Ph.D. Thesis, Wageningen University, Wageningen, The Netherlands.

Tan, N.C.G., Opsteeg, J.L., Lettinga, G. and Field, J.A. (1999) Integrated anaerobic/aerobic EGSB bioreactor for azo dye degradation. In *Bioremediation of nitroaromatics and haloaromatic compounds*. Battelle Press, Columbus, Richland, USA. pp. 253–258.

Tan, N.C.G., Slenders, P., Svitelskaya, A., Lettinga, G. and Field, J.A. (2000) Degradation of azo dye Mordant Yellow 10 in a sequential anaerobic and bioaugmented aerobic bioreactor. *Water Sci. Technol.* **42**(5–6), 337–344.

U.S. Army Environmental Center (2000) *Pollution Prevention & Environmental Technology Division*, SFIM-AEC-ET-TR-200116, U.S. Army Environmental Center. FY 2000-Annual Report.

U.S. Department of Defense (2003) *Environmental Security Technology Certification Program. Mineralization of TNT, RDX and by-products in an anaerobic granular activated carbon-fluidized bed reactor*; U.S. Department of Defense, CP-0004.

Van der Zee, F.P., Villaverde S. (2005) Combined anaerobic-aerobic treatment of azo dyes-A short review of bioreactor studies. *Water Res.* **39**, 1425–1440.

Van der Zee, F.P. (2002) Anaerobic azo dye reduction. PhD Thesis. Wageningen University, Wageningen, The Netherlands.

Van der Zee, F.P., Bisschops, I.A.E., Lettinga, G. and Field, J.A. (2003) Activated carbon as an electron acceptor and redox mediator during the anaerobic biotransformation of azo dyes. *Environ. Sci. Technol.* **37**(2), 402–408.

Van der Zee, F.P., Bouwman, R.H.M., Strik, D.P.B.T.B., Lettinga, G. and Field, J.A. (2001) Application of redox mediators to accelerate the transformation of reactive azo dyes in anaerobic bioreactors. *Biotechnol. Bioeng.* **75**(6), 691–701.

Vandevivere, P.C., Bianchi, R. and W., V. (1998) Treatment and reuse of wastewater from the textile wet-processing industry: Review of emerging technologies. *J. Chem. Technol. and Biotechnol.* **72**(4), 289–302.

Walker, G.M. and Weatherley, L.R. (1997) A simplified predictive model for biologically activated carbon fixed beds. *Process Biochem.* **32**, 327–335.

Wang, Y. (2000) Solar photocatalytic degradation of eight commercial dyes in TiO2 suspension. *Water Res.* **34**, 990–994.

Wiesmann, U., Sosath, F., Borchert, M., Riedel, G., Breithaupt, T., Mohey El-Dein, A. and Libra, J. (2002) Versuche zur Entfärbung und Mineralisierung des Azo-Farbstoffs C.I. Reactive Black 5. *Wasser Abwasser* **143**(4), 329–336.

WTO. (2004) *Background statistical information with respect to trade in textiles and clothing*. World Trade Organization. p. 107.

Wuhrmann, K., Mechsner, K. and Kappeler, T. (1980) Investigation on rate-determining factors in the microbial reduction of azo dyes. *Eur. J. Appl. Microbiol.* **9**, 25–338.

Wujcik, W.J., Lowe, W.L. and Marks, P.J. (1992) Granular activated carbon pilot treatment studies for explosives removal from contaminated groundwater. *Environ. Progr.* **11**(3), 178–189.

Yang, Y., Wyatt, D.T. and Bahorsky, M. (1998) Decolorization of dyes using UV/H2O2 phtochemical oxidation. *Text. Chem. Color Am. D.* **30**, 27–35.

Zaoyan, Y., Ke, S., Guangliang, S., Fan, Y., Jinshan, D. and Huanian, M. (1992) Anaerobic-aerobic treatment of a dye wastewater by combination of RBC with activated sludge. *Water Sci. Technol.* **26**(9–11), 2093–2096.

Zhou, P., ASCE, M., Su, C., Li, B. and Qian, Y. (2006) Treatment of High-Strength Pharmaceutical Wastewater and removal of Antibiotics in Anaerobic and Aerobic Biological Treatment Processes. *J. Environ. Eng.* 129–136.

Zoh, K.D. and Stenstrom, M.K. (2002) Aplication of a membrane bioreactor for treating explosives process wastewater. *Water Res.* **36**, 1018–1024.

8

Environmental technologies to remove nitrogen from contaminated leachates

M. Altınbaş

8.1 INTRODUCTION

Leachate is a liquid, mostly water, which seeps out from storage facilities of the solids, the base of landfills, composting pile, in which the materials are subjected to biodegradation and/or contacted to water. The word "leachate" has always been associated with undesirable situation. However, there remains an inevitable consequence of the storage of the materials; therefore, we have to deal with eliminating or minimizing the impact on the environment. Various monitoring studies showed that even in small scale storage facilities, leachate easily changes the quality of ground water (Kjeldsen and Christophersen 2001; Della Rocca *et al.* 2007). The commonly encountered storage facilities are piles of solid wastes, sewage sludges and manures used for land applications, coals, woody debris, acid mines, peat mine drainages, etc.

The volume and strength of generated leachate is predominantly a function of water availability, waste characteristics, and surface conditions. In general, leachates may contain very high concentrations of particulate and dissolved organic and inorganic substances. Their concentrations may be up to the magnitude of 100–200 times higher than those of municipal wastewaters. While moisture content of wastes and/or rainfall extracts the organic and inorganic substances, various contaminants enter the leachate solution while it moves through the pile. Then from piles, it reaches to the collection ponds in the presence of engineered liners, otherwise it spreads out to the adjacent environment. Leachate from non isolated piles is a major nonpoint source of nutrients causing serious uncontrolled environmental pollution.

Leachate may have high nitrogen concentrations, which leads to eutrophication and decreases the DO (dissolved oxygen) level in receiving watercourse. Inevitably, cost effective and environmentally sustainable treatment solutions should be considered as a priority for nitrogen removal from leachate for the protection of sensitive water bodies from eutrophication. The purpose of this chapter is to give comprehensive review of various nitrogen removal technologies, their applications and limitations supported with several case studies for different leachate characters.

8.2 SOURCE OF NITROGEN CONTAMINATED LEACHATES

The known point sources of nitrogen contaminated leachates are landfills, composting plants, peat mines, and wood piles. Beside these, it should also be considered that nitrogen may be sourced from large mass of solid materials when water percolates through it. As listed in Table 8.1, leachates from landfill and composting contain significant amount of nitrogen. Reuses of the leachates from composting plants are most of the time possible; however, some portion

Table 8.1. Total nitrogen concentrations of leachates from point sources

Source	Total Nitrogen Concentration (mg N/L)	References
Landfill	50–5000	Collective data from literature
Composting	250–1602	Krogmann and Woyczechowski (2000)
Peat mine	3–18	Klove (2001)
Wood piles	1–13	Hedmark and Scholz (2008)

must be disposed of in an environmentally acceptable way. On the contrary, landfill leachate has a primer importance due to its volume and characteristic features. Therefore, much research effort has been directed to this area.

Landfilling of solid wastes is the most acceptable and prevalent method for disposal of them. Around 2 million tones of nitrogen are disposed of annually by landfilling in the EU countries (Jokela and Rintala 2003). Landfills may be challenging to the environment, which may spread over its contaminants by producing the leachate. Leachate emanating from municipal solid waste is the major source of environmental pollution worldwide. Control of leachate, therefore, has become area of concern for most environmental engineers. Control should not be taken as a short term care; it often passes beyond the operational life of the landfills, which may last for centuries (Kjeldsen et al. 2002).

Landfill leachate characterization is a crucial factor for determining the optimal management and treatment strategies regarding the necessity of long term operation of the landfills. Landfill leachates are characterized by high concentration of organics, ammonium, inorganic salts, and heavy metals, some of which contains toxic compounds. Composition of landfill leachate shows spatial and temporal variations depending on many parameters listed as follows: landfilling technology, hydro-geological structure of landfill, capacity of the landfill, solid waste composition, moisture content of the waste, climate and the extent of percolation of rainwater, type of leachate collection system, age of the landfill, biological and physico-chemical processes occurring in the landfill. Among those, type and age of the landfilled materials are primarily determinant factors of the leachate quality.

Since the ammonium level of the leachate remains stable for a long time (Kjeldsen et al. 2002) as seen from Figure 8.1, and taking into account the long

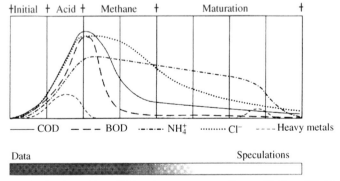

Figure 8.1. Distribution of landfill leachate concentrations with time (Kjeldsen et al. 2002)

term effect of leachate even after the closure of landfill sites, nitrogen content of the leachate should be considered primarily. In another comprehensive review paper, similarly, significant differences between leachate from acidic and methanogenic phases were not observed (Christensen *et al.* 2001). Nitrogen in leachate mainly sourced from the wastes containing proteins such as food wastes, biosolids and yard wastes. The amounts of these waste types present in the landfill affect the concentration and the flow rate of nitrogen in the generated leachate.

The form of the nitrogen species present in the leachate depends on the stabilization phases of the landfill, reflecting the age of landfill. These phases are proceeded in the order of initial, transition, acid, methane and maturation. Initial phase is mainly represented by high organic carbon and nitrogen content. As landfill ages, ammonium ion is the dominant species by ammonification of the organic nitrogen. Further degradation of ammonium ion may not be observed under natural anaerobic landfilling conditions due to the absence of major sink mechanism for ammonium. However, some internal landfill processes may change the form of nitrogen as it will be mentioned in the following parts of this chapter. As a consequence of landfill aging, leachate is represented by low C/N and BOD_5/COD ratios, as well as high contents of ammonium.

Since, there are numerous published data about the nitrogen content of landfill leachate, quite variable values were ascertained for specific cases. Even though the nitrogen content categorized generally under the age of landfill as young, medium and old, this classification published from different authors showed significant differences from each other (Alvarez-Vazquez *et al.* 2004). As each landfill has its own constituents, produced leachate has also its own unique properties. The nitrogen concentration of landfill leachate may be in the range of 50–5000 mg/L according to collective data from literature. If the extreme end point data is to be kept out for further evaluation, the data published from Ehring *et al.* (1984), can be selected as the best representative ones showing the variations by landfill age (Figure 8.2).

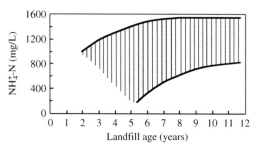

Figure 8.2. Relationships between NH_4^+ and the landfill age (Ehring 1984)

8.3 NITROGEN REMOVAL TECHNOLOGIES

As environmental regulations and requirements became more strict, different technologies have been developed and implemented for the nitrogen removal from leachates. Quantity and quality of leachate in terms of nitrogen species together with the other contaminants are critical parameters for selecting the optimum treatment process chain due to its variability from day to day and year to year. Thus, the treatment chain should provide the maximum amount of long-term flexibility to assure fulfillment with discharge standards and regulations.

Nitrogen can be removed by physical, chemical or biological methods using on-site or in-situ alternatives (Figure 8.3). Sometimes the combination of these methods is necessary to reach the desirable effluent quality considering other components of the contaminants as well. Effluent discharge alternatives, regulatory constraints, technological alternatives and costs should be considered before deciding to apply the best suitable technology.

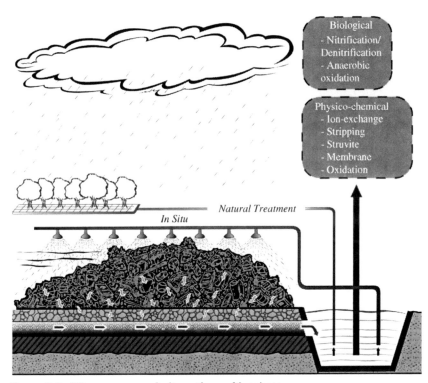

Figure 8.3. Nitrogen removal alternatives of leachate

For small landfill areas, natural leachate treatment systems, lagoons and *in situ* applications are preferable due to its requirements of low cost and low maintenance. However, these alternatives are limited with the environmental and climatic conditions of landfill area. In case of high capacity landfills, controlling of leachate will become more difficult, therefore more advanced treatment processes should be selected.

Due to the changing characteristics of leachate, the initially set up treatment process will require extensive modifications as the landfill ages. For this reason, one of the most convenient and applied methods for leachate control is to treat landfill leachates at central municipal wastewater treatment plants. Addition of leachate to municipal wastewater in appropriate ratios (0.1 and 5% by volume) helps to control the leachate in landfills. For this combination, the municipal wastewater treatment plant should have the capacity to tolerate this additional load. However, due to its high content of toxic organic and inorganic pollutants, this solution may increase the toxicity of municipal wastewaters since the adverse effect of landfill leachate has been reported in several studies. Del Borghi *et al.* (2003) found that leachate greater than 10% by volume has adverse effects on the wastewater treatment plants. It was also reported that less than 2% leachate combination with municipal wastewaters had no significant effect on the performance of treatment plants (Qasim and Chiang 1994). The negative effect will be on the sludge production in terms of both quality and quantity. On the other hand, trial with the 11% leachate addition showed that the overall nitrogen removal was in the range of 35–50% using SBR (Diamadopoulos *et al.* 1997).

8.4 BIOLOGICAL PROCESSES

Leachate contains non-oxidized forms of nitrogen, either organic nitrogen or ammonium. Although other constituents of the leachate dramatically change over time, ammonia concentration is the less affected parameter related to the stabilization phases of landfill. Among these constituents, biological nitrogen removal systems are mainly affected by the readily biodegradable carbon, high salinity and presence of inhibitory compounds.

Simply, leachate in terms of composition can be classified as young and old according to the age of the landfills. Biological processes are very effective and economical, especially in the early stage of the landfills, since they consist of high amount of biodegradable materials. Anaerobic processes, in which the organic carbon is converted to methane, are more preferable; however, it provides only the ammonification of the organic nitrogen, so other conventional or modified biological processes will be necessary for further elimination of nitrogen species from leachate. Therefore, biological nitrogen removal should proceed as a sequence of two different processes. However, as leachates from maturation phase contain less biodegradable carbon and high nitrogen, different treatment concepts have to be considered for the efficient removal of contaminants. In summary, BOD_5/N ratio and concentration of nitrogen species are important criteria in defining what type of treatment system can be used to remove nitrogen. Beside that, phosphorous limitation is generally observed for the biological treatment of landfill leachate. Especially, the need of phosphorus addition for the leachate from young landfill sites will be higher due to the high biodegradable fraction of the leachate.

The biological processes specifically applied for nitrogen removal from leachate will be reviewed in the following sections. Process fundamentals for biological nitrogen removal technologies were mentioned in detail in chapter 6.

8.4.1 Nitrification-denitrification

Nitrification-denitrification processes are the most used and conventional methods for converting the NH_4^+ to N_2 gas. Large number of combinations of aerobic and anoxic processes necessary for nitrification and denitrification, respectively, were proposed together with the different process configurations. Nitrification can be achieved by conventional treatment processes with $>95\%$ ammonium removal with regard to the time variable nature of the leachate. Table 8.2 gives nitrification performances of landfill leachate treated with different process configurations.

Leachate can be treated by the most common process, Modified Ludzak-Ettinger (MLE), consisting of anoxic reactor followed by aerobic reactor for nitrification and denitrification. This single-sludge system is an effective process for the ammonium concentrations of less than 400 mg/L (Elefsiniotis *et al.* 1989). The removal of higher nitrogen concentrations was possible with additional processes (Calli *et al.* 2003; Bae *et al.* 2004). For example, the configuration of two sequential MLE processes was found more efficient for high amount of nitrogen (Ilies and Mavinic 2001). Pre-anoxic and post-anoxic denitrification referred to as a Bardenpho process was examined for the leachate having ammonium concentration of about 2200 mg/L. This system showed stable operation even with fluctuations of over 300 mg/L in the influent ammonium concentrations, achieving a complete nitrification and effluent NO_x concentrations of less than 60 mg/L. At low temperature like Nordic landfills, complete nitrification was achieved by activated sludge (AS) or MLE systems at 7–11°C with a loading rate of 0.01–0.03 kg NH_4^+-N/kg-VSS.day (Pelkonen *et al.* 1999). Another AS system showed maximum ammonium removal rate of 0.13 kg NH_4^+-N/kg-VSS.day at 13°C (Knox 1985).

Table 8.2. Nitrification performances of landfill leachate treated with different process configurations

Process	Operational Conditions for nitrification	BOD_5/COD	BOD_5/N	NH_4^+-N Load		NH_4^+-N (mg/L)		References
				(kg/m³-day)	(kg/kgVSS-day)	Inf.	Eff.	
MLE	HRT: 3.4 h, T: 20°C	0.2	0.2	1.1	0.2	1200	~50	Shiskowski and Mavinic (1998)
BARDENPHO	HRT: 6.8 h, T: 20°C			0.5	0.07	1100	<1	
Two-stage AS	HRT: 0.5–2 d, T: 25°C	0.15	0.5–0.6	0.13–0.25		1400–1800	<1	Bae et al. (1997)
AS	HRT: 3 d, T: 10°C				0.027	53–270	N.D.	
AS + carrier	HRT: 3 d, T: 5°C				0.01		N.D.	
Anoxic + aerobic (bentonite)	HRT: 1–1.8 d, T: 20°C	0.5–0.6	4	0.15–0.28	0.02–0.11	402	<5	Wiszniowski et al. (2007)
Two-stage ND	HRT: 2–2 d, pH: ~7	<0.1	–	0.14–0.18		800–2200	<20	Calli et al. (2003)
Anaerobic + aerobic	HRT: 4 d, T: 23°C, pH: 6.5–8	0.4–0.5	5.4–6	0.8		1635–1810	<10	Im et al. (2001)
SBR	HRT: 6 h, T: 22°C, pH: 6.9–8.2, aerobic cycling	0.06	<0.05	1.8–5.9	0.6–1.1	<900	N.D.	Doyle et al. (2001)
SBR	HRT: 8 h, pH: 7.8, aerobic cycling	0.13	0.25	0.4	0.6	83–336	50	Henderson et al. (1997)
SBR	HRT: 1 d, pH: 7–8, aerobic/anoxic/aerobic cycling	0.6	2.5–7.5	1.1–2.5	0.3–0.7	757–1017	<10	Yilmaz and Ozturk (2003)
MBR	HRT: min 4.3 d, T: 25°C, pH 8.5	–	<3	<0.6		1400	50	An et al. (2006)
MBR, (PN)	T: 33°C, pH: ~8	0.2–0.8	0.5–3.4	<0.9		1000–1500	<15	Canziani et al. (2006)

MBR: Membrane bioreactor, PN: Partial Nitrification, AS: Activated Sludge, ND: Nitrification Denitrification, SBR: Sequencing Batch Reactor, N.D.: Not Detected

Anaerobic-aerobic system can be applied for nitrification/denitrification treating leachate from young or medium age landfill characterized by high amount of biodegradable organic material and nitrogen. In this system, the anaerobic reactor provides simultaneous methane production and denitrification. This will enhance also efficient nitrification in the following aerobic reactor by reducing the biodegradable carbon and increasing the alkalinity. It was found that methane production is ceased only 10–20% from denitrification process (Im *et al.* 2001). Anaerobic-anoxic/aerobic system with a circulation of effluent to the anaerobic process was also able to remove 1200–1800 mg/L of ammonium concentration down to less than 15 mg/L by partial nitrification for leachate from old landfill (Zhang *et al.* 2007). In this system, denitrification of nitrite was achieved with removal efficiencies ranging from 67% to 80% in the anaerobic process.

Accumulation of nitrite in the processes targeted complete nitrification is the most common problem encountered in the leachate. This is mostly caused by the increased level of free ammonia concentration (Ilies and Mavinic 2001; Kim *et al.* 2006). Beside that, chromium (Cr^{3+}) and nickel (Ni^{2+}) were also found to have an inhibitory effect on nitrification processes of leachate from old landfill at the concentrations of 0.3 and 0.7 mg/L, respectively (Harper *et al.* 1996). Although nickel inhibited nitrification at low concentrations; denitrification was not affected up to 2 mg/L Ni concentrations.

8.4.1.1 SBR

The Sequencing Batch Reactor (SBR) has also been widely explored and extensively utilized for the treatment of leachate due to its operational flexibility and effluent stability. Mace and Mata-Alvarez (2002) summarized some SBR cases applied to landfill leachate. It can be said that SBR processes efficiently remove the total nitrogen present in the landfill leachate with sequential anoxic/aerobic cycles with several different alternating operational options. Meanwhile, removal of high content of nitrogen can be achieved by increasing cycles in terms of both number and period of each cycle to reach the desired effluent concentrations.

Aerobic/anoxic operations of the SBR process with the addition of methanol as a carbon source removed nitrogen more than 95% from landfill leachate ranged from 100 mg/L to 330 mg/L (Hosomi *et al.* 1989). The initial average 887 mg NH_4^+-N/L concentration was reduced with 99% efficiency by applying aerobic/anoxic/aerobic/settling sequential operations make up total 24 hours batch cycles (Yilmaz and Ozturk 2003). In this SBR system, acetate and leachate were alternatively used as an external carbon source for denitrification. In case of supplementation of acetate, 99% of nitrate removal was possible with a

COD/NO_3^--N ratio of 5.9. However, raw leachate addition as external carbon source limited the effluent quality due to the introduction of additional ammonium which caused the increase of nitrate in the effluent stream. Nitrogen sourced from composting leachate was also successfully removed with the SBR technology (Laitinen et al. 2006). With only aeration cycling in a 24 h, the average initial 210 mg NH_4^+-N/L was reduced to less than 3 mg NH_4^+-N/L. Effect of feeding pattern on SBR performance was also studied by Klimiuk and Kulikowska (2005) and they found that short filling period of SBR had better performance than the filling during aeration period for the removal of organics and ammonium from leachate.

Although, SBR systems are generally considered as a single sludge system, two-sludge system is also possible for landfill leachate. Two-sludge SBR consisting of two SBRs operating under aerobic and anoxic conditions separately and sequentially for nitrification and denitrification demonstrated complete removal of inorganic nitrogen species for the total nitrogen concentration of 1950–2450 mg/L (Kaczorek and Ledakowicz 2006).

Operation of SBR with leachate from mature landfill (BOD_5/COD < 0.05) showed that complete nitrification with a high loading rate of 5.9 kg NH_4^+-N/m^3 − day was possible within 6 h (Doyle et al. 2001). This rate was at least 4 times higher than the observed in other nitrification systems (Table 8.2). Moreover, in this system, greater suspended solids concentration (up to 13 g/L) could be maintained with Sludge Volume Index (SVI)s less than 50 mL/g. Full scale application of this SBR system was also in agreement with the results explained above.

To achieve the partial nitrification processes, SBR technology stands potentially suitable due to flexibility for optimization of nitrogen removal. Especially, switching of nitritation to denitrification could be easily achieved when the maximum level of nitrite concentrations has been reached.

8.4.1.2 Lagoons

Lagoons operated on a flow-through basis are also mostly used means for the on-site treatment of landfill leachate. The practical application, simplicity in terms of design and operation are important advantages of the lagoon treatment. Beside conventional lagoons, dual powered flow through lagoon systems are also found to be of practical application in many areas. Two-stage anaerobic/facultative lagoon showed stable operation during 10 year monitoring period with a 77% total nitrogen removal efficiency at a HRT of 250 days (Frascari et al. 2004). Nearly complete nitrification was achieved in the aerated lagoon even at 10°C temperature with a HRT of 20 days (Haarstad and Maehlum 1999). In aerated lagoon, 130–175 mg NH_4^+-N/L could be reduced to less than 10 mg/L with a HRT of more than 10 days (Robinson and Grantham 1988). In this system, nitrogen removal was mainly attributed to incorporation in biomass and removing as sludge due to high ratios of BOD_5/N (22–40). In another study, alkalinity limitation in the nitrification process was observed during the removal of average 3480 mg/L ammonium concentration in the aerated lagoon, which could have the effluent ammonium of 630 mg/L (Parkes et al. 2007).

8.4.1.3 Membrane bioreactors

Nitrification is readily achievable by abovementioned biological treatment systems. To increase efficiency and overcome some sludge settling problems encountered during operation of biological nitrogen removal systems, Membrane Bioreactor (MBR)s are started to be used widely. Advantage of this process is keeping high amount of biomass inside the reactor together with the elimination of clarifier unit, allowing intensification of the reactor sizes and processing the concentrated streams with a stable effluent water quality. The location of membrane will be either in the reactor basin or outside of the basin. In the reactor basin, submersed type of membranes are operated with a negative pressure and low fluxes like 8–10 L/m^2-h. However, external flow membranes are operated with a positive pressure with high fluxes of 60–90 L/m^2-h. Application of both types of ultrafiltration membrane into MLE system was successfully treated landfill leachate reducing the 2000–2700 mg/L ammonium to less than 10 mg/L (Wens et al. 2001). In this study, two different types of leachate from young and old landfills were also studied and no differences were observed between them in terms of effluent quality.

Membrane Sequencing Batch Reactor (MSBR) was used for the treatment of leachate from old landfill by applying same kind of operational steps as applied for conventional nitrification/denitrification SBR systems (Tsilogeorgis et al. 2008). The only difference was that instead of applying settling phase, the membrane hollow fiber ultrafiltration was used for discarding the effluent. In this system complete nitrification was observed and the effluent nitrate concentration was less than 20 mg/L succeeded by addition of methanol for denitrification process, with a total cycling time of 12 h.

Generally, integrated membrane processes are applied more to remove organic matters than nitrogenous compounds. The MBR process operated as an alternative to SBR process for composting leachate demonstrated no significant differences in terms of nitrification efficiency. However, it appears that reasonable COD removal efficiency and suspended solids could be attributed to the membrane bioreactor different from the SBR process (Laitinen et al. 2006). In case of more stringent effluent criteria, membrane systems are mainly designed in the order of biological degradation and then physical separation. NH_4^+-N can be removed from 1400 mg/L to the level of 50 mg/L by sequential MBR and Reverse Osmosis (RO) processes (Ahn *et al.* 2002b).

8.4.1.4 Denitrification

Denitrification can take place in an anoxic environment with either autotrophic or heterotrophic microorganisms. Organic carbon in landfill leachate can be successfully used as a carbon source for heterotrophic denitrification at a COD/N ratio of equal or higher than 4 (Yilmaz and Ozturk 2003; Wiszniowski *et al.* 2007). As landfill ages, the reduction of nitrate becomes difficult due to the absence of biodegradable carbon. Denitrification performances of landfill leachate treated with different process configurations with different electron donors are presented in Table 8.3. As can be seen, the necessity of external electron donor addition is generally supported by several sources such as methanol, acetate, methane, ethylene glycol and sulfur. Denitrification of leachate with biogas recovered from landfill site was found also feasible using fluidized bed reactor (Werner and Kayser 1991). However, presence of oxygen and high amount of CO_2 in the gas stream adversely affected the process efficiency.

The treatment cost of the leachate will increase remarkably due to the high theoretical need of carbon to reduce nitrate to nitrogen. To reduce this operational cost, sulfur-utilizing denitrification processes can be successfully used instead of heterotrophic denitrification. In these processes, inorganic sulfur compounds such as sulfide (S^{2-}), elemental sulfur (S^0), thiosulfate ($S_2O_3^{2-}$), or sulfite (SO_3^{2-}) are used as electron donor. Sulfur-utilizing denitrification was proposed as a post denitrification after MLE process (Bae *et al.* 2004). The performance of this denitrification process seems more effective by reducing the effluent concentration of NO_3^--N less than 50 mg/L. Denitrification performance is dependent on the NO_3^-/S^{2-} ratio, which is decisive on the product formation and the rate (Beristain-Cardoso *et al.* 2006). However, although its usage is economical compared to other carbon sources used for denitrification, the effluent SO_4^{2-} concentration will be more than 2500 mg/L, which should be considered while discharging to the receiving waters. Another limitation is the necessity of alkalinity to perform this sulfur-utilizing denitrification process. To overcome this alkalinity requirement for autotrophic denitrification, heterotrophic denitrification was also stimulated by addition of acetate to sulfur bed reactors (Lee *et al.* 2001). Complete denitrification was possible without alkalinity addition by applying the acetate with a 44% of the total theoretical electron requirement.

Table 8.3. Denitrification performances of landfill leachate treated with different process configurations

Process	Operational Conditions	BOD5/N	Electron Donor	NO3-N load, (kg/m3-day)	NO3-N (mg/L) Inf.	NO3-N (mg/L) Eff.	NO3-N removal (%)	References
MLE		<4	Leachate	1.2	–	400		
MLE + SPBR (post DN)	T: 35°C, pH: ~7	<2	Leachate + Sulfur	1.2	–	<50	>99	Bae, et al. (2004)
Two-stage ND AS	HRT: 8–12 h	–	Acetate	5.4 0.06	–	<40	>98	Calli, et al. (2003)
TF FBR	T: 18–21°C	~0.01	Methane	0.15 0.55	~850	<1	>99	Werner and Kayser (1991)
SPBR	HRT: 6.8 h T: 35°C	–	Sulfur + acetate	0.5	900	N.D.	>99	Lee et al. (2001)
MBR	T: 35°C HRT: 24 h submersed membrane	<1	ethylene glycol	0.3–0.4	~750	~30	~96	Praet et al. (2001)
Two-stage SBR (PD)	HRT: 24 h anoxic cycling	0.3	Methanol	~0.4	320	<0.1	>99	Klimiuk and Kulikowska (2004)

MLE: Modified Ludzack-Ettinger, SPBR: Sulfur Packed Bed Reactor, AS: Activated Sludge, TF: Trickling Filter, FBR: Fluidized Bed Reactor, SPBR: Sulfur Packed Bed Reactor, MBR: Membrane Bioreactor, SBR: Sequencing Batch Reactor, PD: Post Denitrification

Denitrification is limited with the recycle ratio. The effect of recycle ratio of effluent to the inflow leachate stream was also revealed by several studies. Bae *et al.* (1997) reported that recycle ratio greater than 4 resulted in better nitrification. This high ratio can also prevent the sludge rising problem (Elefsiniotis *et al.* 1989; Bae *et al.* 1997).

8.4.1.5 Attached growth

The immobilization and accumulation of high amount of nitrifying bacteria population on the support material may enhance the nitrogen removal processes. Table 8.4 gives nitrification performances of landfill leachate treated with different biofilm process configurations.

Biomass carrier material could be selected as stationary or suspended type which could significantly affect the nitrogen removal efficiency. The generally used suspended carries are sand, activated carbon or lighter polyethylene materials. Welander *et al.* (1997) studied with three different suspended carriers including polyethylene and cellulose type and the carrier media consisting of small cubes of macroporous cellulose showed significant nitrogen removal efficiency for landfill leachate due to the macro surface area of 1700 m^3/m^2.

Especially at low temperatures, attached growth systems are more preferable in terms of increasing the density of ammonium oxidizing bacteria. Therefore, the nitrification rate is less influenced by low temperatures. It was found that operation of biofilm systems showed only 33% reduction of nitrification rate at 5°C compared to that of at 20°C (Welander *et al.* 1997). At 5°C temperatures, addition of plastic bio-carrier materials into the activated sludge system resulted in the complete nitrification with a rate of 10 mg NH_4^+-N/VSS-day, while the removal rate was only 6 mg NH_4^+-N/VSS-day with the conventional activated sludge system (Hoilijoki *et al.* 2000).

Complete nitrification can be achieved by Rotating Biological Contactor (RBC)s, which is the most common biofilm based process. Ehring (1984) reported that less than 2 g N/m^2-day loadings are convenient for the complete nitrification. The advantage of this system as well as other biofilm systems is the simultaneous nitrification and denitrification processes due to the biofilm formation which creates the aerobic and anoxic zones itself.

The Sequencing Batch Biofilm Reactor (SBBR) was also applied for the treatment of landfill leachate (Kalyuzhnyi and Gladchenko 2004; Tengrui *et al.* 2007a). In these processes, biofilm were formed on either stationary filter or lighter carrier materials. In laboratory scale applications, it showed nitrification rates higher than 95% with aerobic/anoxic sequential phases.

Table 8.4. Nitrification performances of landfill leachate treated with different biofilm process configurations

Process	Conditions	BOD_5/COD	BOD_5/N	NH_4^+-N Load $(kg/m^2 \cdot day)^a$ $(kg/m^3 \cdot day)^b$	NH_4^+-N (mg/L) Inf.	Eff.	References
Two stage GAC-FBR	Pre-treatment with lime precipitation	0.05–0.09	0.2–0.3	0.71[b]	220–800	<80	Horan et al. (1997)
Air-lift	HRT: 0.6 d, T: 20°C, macroporous cellulose carriers	0.04–0.1	0.04–0.1	0.96[b]	800–1300	<1	Welander et al. (1997)
RBC	T. 18–22°C	~0.4	<1	<0.002[a]	206–1346	N.D.	Ehrig (1984)
RBC	HRT: 0.3 d Temp. 18°C	~0.1	0.25	<0.0045[a]	83–336	26	Henderson et al. (1997)
BC	Temp. 25°C	~0.4	1.6–1.8	0.002[a]	833	<0.2	Hippen et al. (1997)
TF	HRT: 0.5–8 d Temp. 16°C	~0.1	0.2–1.4	0.31[a]	176–520	1–34	Knox (1985)
SBBR	HRT: 0.4 d Fibrous carriers	0.05	0.06	0.51[b]	1100	<10	Tengrui et al. (2007a)
MBBR	HRT: 1.25 d Nano-size carriers	0.4	13–19	<0.32[a]	350–400	6–10	Chen et al. (2008)

MBBR: Moving Bed Biofilm Reactor; RBC: Rotating Biological Contactors; GAC-FBR: Granular Activated Carbon-Fluidized Bed Reactor; TF: Trickling Filter; SBBR: Sequencing Batch Biofilm Reactor
[a] based on surface area, [b] based on volume of the reactor

In Moving Bed Biofilm Reactor (MBBR)s, the biomass is grown on the lighter carrier materials that move along with flow of the reactor. While in operation, it should be kept at the optimum level which can be suspended during aeration. High loading shocks were successfully applied with the bio-carriers consisting of a mixture of high density polyethylene and nano-sized coke powder or zeolite and activated carbon (Chen et al. 2008). These reactors can be also operated in SBR mode, which gives additional advantage in terms of operational flexibility (Loukidou and Zouboulis 2001).

In some cases, ammonia should be reduced to certain levels before introducing the biological process to prevent shock loadings. It was found that the addition of adsorbent alleviate the adverse effects of leachate on biological nitrogen removal processes (Aktas and Cecen 2001; Loukidou and Zouboulis 2001). The positive effect of adsorbent such as Activated Carbon (AC) and zeolite addition into activated sludge process was observed in terms of nitrogen removal performance (Kargi and Pamukoglu 2004). In the presence of 1000 mg/L of powdered AC, the nitrification rate increased from 0 to 50% with reaction period of 96 hours on the mixture of leachate and domestic wastewater (5% leachate, by volume) (Aktas and Cecen 2001). In another study, Aghamohammadi et al. (2007) reported the results of augmentation of activated sludge process for nitrification with powdered AC. NO_2^--N concentration was found as 1666 and 2060 mg/L for powdered AC added and not added to the activated sludge processes, respectively. Beside its ability to adsorb the ammonium, surface of the adsorbent may be used for attached growth of the microorganisms for nitrification. As a result of biological degradation, active sites on the AC will be released and allow further adsorption of ammonium. Therefore, addition of AC in activated sludge processes has attracted a great deal of interest, lately.

8.4.2 Nitritation-denitritation

In particular, in case of low COD/N ratio, partial nitrification represents a promising process for nitrogen removal from leachate. Instead of complete nitrification, ammonium is oxidized to nitrite; then nitrite is reduced directly to nitrogen without passing to the nitrite oxidation step. This partial nitrification process will yield about 25% less oxygen and 40% less carbon source (if used methanol) compared to the conventional nitrification/denitrification process. SHARON (Single Reactor High Activity Ammonium Removal over Nitrite), technology has been widely used to deal with the high nitrogen wastewaters (Van Kempen et al. 2005). In nitritation systems, the need of nitrite oxidizing bacteria, which are more sensitive to the environmental changes than ammonium oxidizing bacteria, will be eliminated. In full scale applications, there are several ways to achieve the partial nitrification. It basically depends on the washing out or inhibition of the nitrite oxidizers by lowering the hydraulic retention times and dissolved oxygen, increasing the temperature to around $35\,^{\circ}C$ and/or inhibiting with free ammonium (Khin and Annachhatre 2004; Ganigue *et al.* 2008). On the contrary, Canziani *et al.* (2006) reported that partial nitrification of ammonium in MBR systems, aiming longer sludge retention times, was successful by keeping the oxygen concentrations below 0.5 mg/L. Beside that, if free ammonium concentration is higher than 1 mg/L, nitrite oxidizers can be easily inhibited (Kim *et al.* 2006; Zhang *et al.* 2007). For leachate, this is very commonly observed due to high nitrogen content and higher pH generally more than the neutral values. Because of low C/N ratio of leachate, partial nitrification seems more promising in terms of economical and practical operation of the biological systems.

8.4.3 Nitritation-Anammox process

Anammox process is the oxidation of ammonium to nitrogen gas using nitrite instead of oxygen. Process details of deammonification were given in Chapter 6. Leachate from old landfills having low BOD_5/COD and low COD/N ratios can take the advantage of ammonium removal with anammox process due to its considerably small amounts of organic carbon and oxygen requirements in contrast to conventional nitrification/denitrification. As observed also for other types of wastewater, nitrate formation is limiting the efficiency of anammox processes due to not being successful in providing the stoichiometric NO_2^-/NH_4^+ ratio of 1.0:1.3. Therefore, effective process selection and operation is necessary for the nitritation step to prevent nitrate formation and concomitantly stimulate the partial nitrite formation. By this way, suitable influent can be obtained for the subsequent anammox reactor for the ammonium removal. One of the ways of ammonium oxidation into nitrite in a stoichiometric ratio can be controlling and limiting the available alkalinity. Beside that, pH, free nitrogen, dissolved oxygen, NH_4^+ loading rate are also critical factors for NH_4^+ removal from leachate in anammox process. To maintain these control parameters in an optimum range, process control by real-time strategies is necessary. SHARON and CANON (Completely Autotrophic Nitrogen Removal over Nitrite) are the most common treatment technologies used for partial nitritation-anammox process.

For the anammox reactor, the important design criterion is the biomass retention time due to the low growth rate of anammox bacteria. Generally it requires long start up period for stable operation of the reactor. Selected process, therefore, should favor the relatively slow growing bacteria. Biofilms and granules are more widely used technologies for anammox reactor. However, SBR was also successfully applied to leachate from old landfill (Ganigue *et al.*

2008). From 1500 mg NH_4^+-N/L containing landfill leachate, equimolar ratio of ammonium and nitrite was obtained by limiting the alkalinity with SBR operated at 8 h working cycles (Ganigue *et al.* 2007). This favorable condition was provided in spite of high solid retention time (3–7 days) and high DO concentration (2 mg O_2/L) which may also stimulates the nitrite oxidizing process. Feed strategies of SBR have found to affect the process performance in terms of obtaining the stable effluent quality. Operation of SBR with step feed strategy showed less being affected from disturbances and consequently less fluctuation in its operation (Ganigue *et al.* 2008).

Fixed bed biofilm reactor was also successfully applied to obtain partial nitritation by controlling the dissolved oxygen and ammonium loading rate. Partial nitrite formation (NO_2^-/NH_4^+:1.0/1.3) was obtained at a temperature of 30°C, load of 0.2–1 kg NH_4^+-N/m^3-day and dissolved oxygen concentration of 0.8–2.3 mg/L (Liang and Liu 2007). In another study with the RBC type of reactor, 58% of the initial 600 mg NH_4^+N was removed by anammox process at 17°C (Cema *et al.* 2007). In this system, low nitrogen removal rate was attributed to the presence of nitrite oxidizers. Nitrate production was accounted for 50%, higher than the acceptable level of 11% for anammox process. Anammox bacteria are sensitive to high nitrite concentrations. At the average NO_2^--N concentration of 1011 mg/L, inhibition effect on the anammox process was observed in the biofilm reactor (Liang and Liu 2008). However, the inhibition effect was observed at concentrations higher than 100 mg/L of nitrite for synthetic wastewater (Strous *et al.* 1999).

8.5 PHYSICO-CHEMICAL PROCESSES

8.5.1 Ion-exchange

The performance of ion-exchange processes is mainly dependent on the thermodynamic equilibriums. Since it has to be determined experimentally for each type of adsorbents, leachates also show significant differences depending on the characterization. It is known that the media type has a direct effect on the efficiency of ammonium removal. Several adsorbents can be used to remove nitrogen from leachate, in some cases zeolites, such as clinoptilonites or sepiolite, wollastonite, bentonite, peat, waste materials, etc. Table 8.5 gives a few examples of ion exchange systems applied on leachate treatment. Presence of dissolved organic and inorganic substances influence the uptake of ammonium ion, since zeolite demonstrates also COD removal of 10–20% together with ammonia removal (Papadopoulos *et al.* 1996; Ahn *et al.* 2002a).

Table 8.5. Removal performances of ion-exchange for leachate

Wastewater	COD (g/L)	Adsorbent Conc. (mg/L)	Adsorption capacity (mg NH_4^+-N/g adsorbent)	NH_4^+-N (mg/L)		References
				From	To	
CL	978	10 (Clinoptilolite)	7.7	200	17	Liu and Lo (2001b)
LL + MWW (%10 leachate by vol)	13570	38 (zeotilized coal fly ash)	0.4	222	132	Otal et al. (2005)
LL	8500	0.1 (Clinoptilolite)	10.4	3257	520	Papadopoulos et al. (1996)
LL + coagulation	2400–3300	Clinoptilolite and mordenite	16.4	1300–1500	110–130	Ahn et al. (2002a)
LL	785–985	Clinoptilolite	–	1041–1605	305–440	Kim et al. (2003)

CL: Composting leachate, LL: Landfill Leachate, MWW: Municipal wastewater

Leachate from composting plant processing the solid wastes of vegetable greenhouse facility treated with clinoptilolite in a packed columns and over 98% ammonium was removed with a six successive loading, which yielded an adsorption capacity of 1.31 mg NH_4^+-N/g zeolite (Liu and Lo 2001a). However, ammonium adsorption capacity was influenced by the presence of competitive ion potassium, for which the clinoptilolite has a greater affinity. The removed potassium concentration was 3770 mg/L, which was much higher than the ammonium concentration of 255 mg/L.

There is a need for the regeneration or replacing the media frequently due to high salt content. The regeneration of the media is generally achieved with sodium chloride solution. It was found that the third regeneration of adsorbent decreased the specific surface area and ion exchange capacity in leachate due to the appearance of macro-pores (Turan et al. 2005). Adsorption capacity of the media depends not only on its total surface area but also on its inner porous structure type and functional groups present on these pores. Therefore, the frequent replacement of adsorbent should be taken into account for leachate due to its higher dissolved organic and inorganic content.

Ion-exchange can also be applied for nitrate removal from leachate after biological nitrification or electrochemical oxidation processes, which are resulted in the production of nitrate. It was reported that the 200 mg NO_3^--N formed after electrochemical oxidation was reduced with macro-porous anionic resin to the concentration of below 10 mg/L (Cabeza et al. 2007b).

8.5.2 Ammonia stripping

Ammonia stripping stands as one of the cost-effective treatment options for nitrogen removal from leachate. At higher pH above 9, the ionic form of ammonia (NH_4^+) turned out to the free nitrogen (NH_3-N) which is in the gas form. Bubbling air through the system causes free nitrogen to move into the atmosphere. pH, ammonium concentration, aeration rate, and reactor configuration are the critical factors for effective designing of stripping processes. High pH values are generally supplied with the lime addition, which also gives additional removal of the other contaminants by coagulation. Higher contact of gas bubbles to the liquid is necessary for efficient removal of NH_3-N. Therefore, the increase of the ratio of reactor surface area to liquid volume yields higher ammonia removal. Application of tower and lagoon type stripping resulted in different removal efficiencies Ammonia can be removed up to 90% with tower stripping depending on the origin of leachate (Table 8.6); however, only 50% ammonia removal was achieved from landfill leachate with a lagoon system (Keenan *et al.* 1983). Stripping can be used either before or after the biological systems regarding the purpose of ammonia removal. It will be the case to overcome possible ammonia toxicity, before biological treatment.

Table 8.6. Ammonia stripping studies applied on landfill leachate

Pretreatment	Stripping conditions	NH_4^+-N, (mg/L) From	To	Time, hour	pH	References
Raw	Fine bubble diffusers	~2200	~100	12	11	Calli *et al.* (2005)
Raw	Perforated PVC pipe, Flow rate: 5 L air/L-min	556	39	24	11	Cheung *et al.* (1997)
Raw	Bubble column Flow-rate: 3.3 L air/L-min	1908	954	6	12	Collivignarelli *et al.* (1998)
Raw	Bubble column with plastic carriers, Flowrate: 0.17 L air/L-min	150	16	24	11	Marttinen *et al.* (2002)
Raw	Fine bubble diffusers Flowrate: 7.6 L air/L-min	1025	287	2	12	Ozturk *et al.* (2003)
Coagulation	Rotating paddle agitator	~1600	100	0.67	12	Andres *et al.* (2007)
Coagulation	Porous air diffuser Flowrate: 0.35 L air/L-min	800	<5	96	11	Silva *et al.* (2004)

Due to high initial alkalinity of landfill leachate, lime addition increases in a substantial ratio. It is also possible to achieve efficient ammonium removal without increasing the pH to around 11–12. Instead of adding high amount of lime, keeping the temperature between 60 and 70°C without pH adjustment was found as economical and efficient way of ammonia removal with a rate of approximately 70% (Collivignarelli *et al.* 1998).

With a longer aeration period, it is possible to remove major parts of the ammonium providing the discharge limitations. However, the main concern is the release of NH_3, which should be absorbed in sulfuric or hydrochloric acid to prevent air pollution. Beside that, calcium carbonate scaling is another important problem that should be taken into account.

8.5.3 Struvite precipitation

Struvite (MAP; magnesium ammonium phosphate) precipitation is a preferable ammonium removal method for landfill leachate having high concentrations of ammonium and low BOD/COD ratios. As mentioned in Chapter 9, the struvite precipitation occurs based on the stoichiometric equation of Mg^{2+}, NH_4^+, PO_4^{2-} ions at alkali pH's. In general, it is not possible to observe that the Mg^{2+} and PO_4^{2-} ions are present in equal molar of NH_4^+ in the leachate. To achieve higher ammonium removal efficiencies, additional phosphorous and magnesium sources are necessary. Besides that, it is an advantageous technology, due to its simplicity, requirement of small footprint and robustness as well as the using the struvite as a slow release fertilizer.

As can be seen from the Table 8.7, residual ammonium is always observed after struvite precipitation due to its solubility in the liquid phase. Therefore, due to this residual ammonium concentrations after precipitation, struvite

Table 8.7. Struvite recovery efficiency of landfill leachate

Pretreat-ment	Mg^{2+}:NH_4^+: PO_4^{3-} ratio	Mg^{2+} Source	PO_4^{2-} Source	pH	NH_4^+-N (mg N/L)		References
					From	To	
Raw	1:1:1.2	$MgCl_2$	NaH_2PO_4	9	2132	36	Barnes *et al.* (2007)
Raw	1:1:1	$MgCl_2$	NaH_2PO_4	8.6	5619	112	Li and Zhao
	1:1:1	MgO	H_3PO_4	9	5619	~1900	(2001)
Anaerobic	1:1:1	$MgCl_2$	NaH_2PO_4	9.2	2240	255	Ozturk, *et al.* (2003)
Raw	1:1:1	$MgCl_2$	Na_2HPO_4	9	1100	28	Tengrui *et al.* (2007b)

process can be used as a pretreatment step for the biological nitrogen removal systems, which could alleviate the possible high nitrogen content effect on the biological systems.

The drawbacks of the process are the high consumption of phosphate and magnesium salts since they provide the stoichiometric ratio of Mg^{2+}: NH_4^+: PO_4^{3-}, resulting in high operational costs. There are some trials to reduce the cost by using repetitively the magnesium and phosphate sources. This process can be managed by decomposition of MAP precipitates at high pH and temperature, and releasing the ammonia. Then magnesium and phosphate can be used for the successive MAP precipitation reaction. He et al. (2007) found that 3 times reuse of the MAP decomposition residues for the removal of ammonium was efficient by 44% reducing the chemical cost.

8.5.4 Membrane separation

There are many membrane separation applications for treating the leachates regardless of origin of young or old landfills. Also, it is becoming more prevalent for leachate applications that require a high quality effluent (Peters 2003). The mechanism is simply blocking the nitrogen species as well as other contaminants in the pressure-driven membranes as a physical barrier. Membrane separation generally involves Reverse Osmosis (RO) for effective separation, although nanofiltration was applied in some cases. All nitrogen species will be removed by RO processes with specialized modules. However, it was found that negatively charged membranes remove negatively charged nitrite and nitrate efficiently compared to the positively charged ammonium (Elimelech et al. 1994). Some examples on performance of RO applied on landfill leachate are summarized in Table 8.8.

Related to the discharge standards, the nitrogen species mostly have to be at very low levels. After biological nitrification/denitrification, to provide discharge limits, two-stage RO plant was successfully applied (Hippen et al. 1997). Within this treatment chain, while the significant part of ammonium was removed with the biological process, the residual nitrate was removed to less than 10 mg/L with RO. In another study, two-stage RO system showed total nitrogen removal of >99% for the initial ammonium concentration of 366 mg/L (Peters 1997). Container mounted a transportable full scale RO system was also applied to very high ammonium concentrations of 2500 mg/L with a removal rate of 88% (Beone et al. 1997).

Application of RO technology can be useful to produce high quality effluent itself; however, depending on the quality of generated leachate, it will be feasible to place a micro-filtration or ultra-filtration unit upstream of RO unit. In the

Table 8.8. Performance of reverse osmosis applied on landfill leachate

Pretreatment	Material/type	Pressure (bar)	Flux (L/h-m²)	NH_4^+-N (mg/L)		NO_3^--N (mg/L)		References
				From	To	From	To	
Nitrification/ Denitrification	Two-stage: 1st tubular, 2nd spiral wound	40	15–30	2–92	0.5–4	129–440	10–23	Baumgarten and Seyfried (1996)
Nitrification/ Denitrification	Polyamide/ spiral wound	–	15–22	40	<1	<30	<2	Cornelissen et al. (2001)
Raw	Polyamide	60	21–29	3052	10–12	–	–	Di Palma et al. (2002)
Raw	Two-stage: Tubular	20–50	35	140	8	–	–	Linde et al. (1995)
Sand filtering	Disc tube	12–20	6–13	1–30	<1.5	–	–	Ushikoshi et al. (2002)

application of sequential ultra-filtration and RO on anaerobically pretreated landfill leachate, NH_4^+-N and conductivity removals of 72 and 82% were obtained, respectively (Ozturk *et al.* 2003). The NH_4^+-N and conductivity values were reduced to 270 mg/L and 4000 mS/cm, respectively, following a single pass from the ultra-filtration membrane at 4 bar pressure and RO membrane at 55 bar pressure. The permeate flux was about 10 L/m^2-h for ultra-filtration membrane and 58 L/m^2-h for RO membrane. Results indicated that post-treatment application of combination of membrane systems after anaerobic treatment was an appropriate treatment alternative for leachates from young landfill.

Nano-filtration can be run at lower pressure than RO which gives advantage to former one for cost and membrane fouling. On the contrary, the removal efficiencies are not as high as the RO does. The treatment of low strength landfill leachates with nano-filtration operated at 6–8 bar showed 27–56% ammonium removal (Marttinen *et al.* 2002). In another study, nano-filtration membrane having 450 Da molecular weight cut off showed 21% ammonium removal at 20 bar with low strength landfill leachate (Trebouet *et al.* 2001).

RO separation has distinct advantage for reducing ammonium as well as other pollutants to below detection levels. However, high operating pressures (30–60 bar), management of concentrate stream, scaling and fouling problems should be taken into considerations. Renou *et al.* (2008) reported a requirement of frequent membrane cleaning to prevent the irreversible membrane fouling. It should be kept in mind that cleaning of RO membranes is more complicated than low pressure systems due to frequent formation of different type of scaling. It was also observed that thin film formation of extra-cellular polymeric substances in leachates is pretreated with biological systems. It is difficult to remove membrane surface; therefore, RO feeding without biological pretreated leachate showed upward tendency (Holweg *et al.* 2003). Hurd *et al.* (2001) reported the possibility of using low pressure (~10 bar) RO treatment for landfill leachate. Within this operation, ammonium removal efficiencies greater than 88% were obtained with a permeate flux of 26–54 L/m^2-h.

8.5.5 Chemical oxidation

Among the many wide oxidation techniques, so far only electrochemical, wet air and photochemical oxidations are known as effective for complete ammonium removal from wastewaters. It should be kept in mind that other chemical oxidations processes may also be effective to remove nitrogen from wastewaters. However, less attention has been directed toward investigating the oxidation processes for nitrogen removal.

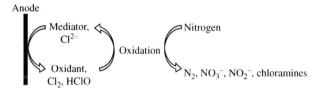

Figure 8.4. Schematic representation of electrochemical oxidation of nitrogen

The principle of nitrogen removal via electrochemical oxidation is the generation of strong oxidants at anode surface by electrical charge and then indirect oxidation of ammonium by means of diffusion of oxidants into the bulk solution (Figure 8.4). The agents generated during indirect oxidation can be chlorine/hypochlorine, hydrogen peroxide, and metal mediators. Chloride may easily be converted to chlorine/hypochlorites, which are strong oxidants due to their low stability. Since presence of reasonable amount of chloride and high electrical conductivity of leachate, optimum conditions are provided for electrochemical oxidation of ammonium. Reactions for electrochemical oxidation with chloride are similar to the breakpoint chlorination as follows:

$$2Cl^- \Leftrightarrow Cl_2 + 2e^- \tag{8.1}$$

$$Cl_2 + H_2O \Leftrightarrow HClO + H^+ + Cl^- \tag{8.2}$$

$$2NH_4^+ + HClO \Leftrightarrow N_2 + 2H_2O + 6H^+ + 2Cl^- \tag{8.3}$$

Various electrochemical oxidation applications on landfill leachate are given in Table 8.9. Ammonium oxidation efficiency is directly related to the production of chlorine/hypochlorites. The high rate production of these oxidants can be succeeded by selecting effective anode materials, keeping high chloride concentrations, and increasing current density. It was found that anode material, current density, and chloride concentration had similar effects on oxidant production efficiency (Chiang *et al.* 1995). The various anode materials applied specifically for landfill leachate, including graphite, lead dioxide coated titanium (PbO$_2$/Ti), ruthenium oxide coated titanium (RuO$_2$/Ti), tin oxide coated titanium (SnO$_2$/Ti), ternary Sn-Pd-Ru oxide coated titanium (SPR) was tested and SPR found as the better current efficiency and electrocatalytic activity (Chiang *et al.* 1995). Boron doped diamond is also effective anode due to its high chemical inertness and longer lifetime (Cabeza *et al.* 2007a). Additional Cl$^-$ increased the

Table 8.9. Electrochemical oxidation applications on landfill leachate

Pretreatment	Electrodes	Time (min)	Cl^- (mg/L)	Current density (mA/cm^2)	NH_4^+-N Inf	NH_4^+-N Eff	NO_3^--N Eff	References
Raw	Anode: Boron doped diamond Cathode: SS	360	3235	90	1934	N.D.	~900	Cabeza et al. (2007a)
Biological and physico-chemical		240	8570	30	860	<15	~200	
Raw	Anode: Sn-Pd-Ru-Ti Cathode: steel	240	10000	150	2100–3000	N.D.	–	Chiang et al. (1995)
Raw	Anode: Ti-PbO$_2$ Cathode:Ti	360	–	80	~780	N.D.	–	Ihara et al. (2004)
Raw	Anode: Pt-Ti Cathode:SS	360	2600	–	1200	50	–	Vlyssides et al. (2003)

*N.D.: Not Detected, SS:Stainless Steel

treatment efficiency (Chiang et al. 1995) and also reduces the NO_3^--N formation (Cabeza et al. 2007a).

Alkaline pHs encourage the formation of chlorine/hypochlorite from chloride, which increases the efficiency of indirect oxidation (Chiang et al. 1995). It was also found that NH_4^+-N oxidized to nitrate at alkali pH; however, in acidic pH it converted to nitrogen gas in wet air oxidation systems (Cao et al. 2003). These studies about the pH effect on ammonium oxidation yielded contradictory results.

The presence of organic contaminants leads to competition between COD and nitrogen removal. Kinetic results demonstrated that the NH_4^+-N was removed primarily when the indirect oxidation takes place. However, COD removal takes priority when direct oxidation became prevalent (Deng and Englehardt 2007). For effective removal of nitrogen, therefore, organic carbon should be removed in advance in order to minimize the loading and energy consumption of the electrochemical oxidation.

Important advantage of electrochemical removal of nitrogen is the operation with simple reactor configuration, which is open to modifications. It can be used as a post-treatment alternative since complete nitrogen removal and effluent COD concentration of less than 5 mg/L is possible. Therefore, an electrochemical unit can be placed at the end of treatment chain for the stronger leachates. The main disadvantages of this oxidation are the possibility of chloramines formation, organic halogens formation and requirement of high energy consumption. In the oxidation study, several chlorinated organic carbons and increased level of absorbable organic halogens (AOX) were detected during electrolysis of pre-treated landfill leachate (Lei et al. 2007).

It was also found that ammonium can be oxidized photocatalitically together with COD. The ammonium was decreased in 12 h from 2000 to less than 100 mg/L at pH 8.8 by photocatalytic oxidation system having 254 nm ultraviolet lamps (Cho et al. 2002).

8.6 WETLANDS AND OTHER NATURAL TREATMENT SYSTEMS

The low-cost and environmentally friendly wetland and natural treatment systems can be used for nitrogen removal from leachate on either old landfill or pretreated young landfills. Wetlands are natural filters with high biodiversity and ability to remove a wide range of pollutants. Constructed wetlands are generally classified as subsurface and flow surface systems, which were both successfully applied to the low strength leachate for ammonia removal. Removal of ammonia in wetlands might be result of a combination of different processes, which are adsorption, ion-exchange, plant uptake, volatilization, nitrification and deni-trification. Vegetation plays key role in nitrogen removal using mechanism of plant uptake and providing oxygen transfer through the roots in which stimulat-ing the nitrification (Wu et al. 2001). It was reported that nitrogen could be removed by only plant uptake with 0.8 g/day nitrogen load without involving other nitrogen removal mechanisms (Polprasert and Sawaittayothin 2006). To increase the nitrification efficiency of wetlands the aeration system may be placed at the bottom of the pools. With this method, the initial ammonium concentrations of 175–253 mg/L was reduced to the values of 4–14 mg/L by operating the wetland throughout the year, at which the observed winter temper-ature was −2°C (Nivala et al. 2007). Beside that, decrement of ammonium removal rates depending on the seasons, especially in winter, will be overcome by insulating the wetlands (Picard et al. 2005). In case of nitrate resulting from the nitrification process, it could be efficiently eliminated by denitrification process adding acetate as external carbon source in the three-stage wetlands (Rustige and Nolde 2007).

Wetland treatment stands as a feasible alternative to conventional treatments especially for low strength or pre-treated landfill leachates (Table 8.10). Effective performance of such systems mainly depends on the selection and maintenance of health. The selected plant should resist to salinity. Leachate from young age landfill may contain the chloride up to 5000 mg/L concentration, which will have detrimental effect on the plants of wetlands. Therefore, the attention should be taken while selecting the plants. Commonly used reed beds in the UK, P. australis, can tolerate 12000 mg/L of NaCl (Barr and Robinson 1999).

One potential option for nitrogen removal from leachate is the irrigation through the lands, especially forest and grassland ecosystems. The application is generally managed by either sprinkling or surface flow. This option is generally applicable to the landfills generating between 1,000 and 10,000 m^3 of leachate annually (Jones et al. 2006). This amount corresponds to the leachate applied to soil at the rates of approximately 250 m^3/ha-year. Beside that generally low strength leachate having the ammonium concentration between 100–200 mg/L found as a minimal effect to the environment (Maurice et al. 1999). Use of landfill leachate as irrigation water demonstrated that the content of nitrogen was readily available for plant uptake and can be replaced with substantial amount of fertilizers (Cheng and Chu 2007). The presence of other macro and micro nutrients in the leachate provides additional advantage for the purpose of irriga-tion. Energy crops such as willows and poplars can be used as a means of landfill leachate treatment by irrigation (Thorneby et al. 2006; Zalesny et al. 2007). The practical applications of willow short rotation coppice as a technology to treat

Table 8.10. Performances of constructed wetlands

| Wetland Type | Pre-treatment | BOD$_5$/N | NH$_4^+$-N | | NO$_3^-$-N | Plantation | References |
			Inf	Eff	Eff		
Vertical SSFCW	Raw	1.2	~611	13.6	3.7	I. Aquatica (Forsk)	Aluko and Sridhar (2005)
Horizontal SSFCW	Raw	1–2.4	13–26	<3	ND	T. Angustifolia (cattail)	Polprasert and Sawaittayothin (2006)
SSFCW	SBR	0.1	~226	1	18.4	P. australis (reed)	Barr and Robinson (1999)
SFCW	Extended Aeration	<1	1–60	0.3–15	0.1–8	Macrophytes, reeds, aquatic plants	Martin and Johnson (1995)
Vertical and Horizontal SSFCW	Raw	0.1	275–753	11–546	7.4–68	P. australis (reed) and T. latifolia (cattail)	Bulc (2006)

*SSFCW: Subsurface flow constructed wetland; SFCW: Surface flow constructed wetland landfill leachate, not only removes the nitrogen efficiently, but also provides valuable resource for the future (Godley *et al.* 2005). The total nitrogen was reduced from 1600 kg N/ha-year to approximately to 100 kg N/ha-year, over a ten year period using willow crops (Hasselgren 1998). Willow-coppice plantation was also found effective for treating annual quantity of 195.00 m^3 of leachate with an average nitrogen content of 24 mg/L and load of 130 kg/ha-year (Rosenqvist and Ness 2004). However, phytotoxic effects of leachate were also observed related to high electrical conductivity and irrigation methods (Duggan 2005; Watzinger *et al.* 2006). The irrigation with three different type of leachate from old landfill was tried on herbaceous species (Hernandez *et al.* 1999). Phytotoxicity symptoms were observed for leachates having high conductivity values (16 and 30 mS/cm). The nutrient composition was also unbalanced for plant growth. Regarding the high N/P ratio of landfill leachate, P was the deficient nutrient and therefore resulted in less harvesting compared to the low N/P ratios.

Climate plays a key role in the irrigation systems and depending on the vegetation type they cannot be applied in the non-growing season. Moreover, there is a potential risk of pollutant migration into groundwater. In general, landfill leachate can be treated without the adverse effect on soil, groundwater and vegetation with careful operational strategies. Nevertheless, ground and surface water quality should be observed periodically to detect and prevent the possible contamination in advance.

Nitrogen removal was also possible with the growth of water hyacinth, a rapidly growing and floating aquatic plant. However, its application is limited to low strength leachate due to the requirement of low salinity (El-Gandy *et al.* 2004). Aquatic plants were also found to be an efficient way of nitrogen removal when the aerated lagoons are covered with them (Martin and Johnson 1995). With this combination, the initial total nitrogen concentrations of 27–690 mg/L were reduced to 60 mg/L with a retention time of around 500 days. Furthermore, removal of nitrogen to a total effluent nitrogen concentration of less than 3 mg/L was succeeded by passing through the surface flow wetland systems with a hydraulic loading of 227 m^3/day.

Biofilters can be used with the purpose of leachate polishing. Nitrogen removal in the filter column containing grained bark was 45–76% (Haarstad and Maehlum 1999). Peat beds were also used as an efficient way of ammonium removal with a rate of 3.4 g/m^2-day, which is comparable with other biofilm reactor systems (Heavey 2003). Infiltration bed comprised sand layer overlying sawdust removed 96% of the 25 mg/L total nitrogen by creating natural aerobic and anoxic zones in the filter. It has been also suggested that sawdust has a sufficient carbon for denitrification process up to 30 years without replenishment

(Robertson and Anderson 1999). Ten years old aged refuse is used for bio-filtration of leachate and initial ammonium concentrations of 500–800 mg/L was reduced to 15–25 mg/L at a hydraulic rate of 80–200 L/m^3 refuse-day (Youcai et al. 2002). However, removal efficiency for total nitrogen remained in the level of 20–30%, which the effluent mainly composed of NO_2^- and NO_3^--N.

8.7 *IN SITU* APPLICATIONS

Concept of regarding the landfill sites as biological reactors has attracted great attention since last two decades. There is not significant difference between conventional and bioreactor landfills in terms of designs and functions. Some small modifications and transient operation modes have to be implemented to consider the landfills as a bioreactor. It should be taken into account that the isolation of the waste body by engineered liners from the surface and the groundwater is a very critical point while operating the bioreactor landfills. Possible leakages from landfill will deteriorate the environment of the landfill sites more than the leachate from conventionally operated landfills. The aim of this bioreactor approach is to stabilize the waste constituents by accelerating the decomposition of solid wastes. Maturation phase is therefore, reached much faster than the conventional ones. This enhancement can be provided by water addition and/or leachate recirculation through the waste body, in which creates the optimum conditions for micro-bial degradation. However, with a recirculation of leachates for a long period, ammonium accumulation to higher levels was observed compared to the conventional single pass of the leachate (Burton and Watson-Craik 1998).

Form of nitrogen in bioreactor landfill may be changed by several physico-chemical and biological processes simultaneously or sequentially. These processes may be nitrification, denitrification, anammox, ion-exchange, sorption, and stripping depending on the operational strategies. Biological reactors are operated as by either aerobic, anaerobic or hybrid. The common operation type of the landfill bioreactor is treating leachate by recirculation through areas of the landfill covered with wastes. Under anaerobic conditions, while biodegradable organic carbon is readily removed, the organic nitrogen is also easily converted to the ammonium nitrogen. Further degradation of ammonium may not be observed due to the absence of major sink mechanism for ammonium. The ammonium present in this stabilized leachate by recirculation through the landfill can be successfully treated further by on-site physico-chemical methods (Diamadopoulos 1994).

In situ aeration of waste body provides rapid and significant ammonium nitrogen consumption through nitrification (Onay and Pohland 1998; Berge *et al.*

2006). In hybrid systems, nitrification and denitrification can be provided by alternating aerobic and anaerobic conditions inside the waste body. However, aerobic bioreactors can be considered also as a hybrid system. The presence of both aerobic and anoxic areas, even with continued aeration inside the waste body, may cause the simultaneous nitrification and denitrification. Berge *et al.* (2006) found that in situ nitrification and denitrification was feasible in a continuously aerated landfill body with a specific nitrogen removal rate of 0.2 mg N/day-g dry waste. The high half saturation constant of 60 mg/L found in the bioreactor compared to the ammonia removal from wastewater treatment systems may be attributed to the inherent heterogeneity and presence of confined areas in the waste body resulted in the less contact efficiency of leachate and the waste.

The initial ammonium concentration of 512 mg/L was removed with almost 99% efficiency by nitrification (Onay and Pohland 1998). It was also reported that leachate circulation reduce the NH_4^+-N from 2110 to 200 mg/L at 250 days in the aerated landfills (Bilgili *et al.* 2007). In a same way, NH_4^+-N in the leachate of aerobic bioreactor decreased from 400 mg/L to 5 mg/L after 175 days of operation (Erses *et al.* 2008). It was also found that ammonium removal was affected by the age of landfilled material where the recirculation was applied. In fresh wastes, the ammonium removal was lower than that of old wastes (Long *et al.* 2008; Shao *et al.* 2008).

Benson *et al.* (2007) reported 5 full-scale landfill applications of bioreactor/recirculation under anaerobic conditions and found that the ammonification rate was significant in all areas. Nitrification was observed in only one landfill, on which the *ex situ* aeration process applied. If the leachate is aerobically treated before recirculation through the waste body by *ex situ* applications, significant part of the NH_4^+-N is converted to nitrate before returning to the landfill (Shou-liang *et al.* 2008). After ex-situ nitrification of leachate, the nitrate is converted to N_2 in the bioreactor by denitrification process (El-Mahrouki and Watson-Craik 2004). He *et al.* (2006) applied SBR for *ex situ* nitrification to the leachate from young landfill. Then nitrified leachate could be efficiently denitrified with fresh waste bodies. Although, nitrification can be carried out by heterotrophic denitrification, autotrophic denitrification via inorganic sulfur compounds can also be observed in low biodegradable waste bodies. It was found that in acidogenic phase of the landfill, heterotrophic denitrification dominated the process due to the availability of easily biodegradable carbon; however, in the methanogenic phase, autotrophic denitrification was detected by sulfate accumulation (Vigneron *et al.* 2007).

In situ applications may also be combined with other natural treatment methods. Alternative to conventional landfill capping, phytocapping in which infiltration of surface applied leachate may be used for the nitrogen removal mainly by the sorption mechanism (Phillips and Sheehan 2005). Then the nitrogen will be removed from this surface by plant uptake.

Climate is a very critical parameter in determining the applicability of the recirculation. In case of the precipitation exceeds the evaporation rate, landfill body may not tolerate the increased flow rate of the leachate (Miller and Clesceri 2003). Other encountered problems are the leachate accumulation and clogging of the waste bodies. Due to the waste heterogeneity, low porosity and compaction level may cause the channeling of the liquid flow. Therefore, uniform liquid and air distribution via horizontal trenches or vertical pipes inside the landfill is challenging.

8.8 CASE STUDIES

In this part, case histories of some full-scale treatment plants of landfill leachate are presented in detail. The experiences about landfill leachate treatment can also be found in other publications (Qasim and Chiang 1994; Mulamoottil *et al.* 1997).

8.8.1 SBR and reed bed combination (Robinson and Olufsen, 2007)

Operation of an SBR and reed bed for Hampshire County council at Efford landfill site in the New Forest, UK is described in detail. The treatment system including SBR and reed bed was designed to treat up to 150 m^3/day of strong leachate from the landfill, typically between 70–110 m^3/day (Figure 8.5). The reactor was a part-buried SBR tank fitted with venturi aeration and mixing, and roofed in order to retain heat, such that winter temperatures would be maintained

Figure 8.5. Efford Leachate Treatment Plant

Table 8.11. Overall performance of Efford Leachate Treatment Plant, (January 2003 to June 2006)

Parameters (mg/L)	Raw Leachate	SBR Effluent	Reed Bed Effluent	Discharge Limits to Sewer
COD	388–2500 (988)	179–756 (374)	121–501 (301)	2500
BOD$_5$	6–298 (76)	2–95 (11)	1–19 (<3)	
NH$_4^+$-N	120–1141 (569)	0–23 (<2)	0–4 (0)	80
Chloride	791–2660 (1354)	932–1970 (1368)	850–2040 (1393)	2000
Alkalinity	2170–6840 (3576)	202–1970 (807)	211–2010 (818)	

*Mean values given in parenthesis

above 20°C. The SBR process was operated on a 24-hour cycle, each day making a single discharge of clarified effluent into a treated leachate balance tank. A reed bed polishing system, planted with *Phragmites australis*, was employed to provide complete removal of any residual suspended solids, treatment of any residual BOD$_5$ or ammonium, and refractory COD associated with suspended solids or colloidal materials (Table 8.11).

Initially, the plant was operated with only leachates from older parts of the landfill for 18 months. This leachate typically had COD and NH$_4^+$-N values of 500–1200 mg/L and 400–700 mg/L, respectively. After that, varying amount of stronger leachate from the new part of the landfill leachate was pumped to the treatment plant. The blending of these resulted in COD and NH$_4^+$-N values of 2500 mg/L and 1200 mg/L, respectively. In spite of higher load, the plant kept a good effluent quality. NH$_4^+$-N removal was by means of complete nitrification to nitrate, typically with about 75–90% appearing as nitrate-N in final effluent. In the reed bed stage, effluent rarely exceeded 1 mg/L, in spite of occasional spikes in the values of the SBR effluent of up to 10 mg/L (Figure 8.6).

The main operational costs for treatment at Efford comprised staff costs for operations, electricity for leachate pumping and treatment, chemical dosing costs, trade effluent charges. Extremely detailed records were kept for all other operating costs on a monthly basis. Electricity usage during the period of two years averaged 18 kWh/m^3 treated leachate, of which about 25 percent related to leachate collection and pumping, rather than treatment. Trade effluent charges were between 70 and 80% of these costs related to fixed charges, not related to effluent quality. Chemical dosing costs were primarily related to supply of alkalinity as NaOH, dosed as a 32% w/v solution, and amounted to dosing of between 2.0 and 2.6 L/m^3. No dislodging of sludge has been required to date at Efford, during 42 months of operation, since the extended aeration of SBR process let the digestion of sludge.

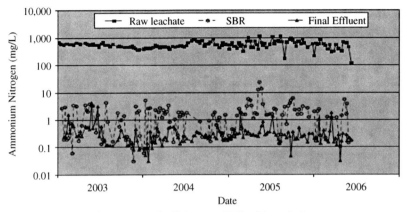

Figure 8.6. Ammonium removal efficiency of Efford Leachate

8.8.2 Upgrading aerated lagoon to nitrification and GAC combination (Hercule *et al.* 2003)

The operation of an aerated lagoon with a granular activated carbon (GAC) system is described; this facility is placed in the Pont-Scorff municipal landfill located in the western part of France. This landfill site operated since 1979, and the engineered cells were in operation since 1996. The amount of municipal solid waste was 50,000 ton/year, including green waste and incineration residues. The former impermeable capping was replaced by the clay cover to reduce the entrance of water in the old landfill. In order to enhance biological efficiency, the aeration volume of lagoon, the floating aerator power and the feeding pattern of aeration basins were modified with an automatic process control during autumn 2002 (Figure 8.7). Therefore, pH, temperature and redox sensors were installed for automatic process control. Vertical sand filter and three granular activated carbon tanks were operated after biological treatment. Each column capacity was 20 m^3, with 7 tons of dry carbon and the feeding flow rate was ranged between 1 and 5 m^3/h. The monthly peak value for flow rate of leachate was 4 m^3/h. The plant design was based on mixing of the leachate from old and young landfill cells (Table 8.12).

The ammonia removal rate was in the range of 83–99%, with an average value of 95% from December 2001 until April 2003 (Figure 8.8). The concentration of nitrites was stable and low in the aerated lagoons showing a good stabilization for nitrification. The efficiency on COD removal was 41% in average. During winter 2001 and 2002, when temperature decreased below 5°C,

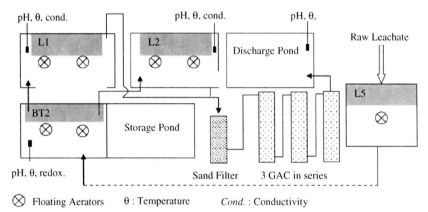

pH, θ, cond.

pH, θ, cond.

pH, θ,

Raw Leachate

L1

L2

Discharge Pond

L5

BT2

Storage Pond

pH, θ, redox.

Sand Filter   3 GAC in series

⊗ Floating Aerators   θ : Temperature   *Cond.* : Conductivity

Figure 8.7. Biological treatment and GAC adsorption scheme after upgrading

the nitrification efficiency slightly decreased, but remained greater than 85% during this period (Figure 8.8). The daily nitrogen load in lagoon was in the range of 14–24 kg N/day. Sulfuric acid and sodium bicarbonate are used for pH adjustment. The phosphorous content of the leachate was sufficient for biomass growth. In L1 lagoon, nitrifying bacteria were added each day to enhance biomass efficiency, especially during winter. As soon as the efficiency was decreased, the effluent of aerated lagoons was diverted to GAC columns. The backwashing of sand-filter and the first GAC column was managed once a week. The second and third GAC columns were backwashed each 2 or 3 weeks. While, the first GAC column was changed each year, this changing period was 1 and a half year for the second column, depending on organic load.

Table 8.12. Overall performance of Pont-Scorff Leachate Treatment Plant

Parameters	Raw (Old cells)	Raw (Young cells)	Biological Effluent	GAC Effluent	Discharge Limits
COD (mg/L)	313–1790	10700	203–1095	23–210	150
BOD$_5$ (mg/L)	–	6840	–	–	30
TKN (mg/L)	122–260	980	–	–	< 30
NH$_4^+$-N (mg/L)	21–598	860	2–32	0.6–13	–
NO$_3^-$-N (mg/L)	–	–	150–190	–	–
pH	7.4–9	7	6.2–8.9	–	6.5–8.5
Conductivity (μS/cm)	1083–9460	15640	3380–7450	–	–

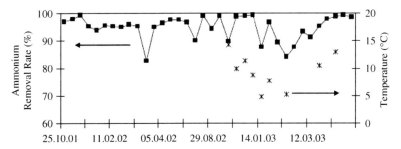

Figure 8.8. Ammonium removal efficiency for modified aerated lagoon

The capital costs for the upgrading of existing biological treatment included the implementation of pumps, aerators, sensors and process automatization. The total capital cost was 3.5 €/m³ of treated leachate for 8 years of depreciation. The operational costs (electric consumption, maintenances of sand filter and GAC columns) depended on the COD concentration of GAC influent (for COD < 600 mg/L and COD > 1000 mg/L, the costs were 6 and 12 €/m³, respectively).

8.8.3 MBR and RO combination (Hercule *et al.* 2003)

A leachate treatment plant in the north of France is the combination of membrane bioreactor (BIOSEP™ - developed by VEOLIA WATER) and reverse osmosis. This landfill site received 800 tons/day of municipal solid waste, with an average composition of 50% household waste and 50% commercial and industrial waste. The leachate was produced by the cells in operation since 1993 and collected through a four storage ponds of 3,000 m³ each. The leachate composition is presented in Table 8.13.

This plant has treated leachate from operational cells at a flow rate of 2.6 m³/h, since June 2000. The BIOSEP™ process combines a biological treatment

Table 8.13. Raw leachate composition (1998 to 2002)

Parameters	1998	1999	2000	2001	2002
COD (mg/L)	2306±869	2130±491	1449±368	1480±418	2563±1032
BOD$_5$ (mg/L)	113±68	82±54	–	–	–
NH$_4^+$-N (mg/L)	459±129	409±196	363±175	366±106	658±331
pH	8.3	8.3	8.2	8.1	8.2
Conductivity (μS/cm)	10230	10740	9638	8804	12290

using activated sludge treatment and physical separation treatment through the use of immersed membranes in the same basin. The membranes were made of polymeric hollow fibers with a breaking threshold of 200,000 Daltons. Thus, the separation was between micro-filtration and ultra-filtration. Bioreactor volume was 60 m^3 and each unit contained 8 membranes modules with a total surface of 372 m^2. While the filtration rate was between 2 and 4.5 m^3/h, the backwashing flux was in the range of 4–8.8 m^3/h. Specific flux on membranes was in the range of 6.3–14.5 L/m^2-h at temperatures between 5 and 26°C. The aspiration induced a differential pressure of 300 mbar. Oxygen was supplied with fine bubble systems located at the base of the basin. The flux showed a decrease since April 2002, thus a complete washing of the membranes by an intensive soak in a hypochlorite solution was done in September and October 2002. The membranes have not been changed since the start up of the plant. A change was planned during the 2nd half of 2003. 10–15 g/L suspended solids was kept in the reactor. The quantity of sludge produced during 2001 and 2002 was 3051 kg and 819 kg of DM (dry matter), respectively. The sludge production, thus represents 0.1 and 0.02 kg DM/kg COD, respectively.

The average efficiency of the membrane bioreactor on ammonia removal was 85% (Figure 8.9). During winter 2001, temperature decreased below 5°C, and then biological treatment performance decreased to less than 40%. During April 2002, ammonium removal rate decreased considerably. In order to reach initial high performance, clear water was filtered through the membranes before feeding with leachate. During winter 2002, the efficiency decreased again due to the increase of ammonia concentration.

The RO plant contained 2 stages of organic tubular membranes, each of them containing 3 modules in series and 8 in parallel. The total surface of filtration was 126 m^2. The applied pressure was 35 bar before a washing cycle and 45 bar

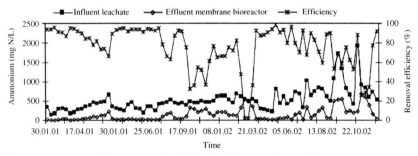

Figure 8.9. Ammonium removal efficiencies for membrane bioreactor

after. The frequency of maintenance was weekly with organic and citric acid washing to prevent fouling and scaling. The average COD removal was greater than 99%. However, when ammonia increased in bioreactor effluent, the RO efficiency also decreased. The average ammonia removal was 93% for the years of 2001 and 2002. The overall performance of the combined treatment was greater than 99% for COD and ammonia removal. The effluent COD, ammonium, and nitrate concentrations were averagely 13, 4 and 24 mg/L, respectively. The RO membranes were changed during 2002, for both treatment stages. The brine production represented 20% of leachate volume. This effluent was reintroduced in the mass of waste by spreading over the surface.

The electrical consumption was 34 kW/m^3 of treated leachate, for the overall process. The capital costs without construction work was 9 €/m^3 of treated leachate, including depreciation over 8 years. The operational costs, including reagents, electric consumption, renewal of membranes every two years and personal requirements were in the range of 9 to 13 €/m^3 of treated leachate.

REFERENCES

Aghamohammadi, N., Aziz, H.b.A., Isa, M.H. and Zinatizadeh, A.A. (2007) Powdered activated carbon augmented activated sludge process for treatment of semi-aerobic landfill leachate using response surface methodology. *Bioresource Technol.* **98**(18), 3570–3578.

Ahn, D.-H., Chung, Y.-C. and Chang, W.-S. (2002a) Use of coagulant and zeolite to enhance the biological treatment efficiency of high ammonia leachate. *J. Environ. Sci. Heal. A* **37**(2), 163–173.

Ahn, W.-Y., Kang, M.-S., Yim, S.-K. and Choi, K.-H. (2002b) Advanced landfill leachate treatment using an integrated membrane process. *Desalination* **149**(1/3), 109–114.

Aktas, O. and Cecen, F. (2001) Addition of activated carbon to batch activated sludge reactors in the treatment of landfill leachate and domestic wastewater. *J. Chem. Technol. Biotechnol.* **76**(8), 793–802.

Aluko, O.O. and Sridhar, M.K.C. (2005) Application of constructed wetlands to the treatment of leachates from a municipal solid waste landfill in Ibadan, Nigeria. *J. Environ. Health* **67**(10), 58–62.

Alvarez-Vazquez, H., Jefferson, B. and Judd, S.J. (2004) Membrane bioreactors vs conventional biological treatment of landfill leachate: a brief review. *J. Chem. Technol. Biotechnol.* **79**(10), 1043–1049.

An, K.J., Tan, J.W. and Meng, L. (2006) Pilot study for the potential application of a shortcut nitrification and denitrification process in landfill leachate treatment with MBR. *Water Sci. Technol. Water Supply* **6**(6), 147–154.

Andres, P., Arturo, D. and Cortijo, M. (2007) Coagulation-flocculation and ammoniacal stripping of leachates from solid waste landfill. *J. Environ. Sci. Health A* **42**(13), 2033–2038.

Bae, J.H., Kim, S.K. and Chang, H.S. (1997) Treatment of landfill leachates: Ammonia removal via nitrification and denitrification and further COD reduction via Fenton's treatment followed by activated sludge. *Water Sci. Technol.* **36**(12), 341–348.

Bae, J.H., Lee, I.S., Jang, M.S., Ahn, K.H. and Lee, S.H. (2004) Treatment of landfill leachate by a pilot-scale modified Ludzack-Ettinger and sulfur-utilizing denitrification process. *Water Sci. Technol.* **50**(6), 141–148.

Barnes, D., Li, X. and Chen, J. (2007) Determination of suitable pretreatment method for old-intermediate landfill leachate. *Environ. Technol.* **28**(2), 195–203.

Barr, M.J. and Robinson, H.D. (1999) Constructed wetlands for landfill leachate treatment. *Waste Manage. Res.* **17**(6), 498–504.

Baumgarten, G. and Seyfried, C.F. (1996) Experiences and new developments in biological pretreatment and physical post-treatment of landfill leachate. *Water Sci. Technol.* **34**(7), 445–453.

Benson, C.H., Barlaz, M.A., Lane, D.T. and Rawe, J.M. (2007) Practice review of five bioreactor/recirculation landfills. *Waste Manage.* **27**(1), 13–29.

Beone, G., Bortone, G., Gastaldello, A. and Tonolo, G. (1997) Treatment of landfill leachate by reverse osmosis. *6th International Landfill Symposium*, Sardinia, Italy, 13–17 October.

Berge, N.D., Reinhart, D.R., Dietz, J. and Townsend, T. (2006) In situ ammonia removal in bioreactor landfill leachate. *Waste Manage.* **26**(4), 334–343.

Beristain-Cardoso, R., Sierra-Alvarez, R., Rowlette, P., Flores, E.R., Gomez, J. and Field, J.A. (2006) Sulfide oxidation under chemolithoauthotrophic denitrifying conditions. *Biotechnol. Bioeng.* **95**(6), 1148–1157.

Bilgili, M.S., Demir, A. and Ozkaya, B. (2007) Influence of leachate recirculation on aerobic and anaerobic decomposition of solid wastes. *J. Hazard. Mater.* **143**(1/2), 177–183.

Bulc, T.G. (2006) Long term performance of a constructed wetland for landfill leachate treatment. *Ecol. Eng.* **26**(4), 365–374.

Burton, S.A.Q. and Watson-Craik, I.A. (1998) Ammonia and nitrogen fluxes in landfill sites: applicability to sustainable landfilling. *Waste Manage. Res.* **16**(1), 41–53.

Cabeza, A., Urtiaga, A., Rivero, M.-J. and Ortiz, I. (2007a) Ammonium removal from landfill leachate by anodic oxidation. *J. Hazard. Mater.* **144**(3), 715–719.

Cabeza, A.A., Primo, O., Urtiaga, A.M. and Ortiz, I. (2007b) Definition of a clean process for the treatment of landfill leachates integration of electrooxidation and ion exchange technologies. *Sep. Sci. Technol.* **42**(7), 1585–1596.

Calli, B., Mertoglu, B. and Inanc, B. (2005) Landfill leachate management in Istanbul: applications and alternatives. *Chemosphere* **59**(6), 819–829.

Calli, B., Tas, N., Mertoglu, B., Inanc, B. and Ozturk, I. (2003) Molecular analysis of microbial communities in nitrification and denitrification reactors treating high ammonia leachate. *J. Environ. Sci. Heal. A* **38**(10), 1997–2007.

Canziani, R., Emondi, V., Garavaglia, M., Malpei, F., Pasinetti, E. and Buttiglieri, G. (2006) Effect of oxygen concentration on biological nitrification and microbial kinetics in a cross-flow membrane bioreactor (MBR) and moving-bed biofilm reactor (MBBR) treating old landfill leachate. *J. Membrane Sci.* **286**(1/2), 202–212.

Cao, S., Chen, G., Hu, X. and Yue, P.L. (2003) Catalytic wet air oxidation of wastewater containing ammonia and phenol over activated carbon supported Pt catalysts. *Catal. Today* **88**(1/2), 37–47.

Cema, G., Wiszniowski, J., Zabczynski, S., Zablocka-Godlewska, E., Raszka, A. and Surmacz-Gorska, J. (2007) Biological nitrogen removal from landfill leachate by deammonification assisted by heterotrophic denitrification in a rotating biological contactor (RBC). *Water Sci. Technol.* **55**(8/9), 35–42.

Chen, S., Sun, D. and Chung, J.-S. (2008) Simultaneous removal of COD and ammonium from landfill leachate using an anaerobic-aerobic moving-bed biofilm reactor system. *Waste Manage.* **28**(2), 339–346.

Cheng, C.Y. and Chu, L.M. (2007) Phytotoxicity data safeguard the performance of the recipient plants in leachate irrigation. *Environ. Pollut.* **145**(1), 195–202.

Cheung, K., Chu, L. and Wong, M. (1997) Ammonia stripping as a pretreatment for landfill leachate. *Water Air Soil Poll.* **94**(1), 209–221.

Chiang, L.-C., Chang, J.-E. and Wen, T.-C. (1995) Indirect oxidation effect in electrochemical oxidation treatment of landfill leachate. *Water Res.* **29**(2), 671–678.

Cho, S.P., Hong, S.C. and Hong, S.-I. (2002) Photocatalytic degradation of the landfill leachate containing refractory matters and nitrogen compounds. *Appl. Catal. B-Environ.* **39**(2), 125–133.

Christensen, T.H., Kjeldsen, P., Bjerg, P.L., Jensen, D.L., Christensen, J.B., Baun, A., Albrechtsen, H.-J. and Heron, G. (2001) Biogeochemistry of landfill leachate plumes. *App. Geochem.* **16**(7/8), 659–718.

Collivignarelli, C., Bertanza, G., Baldi, M. and Avezzu, F. (1998) Ammonia stripping from MSW landfill leachate in bubble reactors: Process modeling and optimization. *Waste Manage. Res.* **16**(5), 455–466.

Cornelissen, E.R., Sijbers, P., van den Berkmortel, H., Koning, J., De Wit, A., De Nil, F. and Van Impe, J.F. (2001) Reuse of leachate waste-water using MEMBIOR technology and reverse osmosis. *Membrane Technol.* **136**, 6–9.

Del Borghi, A., Binaghi, A., Converti, A. and Del Borghi, M. (2003) Combined treatment of leachate from sanitary landfill and municipal wastewater by activated sludge. *Chem. Biochem. Eng. Q* **17**(4), 277–283.

Della Rocca, C., Belgiorno, V. and Meric, S. (2007) Overview of in-situ applicable nitrate removal processes. *Desalination* **204**(1/3), 46–62.

Deng, Y. and Englehardt, J.D. (2007) Electrochemical oxidation for landfill leachate treatment. *Waste Manage.* **27**(3), 380–388.

Di Palma, L., Ferrantelli, P., Merli, C. and Petrucci, E. (2002) Treatment of industrial landfill leachate by means of evaporation and reverse osmosis. *Waste Manage.* **22**(8), 951–955.

Diamadopoulos, E. (1994) Characterization and treatment of recirculation-stabilized leachate. *Water Res.* **28**(12), 2439–2445.

Diamadopoulos, E., Samaras, P., Dabou, X. and Sakellaroppoulos, G.P. (1997) Combined treatment of landfill leachate and domestic sewage in a sequencing batch reactor. *Water Sci. Technol.* **36**(2/3), 61–68.

Doyle, J., Watts, S., Solley, D. and Keller, J. (2001) Exceptionally high-rate nitrification in sequencing batch reactors treating high ammonia landfill leachate. *Water Sci. Technol.* **43**(3), 315–322.

Duggan, J. (2005) The potential for landfill leachate treatment using willows in the UK-A critical review. *Resour. Conserv. Recycling* **45**(2), 97–113.

Ehrig, H.J. (1984) Treatment of sanitary landfill leachate: Biological treatment. *Waste Manage. Res.* **2**(2), 131–152.

Elefsiniotis, P., Manoharan, R. and Mavinic, D.S. (1989) The effects of sludge recycle ratio on nitrification-denitrification performance in biological treatment of leachate. *Environ Tech. Lett.* **10**(12), 1041–1050.

El-Gandy, A.S., Biswas, N. and Bewtra, J.K. (2004) Growth of water hyacinth in municipal landfill leachate with different pH. *Environ. Technol.* **25**(7), 833–840.

Elimelech, M., Chen, W.H. and Waypa, J.J. (1994) Measuring the zeta (electrokinetic) potential of reverse osmosis membranes by a streaming potential analyzer. *Desalination* **95**(3), 269–286.

El-Mahrouki, I.M.L. and Watson-Craik, I.A. (2004) The effects of nitrate and nitrate-supplemented leachate addition on methanogenesis from municipal solid wastes. *J. Chem. Technol. Biotechnol.* **79**(8), 842–850.

Erses, A.S., Onay, T.T. and Yenigun, O. (2008) Comparison of aerobic and anaerobic degradation of municipal solid waste in bioreactor landfills. *Bioresource Technol.* **99**(13), 5418–5426.

Frascari, D., Bronzini, F., Giordano, G., Tedioli, G. and Nocentini, M. (2004) Long-term characterization, lagoon treatment and migration potential of landfill leachate: a case study in an active Italian landfill. *Chemosphere* **54**(3), 335–343.

Ganigue, R., Lopez, H., Balaguer, M.D. and Colprim, J. (2007) Partial ammonium oxidation to nitrite of high ammonium content urban landfill leachates. *Water Res.* **41**(15), 3317–3326.

Ganigue, R., Lopez, H., Ruscalleda, M., Balaguer, M.D. and Colprim, J. (2008) Operational strategy for a partial nitritation-sequencing batch reactor treating urban landfill leachate to achieve a stable influent for an anammox reactor. *J. Chem. Technol. Biotechnol.* **83**(3), 365–371.

Godley, A., Alker, G., Hallett, J., Marshall, R. and Riddell-Balcak, D. (2005) Landfill leachate nutrient recovery by willow short rotation coppice III.Soil water quality. *Arboricultural J.* **28**(1/2), 253–279.

Haarstad, K. and Maehlum, T. (1999) Important aspects of long-term production and treatment of municipal solid waste leachate. *Waste Manage. Res.* 17, 470–477.

Harper, S.C., Manoharan, R., Mavinic, D.S. and Randall, C.W. (1996) Chromium and nickel toxicity during the biotreatment of high ammonia landfill leachate. *Water Environ. Res.* **68**(1), 19–24.

Hasselgren, K. (1998) Use of municipal waste products in energy forestry: highlights from 15 years of experience. *Biomass Bioenerg.* **15**(1), 71–74.

He, P.J., Shao, L.M., Guo, H.D., Li, G.J. and Lee, D.J. (2006) Nitrogen removal from recycled landfill leachate by ex situ nitrification and in situ denitrification. *Waste Manage.* **26**(8), 838–845.

He, S., Zhang, Y., Yang, M., Du, W. and Harada, H. (2007) Repeated use of MAP decomposition residues for the removal of high ammonium concentration from landfill leachate. *Chemosphere* **66**(11), 2233–2238.

Heavey, M. (2003) Low-cost treatment of landfill leachate using peat. *Waste Manage.* **23**(5), 447–454.

Hedmark, A. and Scholz, M. (2008) Review of environmental effects and treatment of runoff from storage and handling of wood. *Bioresource Technol.* **99**(14), 5997–6009.

Henderson, J.P., Besler, D.A., Atwater, J.A. and Mavinic, D.S. (1997) Treatment of methanogenic landfill leachate to remove ammonia using a rotating biological contactor (RBC) and a sequencing batch reactor (SBR). *Environ. Technol.* **18**(7), 687–698.

Hercule, S., Cornu, E., Ballot, J., Coquant, C., Dumensil, C. and Lebourhis, D. (2003) Review of in-istu leachate treatment plant: 3 case studies in France. *9th International Waste Management and Landfill Symposium,* Cagliari, Italy, 6–10 October.

Hernandez, A.J., Adarve, M.J., Gil, A. and Pastor, J. (1999) Soil salination from landfill leachates: Effects on the macronutrient content and plant growth of four grassland species. *Chemosphere* **38**(7), 1693–1711.

Hippen, A., Rosenwinkel, K.-H., Baumgarten, G. and Seyfried, C.F. (1997) Aerobic deammonification: A new experience in the treatment of wastewaters. *Water Sci. Technol.* **35**(10), 111–120.

Hoilijoki, T.H., Kettunen, R.H. and Rintala, J.A. (2000) Nitrification of anaerobically pretreated municipal landfill leachate at low temperature. *Water Res.* **34**(5), 1435–1446.

Holweg, P., Cord-Landwehr, K. and Albers, H. (2003) MSW landfill leachate treatment techniques in Gemny-expeiences of a MSW landfill leachate neighbourhood group. *9th International Waste Management and Landfill Symposium,* Cagliari, Italy, 6–10 October 2003.

Horan, N.J., Gohar, H. and Hill, B. (1997) Application of a granular activated carbon-biological fluidised bed for the treatment of landfill leachates containing high concentrations of ammonia. *Water Sci. Technol.* **36**(2/3), 369–375.

Hosomi, M., Matsusige, K., Inamori, Y., Sudo, R., Yamada, K. and Yoshino, Z. (1989) Sequencing batch reactor activated sludge processes for the treatment of municipal landfill leachate: Removal of nitrogen and refractory organic compounds. *Water Sci. Technol.* **21**, 1651–1654.

Hurd, S., Kennedy, K.J., Droste, R.L. and Kumar, A. (2001) Low-pressure reverse osmosis treatment of landfill leachate. *J. Solid Waste Technol. Manage.* **27**(1), 1–14.

Ihara, I., Kanamura, K., Shimada, E. and Watanabe, T. (2004) High gradient magnetic seperation combined with electrocoagulation and electrochemical oxidation for the treatment of landfill leachate. *IEEE T. Appl. Supercon.* **14**(2), 1558–1560.

Ilies, P. and Mavinic, D.S. (2001) A pre- and post-denitrification system treating a very high ammonia landfill leachate: Effects of pH change on process performance. *Environ. Technol.* **22**(3), 289–300.

Im, J.-h., Woo, H.-j., Choi, M.-w., Han, K.-b. and Kim, C.-w. (2001) Simultaneous organic and nitrogen removal from municipal landfill leachate using an anaerobic-aerobic system. *Water Res.* **35**(10), 2403–2410.

Jokela, J.P.Y. and Rintala, J.A. (2003) Anaerobic solubilization of nitrogen from municipal solid waste (MSW). *Rev. Environ. Sci. Biotechnol.* **2**(1), 67–77.

Jones, D.L., Williamson, K.L. and Owen, A.G. (2006) Phytoremediation of landfill leachate. *Waste Manage.* **26**(8), 825–837.

Kaczorek, K. and Ledakowicz, S. (2006) Kinetics of nitrogen removal from sanitary landfill leachate. *Bioproc. Biosyst. Eng.* **29**(5), 291–304.

Kalyuzhnyi, S.V. and Gladchenko, M.A. (2004) Sequenced anaerobic-aerobic treatment of high strength, strong nitrogenous landfill leachates. *Water Sci. Technol.* **49**(5/6), 301–312.

Kargi, F. and Pamukoglu, M.Y. (2004) Adsorbent supplemented biological treatment of pre-treated landfill leachate by fed-batch operation. *Bioresource Technol.* **94**(3), 285–291.

Keenan, J.D., Steiner, R.L. and Fungaroli, A.A. (1983) Chemical-physical leachate treatment. *J. Environ. Eng. ASCE* **109**(6), 1371–1384.

Khin, T. and Annachhatre, A.P. (2004) Novel microbial nitrogen removal processes. *Biotechnol. Adv.* **22**(7), 519–532.

Kim, D.-J., Lee, D.-I. and Keller, J. (2006) Effect of temperature and free ammonia on nitrification and nitrite accumulation in landfill leachate and analysis of its nitrifying bacterial community by FISH. *Bioresource Technol.* **97**(3), 459–468.

Kim, Y.-K., Park, S.-E. and Kim, S.-D. (2003) Treatment of landfill leachate by white rot fungus in combination with zeolite filters. *J. Environ. Sci. Heal. A* **8**(4), 671–683.

Kjeldsen, P. and Christophersen, M. (2001) Composition of leachate from old landfills in Denmark. *Waste Manage. Res.* **19**(3), 249–256.

Kjeldsen, P., Barlaz, M.A., Rooker, A.P., Baun, A., Ledin, A. and Christensen, T.H. (2002) Present and long-term composition of MSW landfill leachate: A review. *Crit. Rev. Env. Sci. Tec.* **32**(4), 297–336.

Klimiuk, E. and Kulikowska, D. (2004) Effectiveness of organics and nitrogen removal from municipal landfill leachate in single and two-stage SBR systems. *Pol. J. Environ. Stud.* **13**(5), 525–532.

Klimiuk, E. and Kulikowska, D. (2005) The influence of operational conditions in sequencing batch reactors on removal of nitrogen and organics from municipal landfill leachate. *Waste Manage. Res.* **23**(5), 429–438.

Klove, B. (2001) Characteristics of nitrogen and phosphorus loads in peat mining wastewater. *Water Res.* **35**(10), 2353–2362.

Knox, K. (1985) Leachate treatment with nitrification of ammonia. *Water Res.* **19**(7), 895–904.

Krogmann, U. and Woyczechowski, H. (2000) Selected characteristic of leachate, condansate and runoff relesed during composting of biogenic waste. *Waste Manage. Res.* **18**(3), 235–248.

Laitinen, N., Luonsi, A. and Vilen, J. (2006) Landfill leachate treatment with sequencing batch reactor and membrane bioreactor. *Desalination* **191**(1/3), 86–91.

Lee, D.-U., Lee, I.-S., Choi, Y.-D. and Bae, J.-H. (2001) Effects of external carbon source and empty bed contact time on simultaneous heterotrophic and sulfur-utilizing autotrophic denitrification. *Process Biochem.* **36**(12), 1215–1224.

Lei, Y., Shen, Z., Huang, R. and Wang, W. (2007) Treatment of landfill leachate by combined aged-refuse bioreactor and electro-oxidation. *Water Res.* **41**(11), 2417–2426.

Li, X.Z. and Zhao, Q.L. (2001) Efficiency of biological treatment affected by high strength of ammonium-nitrogen in leachate and chemical precipitation of ammonium-nitrogen as pretreatment. *Chemosphere* **44**(1), 37–43.

Liang, Z. and Liu, J. (2008) Landfill leachate treatment with a novel process: Anaerobic ammonium oxidation (Anammox) combined with soil infiltration system. *J. Hazard. Mater.* **151**(1), 202–212.

Liang, Z. and Liu, J.-x. (2007) Control factors of partial nitritation for landfill leachate treatment. *J. Environ. Sci.* **19**(5), 523–529.

Linde, K., Jönsson, A.-s. and Wimmerstedt, R. (1995) Treatment of three types of landfill leachate with reverse osmosis. *Desalination* **101**(1), 21–30.

Liu, C.-H. and Lo, K.V. (2001a) Ammonia removal from compost leachate using zeolite. II. A study using continues flow packed columns. *Environ. Sci. Heal. B* **36**(95), 667–675.

Liu, C.-H. and Lo, K.V. (2001b) Ammonia removal from compost leachate using zeolite. I. Characterization of the zeolite. *Environ. Sci. Heal. A* **36**(9), 1671–1688.

Long, Y., Q.-F., G., Fang, C.-R., Zhu, Y.-M. and Shen, D.-s. (2008) In situ nitrogen removal in phase-seperate bioreactor lanfill. *Bioresource Technol.* **99**(13), 5352–5361.

Loukidou, M.X. and Zouboulis, A.I. (2001) Comparison of two biological treatment processes using attached-growth biomass for sanitary landfill leachate treatment. *Environ. Pollut.* **111**(2), 273–281.

Mace, S. and Mata-Alvarez, J. (2002) Utilization of SBR technology for wastewater treatment: An overview. *Ind. Eng. Chem. Res.* **41**(23), 5539–5553.

Martin, C.D. and Johnson, K.D. (1995) The use of extended aeration and in-series surface-flow wetlands for landfill leachate treatment. *Water Sci. Technol.* **32**(3), 119–128.

Marttinen, S.K., Kettunen, R.H., Sormunen, K.M., Soimasuo, R.M. and Rintala, J.A. (2002) Screening of physical-chemical methods for removal of organic material, nitrogen and toxicity from low strength landfill leachates. *Chemosphere* **46**(6), 851–858.

Maurice, C., Ettala, M. and Lagerkvist, A. (1999) Effects of leachate irrigation on landfill vegetation and subsequent methane emmisions. *Water Air Soil Poll.* **113**(1/4), 203–216.

Miller, P.A. and Clesceri, N.L. (2003) *Waste sites as biological reactors.* Lewis Publishers, Washington.

Mulamoottil, G., McBean, E.A. and Rovers, F. (1997) *Constructed wetlands for the treatment of landfill leachates.* Lewis Publishers, Florida.

Nivala, J., Hoos, M.B., Cross, C., Wallace, S. and Parkin, G. (2007) Treatment of landfill leachate using an aerated, horizontal subsurface-flow constructed wetland. *Sci. Total Environ.* **380**(1/3), 19–27.

Onay, T.T. and Pohland, F.G. (1998) In situ nitrogen management in controlled bioreactor landfills. *Water Res.* **32**(5), 1383–1392.

Otal, E., Vilches, L.F., Moreno, N., Querol, X., Vale, J. and Fernandez-Pereira, C. (2005) Application of zeolitised coal fly ashes to the depuration of liquid wastes. *Fuel* **84**(11), 1440–1446.

Ozturk, I., Altinbas, M., Koyuncu, I., Arikan, O. and Gomec-Yangin, C. (2003) Advanced physico-chemical treatment experiences on young municipal landfill leachates. *Waste Manage.* **23**(5), 441–446.

Papadopoulos, A., Kapetanios, E.G. and Loizidou, M. (1996) Studies on the use of clinoptilolite for ammonia removal from leachates. *J. Environ. Sci. Heal. A* **31**(1), 211–220.

Parkes, S.D., Jolley, D.F. and Wilson, S.R. (2007) Inorganic nitrogen transformation in the treatment of landfill leachate with a high ammonium load. *Environ. Monit. Assess.* **124**(1/3), 51–61.

Pelkonen, M., Kotro, M. and Rintala, J. (1999) Biological nitrogen removal from landfill leachate: a pilot-scale study. *Waste Manage. Res.* **17**(6), 493–497.

Peters, T. (1997) Treatment of landfill leachate by reverse osmosis. *6th International Landfill Symposium* (pp. 395–402), Sardinia, Italy, 13–17 October.

Peters, T.A. (2003) 20 years experience with reverse osmosis in leachate treatment. *9th International Waste Management and Landfill Symposium,* Cagliari, Italy, 6–10 October.

Phillips, I.R. and Sheehan, K.J. (2005) Use of phtocaps in remediation of closed landfills-suitability of selected soils to remove organic nitrogen and carbon from leachate. *Land Contam. Reclam.* **13**(4), 339–348.

Picard, C.R., Fraser, L.H. and Steer, D. (2005) The interacting effects of temperature and plant community type on nutrient removal in wetland microcosms. *Bioresource Technol.* **96**(9), 1039–1047.

Polprasert, C. and Sawaittayothin, V. (2006) Nitrogen mass balance and microbial analysis of constructed wetlands treating municipal landfill leachate. *Water Sci. Technol.* **54**(11/12), 147–154.

Praet, E., Jupsin, H., El Mossaoui, M., Rouxhet, V. and Vasel, J.L. (2001) Use of a membrane bioreactor and activated carbon adsorber for the treatment of MSW landfill leachates. *8th International Waste Management and Landfill Symposium*, Cagliari, Italy, 1–5 October.

Qasim, S.R. and Chiang, W. (1994) *Sanitary Landfill Leachate Generation, Control and Treatment*. Technomic Publishing, Basel.

Robertson, W.D. and Anderson, M.R. (1999) Nitrogen removal from landfill leachate using an infiltration bed coupled with a denitrification barrier. *Ground Water Monit. R.* **19**(4), 73–80.

Robinson, H.D. and Grantham, G. (1988) The treatment of landfill leachates in on-site aerated lagoon plants: Experience in Britain and Ireland. *Water Res.* **22**(6), 733–747.

Robinson, H.D. and Olufsen, J. (2007) Full biological tretament of landfill leachate: a detailed case study at Efford landfill. In the New Forest, Hampshire, UK. *11th International Waste Management and Landfill Symposium*, Cagliari, Italy, 1–5 October.

Rosenqvist, H. and Ness, B. (2004) An economic analysis of leachate purification through willow-coppice vegetation filters. *Bioresource Technol.* **94**(3), 321–329.

Rustige, H. and Nolde, E. (2007) Nitrogen elimination from landfill leachates using an extra carbon source in subsurface flow constructed wetlands. *Water Sci. Technol.* **56**(3), 125–133.

Shao, L.-M., He, P.-J. and Li, G.-J. (2008) In situ nitrogen removal from leachate by bioreactor landfill with limited aeration. *Waste Manage.* **28**(6), 1000–1007.

Shiskowski, D.M. and Mavinic, D.S. (1998) Pre-denitrification and pre- and post denitrification treatment of high ammonia landfill leachate. *Can. J. Civil Eng.* **25**(5), 854–863.

Shou-liang, H., Bei-dou, X., Hai-chan, Y., Shi-lei, F., Jing, S. and Hong-liang, L. (2008) In situ simultaneous organics and nitrogen removal from recycled landfill leachate using an anaerobic-aerobic process. *Bioresource Technol.* **99**(14), 6456–6453.

Silva, A.C., Dezotti, M. and Sant'Anna, G.L. (2004) Treatment and detoxification of a sanitary landfill leachate. *Chemosphere* **55**(2), 207–214.

Strous, M., Kuenen, J.G., and Jetten, M.S.M. (1999) Key physiology of anaerobic ammonium oxidation, *Appl. Environ. Microbiol.* **65**(7), 3248–3250.

Tengrui, L., Al-Harbawi, A.F., L.M., B., Jun, Z. and Long, X.Y. (2007a) Characteristics of nitrogen removal from old landfill leachate by sequencing batch biofilm reactor. *Am. J. Applied Sci.* **4**(4), 211–214.

Tengrui, L., Al-Harbawi, A.F., Qiang, H. and Jun, Z. (2007b) Comparison between biological treatment and chemical precipitation for nitrogen removal from old landfill leachate. *Am. J. Applied Sci.* **3**(4), 183–187.

Thorneby, L., Mathiasson, L., Martensson, L. and Hogland, W. (2006) The performance of a natural treatment system for landfill leachate with special emphasis on the fate of organic pollutants. *Waste Manage. Res.* **24**(2), 183–194.

Trebouet, D., Schlumpf, J.P., Jaouen, P. and Quemeneur, F. (2001) Stabilized landfill leachate treatment by combined physicochemical-nanofiltration processes. *Water Res.* **35**(12), 2935–2942.

Tsilogeorgis, J., Zouboulis, A., Samaras, P. and Zamboulis, D. (2008) Application of a membrane sequencing batch reactor for landfill leachate treatment. *Desalination* **221**(1/3), 483–493.

Turan, M., Gulsen, H. and Celik, M.S. (2005) Treatment of landfill leachate by a combined anaerobic fluidized bed and zeolite column system. *J. Environ. Eng. ASCE* **131**(5), 815–819.

Ushikoshi, K., Kobayashi, T., Uematsu, K., Toji, A., Kojima, D. and Matsumoto, K. (2002) Leachate treatment by the reverse osmosis system. *Desalination* **150**(2), 121–129.

van Kempen, R., ten Have, C.C.R., Meijer, S.C.F., Mulder, J.W., Duin, J.O.J., Uijterlinde, C.A. and van Loosdrecht, M.C.M. (2005) SHARON process evaluated for improved wastewater treatment plant effluent quality. *Water Sci. Technol.* **52**(4), 55–62.

Vigneron, V., Ponthieu, M., Barina, G., Audic, J.-M., Duquennoi, C., Mazeas, L., Bernet, N. and Bouchez, T. (2007) Nitrate and nitrite injection during municipal solid waste anaerobic biodegradation. *Waste Manage.* **27**(6), 778–791.

Vlyssides, A.G., Israilides, C., Loizidou, M., Karlis, P.K., Arapoglou, D. and Fatta, D. (2003) Electrochemical oxidation treatment of leachates from a new municipal landfill site. *9th International Waste Management and Landfill Symposium*, Cagliari, Italy, 6–10 October.

Watzinger, A., Reichenauer, T.G., Gerzabek, M.H. and Blum, W.E.H. (2006) Treatment of landfill leachate by irrigation and interaction with landfill gas. *Environ. Technol.* **27**(4), 447–457.

Welander, U., Henrysson, T. and Welander, T. (1997) Nitrification of landfill leachate using suspended-carrier biofilm technology. *Water Res.* **31**(9), 2351–2355.

Wens, P., De Langhe, P. and Staelens, B. (2001) Biological treatment of leachate by means of biomembrane reactor. *8th International Waste Management and Landfill Symposium*, Cagliari, Italy, 1–5 October.

Werner, M. and Kayser, R. (1991) Denitrification with biogas as external carbon source. *Water Sci. Technol.* **23**, 701–708.

Wiszniowski, J., Surmacz-Gorska, J., Robert, D. and Weber, J.V. (2007) The effect of landfill leachate composition on organics and nitrogen removal in an activated sludge system with bentonite additive. *J. Environ. Manage.* **85**(1), 59–68.

Wu, M.-Y., Franz, E.H. and Chen, S. (2001) Oxygen fluxes and ammonia removal efficiencies in constructed treatment wetlands. *Water Environ. Res.* **73**(6), 661–666.

Yilmaz, G. and Ozturk, I. (2003) Nutrient removal of ammonia rich effluent in a sequencing batch reactor. *Water Sci. Technol.* **48**(11/12), 377–383.

Youcai, Z., Hua, L., Jun, W. and Guowei, G. (2002) Treatment of Leachate by Aged-Refuse-based Biofilter. *J. Environ. Eng. ASCE* **128**(7), 662–668.

Zalesny, J.A., Zalesny Jr, R.S., Coyle, D.R. and Hall, R.B. (2007) Growth and biomass of *Populus* irrigated with landfill leachate. *Forest Ecol. Manage.* **248**(3), 143–152.

Zhang, S.-j., Peng, Y.-z., Wang, S.-y., Zheng, S.-w. and Guo, J. (2007) Organic matter and concentrated nitrogen removal by shortcut nitrification and denitrification from mature municipal landfill leachate. *J. Environ. Sci.* **19**(6), 647–651.

9

Nitrogen recovery via struvite production

M. Altınbaş

9.1 INTRODUCTION

For the removal of the nitrogen content of wastewaters, numerous biological and physicochemical methods have been developed as well as alternative technologies like struvite precipitation. The removal of ammonium through the formation of struvite has extensively been studied because of the advantages of struvite precipitation, such as recovering ammonia as a reusable product and achievement of high nitrogen removal efficiency. This chapter will concentrate on the principles, process parameters and application of struvite precipitation on different types of wastewater together with a case study.

9.2 STRUVITE FORMATION

9.2.1 Brief history

First evidence of the struvite formation by reaction of magnesium with phosphate in the presence of ammonium was found on a grave mound in Hungary originated from the late Bronze Age (Schneider 1985). After described as component of urinary stones (Wollaston 1797), struvite was discovered in guano by Ulex, a Swedish geologist, and called as struvite on behalf of his friend and mentor Heinrich G. von Struve (1772–1851), a Russian diplomat and natural scientist (Griffith and Osborne 1987). In the light of extensive studies, struvite was observed as an important component of infection stones in humans. As a consequence of metabolic disorders, the hydrolysis of urea to ammonia and carbon dioxide together with increase in the urine pH, and the content of NH_4^+, struvite precipitates from human urine (Da Silva *et al.* 2000).

In 1939, the deposition of struvite was reported in a wastewater treatment plant (WWTP) in pipes and other inner surfaces of the treatment process where it caused considerable troubles and expenses (Rawn *et al.* 1939). In this study, hard crystalline deposits found in the supernatant lines of a multiple stage sludge digestion system were identified as struvite. Subsequent finding about the presence of struvite deposits was at the sludge digester of Hyperion Wastewater Treatment Plant, California, where reduction of half percentage of pipeline diameter was observed in the 1960s (Borgerding 1972). Since the last couple of decades, the depositions of the struvite have been reported in the pipe fittings where the turbulence increased; for example, pipe bends, valves, pumps, and screens (Borgerding 1972; Mohajit *et al.* 1989). Struvite, hard, crystalline mineral, can accumulate rapidly, and then limit the capacity and efficiency of pipes and other related equipments earlier than its design life. Field accumulation rates and scale height accumulation of struvite on smooth surfaces was measured as 22.4 g/m^2-day and 1.2 mm/month, respectively (Ohlinger *et al.* 1999). In this study, the possibility of 50% reduction in the flow capacity of 3000 mm pipe within 3 years was assumed with this accumulation rate.

Since the struvite precipitation adversely affects the efficiency of WWTPs performance and causes maintenance problems, different strategies have been investigated to control struvite deposition. The aim of these strategies was to prevent the development of struvite depositions, which lowered the functionality of equipment and facility operation. The struvite deposition can be controlled during the WWTP design either by minimizing phosphorus and magnesium transfer from the liquid to the solids stream or protecting key components, such as dewatering equipment and heat exchangers (Neethling and Benisch 2004).

Besides that, the deposition of struvite crystals has been attempted to control by dilution with water effluents, thermal treatment, jet washing (Borgerding 1972), chemical dosing of iron salts (Mamais *et al.* 1994) or addition of chemical inhibitors (Buchanan *et al.* 1994). Although these strategies have been successfully applied, economical burden to the operational costs was considerably high.

On the other hand, the recovery of struvite as a nutrient source, which has already been proposed by Murray (1858), is more efficient instead of giving effort for controlling the struvite formation in the WWTPs. Due to the possibility of effectively recovering ammonium from wastewaters as easily settleable insoluble compounds with a simple processes, the struvite formation started to be used as a valuable method. Therefore, instead of allowing spontaneous formation, struvite precipitation under controlled conditions and location has then been widely investigated by optimizing the process parameters. Beside achievement of ammonium removal, the produced struvite was found to be the potential area, which could be utilized as marketable product for the fertilizer industry owing to its composition (magnesium ammonium phosphate ($MgNH_4PO_3 \cdot 6H_2O$)).

9.2.2 Applicability of struvite precipitation to N-rich wastewaters

Struvite precipitation can be applied to a wide range of wastewaters as long as the nitrogen present in the form of ammonium. Since optimum conditions should be provided to realize the struvite precipitation, concentration of the constituent ions (Mg, N, and P), pH, and temperature were the main parameters, which were primarily considered for the process design (Borgerding 1972).

The wastewaters pre-treated with anaerobic systems are the most suitable for the struvite crystallization as the organic nitrogen converted into ammonium by ammonification and the pH rises by degradation of the organic acids provided the optimum conditions for struvite precipitation (Ohlinger *et al.* 1998; Munch and Barr 2001). The struvite formation has successfully been performed in several anaerobically digested wastewaters such as swine lagoon liquid (Nelson *et al.* 2003) calf wastewater (Schuiling and Andrade 1999), landfill leachate (Ozturk *et al.* 2003) and dairy manure (Uludag-Demirer *et al.* 2005). In addition to these, biological processes treating high strength of ammonium-nitrogen are also suitable for application of struvite precipitation with the aim of minimizing ammonia toxicity, which is likely to lead the failure of biological treatment systems (Barnes *et al.* 2007). The results of the studies during the treatment of wastewaters containing high strength of ammonia-nitrogen, such as animal manure (Burns *et al.* 2001), baker's yeast and opium alkaloid processing plant effluents (Altinbas *et al.* 2002a), leather tanning wastewaters (Tünay *et al.* 1997), wheat and potato starch factories (Austermann-Haun and Seyfried 1994) have shown the applicability of the struvite precipitation effectively.

Struvite formation has also been applied for improving the quality of compost and urine as a fertilizer. In the case of urine, the hydrolysis of urea leads to nitrogen loss as ammonia (NH_3) during the storage and handling. The formation of struvite crystals could help the recovery of ammonium as in a reusable form and improve the quality of urine as a fertilizer (Ganrot 2008). On the other hand, ammonia emission is a common problem at composting facilities. The formation of struvite crystals in composting process have been found to facilitate the reduction of ammonia loss and a substantial increase in total ammonia nitrogen content of compost product. With the aid of the struvite formation, the nutrient value of the compost can be increased and this is led to enlarge the market of compost as an organic fertilizer (Jeong and Kim 2001).

9.3 PROCESS FUNDAMENTALS

Struvite is a white crystalline substance, which has a distinctive orthorhombic structure consisting of magnesium, ammonium and phosphorus in equal molar concentrations ($MgNH_4PO_4 \cdot 6H_2O$). The structure of struvite crystals varies depending on the growth conditions during reaction. The X-ray diffraction has usually been used for identification of struvite crystals (Doyle *et al.* 2002). The structures may be as equant, wedge shaped, short prismatic and thick tubular (Durrant *et al.* 1999). A crystal structure formed in sludge liquors is given in Figure 9.1 as an example.

Figure 9.1 Crystal of struvite precipitated from sludge liquors (Doyle *et al.* 2002)

9.3.1 Struvite solubility

Supersaturation state of the reaction, which could be evaluated through thermodynamic modeling, has been used for determining whether the reaction will proceed or not. The struvite crystals precipitate when the combined concentrations of Mg^{2+}, NH_4^+, and PO_4^{3-} ions exceed the struvite solubility limit or, in other words, solution becomes supersaturated. Supersaturation is a measure of the deviation of a dissolved salt from its equilibrium value. In supersaturated solution, as a result of the chemical reaction of constituent ions, struvite forms as follows:

$$Mg^{2+} + NH_4^+ + PO_4^{3-} + 6H_2O \rightarrow MgNH_4PO_4 \cdot 6H_2O \qquad (9.1)$$

For dealing with the struvite precipitation problems, equilibrium chemistry has firstly been developed by Stumm and Morgan (1970) and improved with including the effects of ionic strength on the activities of the struvite component ions by Snoeyink and Jenkins (1980). These authors have formulated the *thermodynamic solubility product* (K_{SP}) equation taking into account total molar concentrations of ions in solution of magnesium, free and saline ammonia (N_T) and orthophosphate (P_T) and the *conditional solubility product* (P_S) equation taking into account the ionic strength and the ion activity in addition to total species products as follows:

$$K_{SP} = [Mg^{2+}][NH_4^+][PO_4^{3-}] \qquad (9.2)$$

$$P_S = C_{T,Mg} C_{T,NH_3} C_{T,PO_4} \qquad (9.3)$$

Where $[Mg^{2+}]$, $[NH_4^+]$ and $[PO_4^{3-}]$ are the molar concentrations; $C_{T,Mg}$, $C_{T,NH3}$ and $C_{T,PO4}$ are the total analytic concentrations of magnesium, ammonium and orthophosphate, respectively.

When the conditional solubility product of the constituting ions exceeds the thermodynamic solubility product ($P_S > K_{SP}$), the solution is supersaturated and struvite precipitation will take place, otherwise ($K_{SP} > P_S$), the solution is under-saturated. The differences between these two products can be calculated under certain process conditions and the feasibility of the crystallization process of struvite can be evaluated.

For the calculation of the conditional struvite solubility, most commonly considered ionic species are H_3PO_4, $H_2PO_4^-$, HPO_4^{2-}, PO_4^{3-}, MgH_2PO_4, $MgHPO_4^-$, $MgPO_4^-$, $MgOH^+$, Mg^{2+}, NH_4^+ and $NH_3(aq)$. Besides these ionic species, equilibrium constant of water is also taken into account for the determination of the conditional solubility, as the concentrations of these species are affected by solution pH due to the presence of H^+ and OH^-. Thermodynamic relationships and equilibrium constants (K) of these species and water are given in Table 9.1 at different temperatures.

With respect to these species, mass balance equations for Mg^{2+}, NH_4^+ and PO_4^{3-} can be written as follows:

$$C_{T,Mg} = [Mg^{2+}] + [MgOH^+] + [MgH_2PO_4^+] + [MgHPO_4] + [MgPO_4^-] \quad (9.4)$$

$$C_{T,NH_3} = NH_3(aq) + [NH_4^+] \quad (9.5)$$

$$C_{T,PO_4} = [PO_4^{3-}] + [H_3PO_4] + [H_2PO_4^-] + [MgH_2PO_4^+] + [MgHPO_4]$$
$$+ [MgPO_4^-] \quad (9.6)$$

where [X] shows the molar concentration of each species. The ionization fraction, α_X , can be expressed as a function of bulk fluid pH for each component, using the mass balance equations and equilibrium equations for Mg^{2+}, NH_4^+ and PO_4^{3-} as follows:

$$\alpha_{Mg^{2+}} = \frac{[Mg^{2+}]}{C_{T,Mg}} = \frac{1}{\left(1 + \frac{K_w}{[H^+] \times K_7}\right) + PO_4^{3-}\left(\frac{[H^+]^2}{K_2 \times K_3 \times K_4} + \frac{[H^+]}{K_3 \times K_5} + \frac{1}{K_6}\right)} \quad (9.7)$$

$$\alpha_{NH_4^+} = \frac{[NH_4^+]}{C_{T,NH_3}} = \frac{1}{1 + \frac{K_b}{[H^+]}} \quad (9.8)$$

$$\alpha_{PO_4^{3-}} = \frac{[PO_4^{3-}]}{C_{T,PO_4}}$$

$$= \frac{1}{\left(1 + \frac{[H^+]^3}{K_1 \times K_2 \times K_3} + \frac{[H^+]^2}{K_2 \times K_3} + \frac{[H^+]}{K_3}\right) + Mg^{2+}\left(\frac{[H^+]^2}{K_2 \times K_3 \times K_4} + \frac{[H^+]}{K_3 \times K_5} + \frac{1}{K_6}\right)} \quad (9.9)$$

The struvite solubility product (K_{SO}) can be written as given in Equation 9.10:

$$K_{SO} = \{Mg^{2+}\}\{NH_4^+\}\{PO_4^{3-}\} \quad (9.10)$$

where {X} shows the activity of each species. Expressing the activities in terms of activity coefficient, γ_X ({X} $= \gamma_X [X]$) and molar concentration of

Table 9.1 Major equilibriums for the species involved in struvite precipitation and their equilibrium constants (adapted from Rahaman et al. (2006))

Equilibrium	Equilibrium Constants	pK (25°C)	pK (20°C)	pK (15°C)	pK (10°C)	Reference
$H^+ + H_2PO_4^- \Leftrightarrow H_3PO_4$	K_1	2.148	2.11	2.09	2.07	Smith and Martell (1989)
$H^+ + HPO_4^{2-} \Leftrightarrow H_2PO_4^-$	K_2	7.198	7.21	7.23	7.25	Smith and Martell (1989)
$H^+ + PO_4^{3-} \Leftrightarrow HPO_4^{2-}$	K_3	12.375	12.27	12.43	12.49	Smith and Martell (1989)
$Mg^{2+} + H_2PO_4^- \Leftrightarrow MgH_2PO_4$	K_4	1.207	0.45	0.36	0.32	Childs (1970)
$Mg^{2+} + HPO_4^{3-} \Leftrightarrow MgHPO_4^-$	K_5	2.428	2.77	2.83	2.91	Childs (1970)
$Mg^{2+} + PO_4^{3-} \Leftrightarrow MgPO_4^-$	K_6	4.92	4.8	4.72	4.68	Childs (1970)
$Mg^{2+} + OH^- \Leftrightarrow MgOH^+$	K_7	2.56	2.46	2.46	2.46	Morel and Hering (1993)
$NH_3 + H^+ \Leftrightarrow NH_4^+$	K_b	9.24	9.71	9.62	9.78	Martell et al. (1998)
$H^+ + OH^- \Leftrightarrow H_2O$	K_w	13.997	14.38	14.34	14.52	Smith and Martell (1989)

$pK = -\log_{10}(K)$

species in terms ionization fraction ($[X] = \alpha_X C_X$), K_{SO} becomes:

$$K_{SO} = \gamma_{Mg^{2+}} [Mg^{2+}] \gamma_{NH_4^+} [NH_4^+] \gamma_{PO_4^{3-}} [PO_4^{3-}]$$

$$K_{SO} = \gamma_{Mg^{2+}} \alpha_{Mg^{2+}} C_{T,Mg^{2+}} \gamma_{NH_4^+} \alpha_{NH_4^+} C_{T,NH_4^+} \gamma_{PO_4^{3-}} \alpha_{PO_4^{3-}} C_{T,PO_4^{3-}} \qquad (9.11)$$

The conditional solubility product can then be defined replacing the total component concentration values from equation 9.11.

$$P_S = C_{T,Mg^{2+}} C_{T,NH_4^+} C_{T,PO_4^{3-}} = \frac{K_{SO}}{\alpha_{Mg^{2+}} \alpha_{NH_4^+} \alpha_{PO_4^{3-}} \gamma_{Mg^{2+}} \gamma_{NH_4^+} \gamma_{PO_4^{3-}}} \qquad (9.12)$$

Although numerous values for K_{SO} have been reported in the range of $1 \times 10^{-12.6}$ to $1 \times 10^{-13.15}$, the most commonly used value is $1 \times 10^{-12.6}$ in engineering (Stumm and Morgan 1996; Snoeyink and Jenkins 1980).

As the activity of each ion is affected by the ionic strength of a solution, bulk liquid ionic strength, I, should be determined for the calculation of the activity coefficients of each ion. This ionic strength can be determined indirectly by the measurement of the electrical conductivity of the solution or directly by using following equation:

$$I = \frac{1}{2} \sum_i (C_i Z_i^2) \qquad (9.13)$$

where C_i is the concentration of ionic species, and Z_i is the charge of species.

Guntelberg approximation of the Debye–Huckel theory has frequently been used in wastewater treatment for the estimation of the activity coefficient of ions with ionic strengths smaller than about 0.5 (Snoeyink and Jenkins 1980).

$$-\log \gamma_i = \frac{A Z_i^2 \sqrt{I}}{1 + \sqrt{I}} \qquad (9.14)$$

where, A is the Debye–Huckel constant (0.509 at 25°C).

Correction of ionic strength is difficult for wastewater treatment. With the establishment of the effect of ionic strength, it is possible to interpret the changes in the concentration of the constituent ions or the solution pH in order to eliminate the negative influence of this factor on struvite precipitation in wastewater. The conditional solubility can be plotted as given in Figure 9.2 using the above equilibrium calculations and experimentally determined solubility constants.

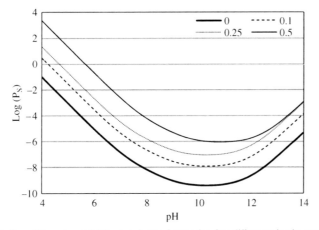

Figure 9.2 Conditional solubility product of struvite for different ionic strengths

In Figure 9.2, the upper part and the lower part of the curve represent the supersaturation area and the undersaturation area, respectively. As seen from Figure 9.2, struvite solubility increases as a result of decrease in the effective concentration of the constituents ions within the operating ionic strength range.

9.3.2 Struvite precipitation

The struvite crystallization kinetic is usually separated into two phases: *nucleation* and *growth*. In the nucleation phase, constituent ions combine to form crystal embryos spontaneously (homogeneous nucleation) or induced by the presence of suitable seed material (heterogeneous nucleation). The time that elapses for nucleation and appearance of crystals is termed as *induction time*. It has generally been used as control parameter for struvite formation. The relationship between induction time and supersaturation ratio was established by Ohlinger *et al.* (1999) and induction time (t_{ind}) was formulated as follow:

$$t_{ind} = \frac{1}{J} \tag{9.15}$$

where, J indicates the homogenous nucleation rate and is formed regarding the statistical nucleation concept and the assumption of the duration of nucleation greater than growth ($t_n \gg t_g$) (Toshev 1973). J is formulated as below (Ohlinger *et al.* 1999):

$$J = \Omega \exp\left[\frac{-\beta v^2 (\gamma^S)^3}{k_B T (-\Delta \mu)^2}\right] \tag{9.16}$$

where Ω is pre-exponential factor, β is the geometric factor, v is the molecular volume, γ^S is the surface energy, k_B is the Boltzman's constant, T is the absolute temperature, and $\Delta \mu$ is the driving force for nucleation, which can be defined with the chemical potential:

$$\frac{-\Delta \mu}{RT} = v \ln S_a \tag{9.17}$$

where R is the universal gas constant, v is the number of ions into which a molecule dissociates and S_a is the activity based supersaturation ratio which can written for struvite as follows:

$$S_a = \left[\frac{[\{Mg^{2+}\}\{NH_4^+\}\{PO_4^{3-}\}]_{initial}}{K_{SO}}\right]^{1/3} \tag{9.18}$$

Solution supersaturation can be included into the induction time equation by substitution of Equation 9.17 and Equation 9.18 into Equation 9.16:

$$\log t_{ind} = \frac{A}{(\log S_a)^2} - B \tag{9.19}$$

where $A = \beta(\gamma^S)^3 v^2 / [(2.3 k_B T)^3 v^2]$ and $B = \log \Omega$ for homogenous nucleation.

After the nucleation, crystals are continuously grown on existing nucleation sites until chemical equilibrium is reached to the growth phase (Ohlinger *et al.* 1999). If the struvite precipitation system is fed continuously with magnesium, ammonium and phosphate in equal molar quantities, crystal growth may continue for an indefinite period (Doyle *et al.* 2002). Growth kinetic can be expressed as simple rate law given by Snoeyink and Jenkins (1980). Ohlinger *et al.* (1999) and Nelson *et al.* (2003) reported that crystal growth rate obeyed the first order reaction rate with a constant changing between 4 and 12 h^{-1}. In experimental studies, the struvite precipitation period has generally been defined as the crystal retention time, which is the sum up of induction and growth times. For different reactor configurations, the maximum crystal retention times are given in the following sections and summarized in Table 9.8.

9.4 PROCESS PARAMETERS

In addition to ionic strength, pH, temperature, total suspended solids (TSS) and dissolved components, such as magnesium, ammonia, and phosphorus ions, presence of other counter-reacting ions (impurities such as calcium) are important process parameters affecting the morphology and size of struvite crystals.

9.4.1 Effect of pH

pH is the most critical parameter to realize the effective struvite precipitation. The optimum pH value for struvite precipitation should be determined depending on the wastewater characteristics. For illustration of the effect of pH on the struvite solubility and constituent ions, conditional solubility product and ionization fraction for each ion is plotted in Figure 9.3 using the equilibrium calculations for 25°C and assuming $I = 0$.

As seen from Figure 9.3, the struvite solubility decreases with increasing pH. The pH increases in solution lead to changes in the concentrations of ionic constituents: ammonium ion concentration will decrease depending on the equilibrium with ammonia (NH_3); the phosphate ion concentration will increase, because of increases in the fraction of total PO_4-P present as the PO_4^{3-} ion; and magnesium ion concentration will decrease, since the $MgOH^+$ is formed as a result of magnesium ion hydrolysis at higher pH values (Snoeyink and Jenkins 1980).

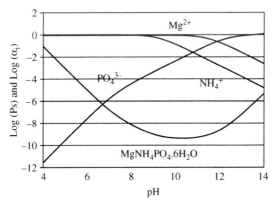

Figure 9.3 Conditional solubility product and ionization fractions for Struvite

Moreover, PO_4^{3-} activity has a greater influence on struvite precipitation than does the pH effect on NH_4^+ activity, due to the pH shifts on phosphate equilibrium towards PO_4^{3-}, although most of the phosphate is in the form of $H_2PO_4^-$ and HPO_4^{2-} at a pH between 4.8 and 6.6 (Ban and Dave 2004).

However, increasing the pH value above 10 leads to an increase in the threshold value of struvite solubility as concentration changes occurred in constituent ions as a function of pH. The alkaline medium causes a low solubility of the precipitate and limits the achievement of supersaturation.

The optimum pH value for minimum struvite solubility may vary depending on the ionic strength and composition of wastewaters. The optimum pH values have been reported to be between 8 and 11 by several researchers depending on the type of wastewater. Although pH of 10.8 has been calculated as a maximum value for minimum solubility, optimum pH values have been observed as 10.3 in synthetic wastewater (Ohlinger et al. 1998). Optimum pH values were also found as 8.5–9.0 in anaerobic digestion wastewater (Celen and Turker 2001), 9.0 in industrial wastewater (Hoffmann et al. 2004), 8.9 and 9.3 in anaerobic swine lagoon liquid (Nelson et al. 2003).

9.4.2 Effect of temperature

The effect of temperature on struvite solubility has widely been investigated. A study on an anaerobic digestion system proposed that as the temperature increases from 0°C to 20°C, struvite solubility increases to a maximum although the precipitation potential steadily decreases above 20°C with increasing temperature (Borgerding 1972). However, in recent studies higher temperature values have been determined for maximum struvite solubility as 50°C (Aage et al. 1997) and 35°C (Bhuiyan et al. 2007) for synthetic wastewater. The different effective temperature values have been found in different waste/wastewaters for the optimization of the ammonium removal as summarized in Table 9.2.

The struvite precipitation has also been prompted by direct heating at temperatures as high as 300°C (Abdelrazig and Sharp 1998). However, an experimental study conducted by He et al. (2007) has shown that the struvite obtained at such a high temperature resulted in a low ammonium removal.

9.4.3 Effect of TSS

In general, struvite precipitation was applied on the pre-treated or the wastewater having less total suspended solids. Meanwhile, it may be expected that the presence of high TSS, reduces the efficiency of ammonium removal. TSS may

Table 9.2 Optimum temperature values for struvite precipitation

Wastewater Type	Temperature (°C)	Initial NH_4^+ Concentration (mg N/L)	NH_4^+ removal Efficiency (%)	References
Swine wastewater	25	2500–2600	90	Maekawa et al. (1995)
Anaerobically digested molasses based industrial wastewater	25–40	1400	90	Celen and Turker (2001)
Landfill leachate	90	2000	96	He et al. (2007)
Poultry manure composting system	65	–	84	Zhang and Lau (2007)

reduce the purity and value of struvite as a fertilizer; as well as inhibit the formation of struvite.

The adverse effect of TSS concentration on struvite precipitation has been reported at concentrations above 1000 mg/L in calf manure (Schuiling and Andrade 1999). In another study on landfill leachate, Barnes et al. (2007) has found that the NH_4^+ removal efficiency decreases from 98% to 78% with increasing TSS concentration from 320 mg/L to 3600 mg/L.

9.4.4 Effect of chemical composition of the wastewater

9.4.4.1 Concentrations of constituent ions

The studies have shown that the optimization of molar concentration ratio of Mg^{2+}, NH_4^+, and PO_4^{3-} (Mg:N:P) in combination with pH is a very common way to find out the efficiency of the removal of the nutrients via struvite formation. In most cases, imbalanced supply of Mg^{2+} and/or PO_4^{3-} has externally been added to induce struvite formation in wastewaters, with high ammonium concentrations. To maximize the ammonium removal efficiency, the optimum molar ratios of Mg:N:P together with pH values was determined for different types of wastewater (Table 9.3).

Magnesium chloride ($MgCl_2$), magnesium hydroxide ($Mg(OH)_2$) and magnesium oxide (MgO) have frequently been used as a magnesium source for struvite precipitation (Table 9.4). In terms of nitrogen removal efficiencies, $MgCl_2$ was found the most effective Mg source compared to other Mg source in

Table 9.3 Optimum Mg:N:P molar ratios and pH for struvite precipitation

Wastewater Type	Working Volume (mL)	Mg:N:P Mol Ratio	pH	Initial NH_4^+ Concentration (mg N/L)	NH_4^+ removal efficiency (%)	References
Synthetic wastewater	500	1.5:1.0:1.5	9	600	97.2	Demeester et al. (2001)
Landfill Leachate	500	1:1:1.2	9	2132	98.3	Barnes et al. (2007)
Leather Tanning Wastewater	500	1.0:1.0:1.0	9.5	6785	94	Zengin et al. (2005)
Anaerobically digested swine wastes	–	1.25:1.0:1.0	9.5	3500	88	Miles and Ellis (2001)
Hydrolyzed Urine	500	1.0:1.0:1.0	9.5	5870	95.8	Kabdasli et al. 2006
Rare-earth smelting company wastewaters	500	1.2:1.0:1.3	11	1540	95	Li et al. (2007)
Textile wastewater	5000	1.0:1.0:1.34	8	450	87	Kabdasli et al. (2000)
Semiconductor wastewater	12 m³	1.0:1.0:1.0	9	155	89	Ryu et al. (2008)
Anaerobically digested molasses based industrial wastewater	200	1.2:1.0:1.2	7.5	1400	90	Celen and Turker (2001)
Composted poultry manure	6000	1.0:1.0:1.0	9	–	84	Zhang and Lau (2007)

Table 9.4 Optimum ammonium removal efficiencies depending on Mg Source for struvite precipitation

Wastewater Type	Working Volume	pH	Mg:N:P Molar ratios	Mg salts	Initial NH_4^+ Concentration (mg N/L)	NH_4^+ removal (%)	Reference
Synthetic wastewater	10 L	9.1	1.3:1:1:1.0	$MgCl_2$	–	53	Lee et al. (2003)
Synthetic wastewater	10 L	10	1.3:1:1:1.0	Seawater	–	54	Lee et al. (2003)
Synthetic wastewater	10 L	9.6	1.3:1:1:1.0	Bittern	–	39	Lee et al. (2003)
Landfill Leachate	30 m^3	8.5	1.3:1:1.5	$MgCl_2 \cdot 6H_2O$	–	70–80	Battistoni et al. (2001)
Landfill Leachate	30 m^3	8.5	2.0:1:1.5	MgO (87%)	–	80–90	Battistoni et al. (2001)
Landfill Leachate	1 L	9	1.0:1.0:1.0	$MgCl_2 \cdot 6H_2O$	2750	92	Li and Zhao (2002)
Landfill Leachate	1 L	9	1.0:1.0:1.0	$MgSO_4 \cdot 7H_2O$	2750	70	Li and Zhao (2002)
Landfill Leachate	1 L	8.38	1.0:1.0:1.0	Bittern	2900	80	Li and Zhao (2002)
Anaerobically digested Dairy manure	2 L	8.5	2.2:1.0 (Mg:N)	$Mg(OH)_2$	480–520	90	Uludağ-Demirer et al. (2005)
Anaerobically digested Dairy manure	2 L	8.5	1.3:1.0 (Mg:N)	$MgCl_2 \cdot 6H_2O$	255	90	Uludağ-Demirer et al. (2005)
Slurry-type swine wastewater	n.a.	8.0–10.0	3.0:1.0:1.5	MgO	2125	98	Kim et al. (2004)
Sludge Fermentation effluent	1.5 L	9.0	n.a.	Waste Lime	88	80	Ahn and Speece (2006)
Cochineal insects processing	400 mL	8.5–9	24 g/L	Low grade MgO	2320	89	Chimenos et al. (2003)
Leachate	250 mL	8.6	1.0:1.0:1.0	MgCO3	2700	91	Gunay et al. (2008)

n.a.: not available

the same M:N:P molar ratios. However, when the Mg:N molar ratio was increased, nearly 90% of nitrogen removal efficiency was possible with all Mg sources. The high nitrogen removal efficiency has also been achieved with the addition of low grade MgO, bittern and seawater used as magnesium source (Chimenos et al. 2003, Lee et al. 2003, Ouintana et al. 2005; Ahn and Speece 2006).

9.4.4.2 The phases for struvite precipitation

During struvite crystallization or dissolution, different phases including magnesium ammonium monohydrate (dittmarite, $NH_4MgPO_4 \cdot H_2O$), magnesium hydrogen phosphate trihydrate (newberyite, $MgHPO_4 \cdot 3H_2O$) and trimagnesium phosphate (bobierrite, $Mg_3(PO_4)_2 \cdot 8H_2O$) can be formed depending on the concentration of constituents ions, pH (Taylor et al. 1963a) and temperature (Sarkar 1991) of the solution. Dissolution of struvite to these phases leads to the release of ammonium and then, the nitrogen component of struvite is missing in the new-formed phases. Therefore, formation of different precipitation products other than struvite causes the decrease in ammonium removal efficiency. pH increases from slightly acidic to slightly basic values leads to change in predominant solid species from newberyite to struvite. In a pH range of 6.4–7.7, newberyite and struvite are thermodynamically stable (Dempsy 1997). On the contrary, struvite and bobierrite are stable phases in alkaline pH (Taylor et al. 1963a). As a result of the exposure of struvite to temperatures between 50–80°C, structure changes were observed from struvite to bobierrite due to loss of ammonium ions (Taylor et al. 1963b). On the other hand, when the struvite was boiled in excess water, it only lost its water to form the monohydrate struvite, dittmarite. Then, it could be transformed back to more stable struvite applying slowly hydration at room temperature (Sarkar, 1991). Relationships between the possible struvite phases were schematically represented in Figure 9.4.

9.4.4.3 Presence of other counter-reacting ions

Because of the presence of other counter-reacting ions in solution, such as calcium and potassium, struvite formation may be inhibited, although the pH and component-ion-molar ratios are appropriate for precipitation. The presence of high concentrations of calcium ion in wastewaters has been found as inhibitory either by competing for phosphate ions or by interfering through the struvite crystallization in terms of size, shape, and purity of the recovered product. The most commonly observed solid phases containing calcium might be $CaHPO_4 \cdot 2H_2O$ (brushite), $CaHPO_4$ (monetite), $Ca_3(PO_4)_2 \cdot (H_2O)$ (amorphous

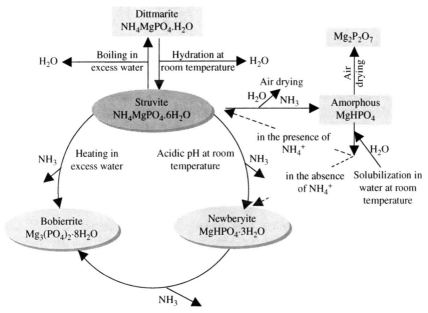

Figure 9.4 Relationship between struvite and other possible phases

calcium phosphate) as well as the crystalline phases such as hydroxyapatite ($Ca_5(PO_4)_3OH$) and whitlockite ($Ca_3(PO_4)_2$). The Mg:Ca ratio is the main factor for controlling the precipitate composition. Efficient struvite formation could be achieved by increasing the Mg:Ca concentrations, since the inhibition of struvite precipitation was observed at Mg:Ca molar ratios lower than 1.0 (Le Corre 2005). Mustovo et al. (2000) reported that Mg:Ca molar ratios greater than 0.6 should be provided for the effective struvite precipitation. It was also recognized that the crystallization of calcium phosphates was affected by the presence of high magnesium concentrations in the Mg:Ca ratio higher than 1.3 (Battistoni et al. 1998). In accordance with these observations, the presence of calcium in wastewater streams should be taken into consideration when determining an ideal magnesium dose for achievement of sufficient ammonia removal.

Potassium ions can also incorporate into struvite precipitate, which results in the formation of K-struvite ($KMgPO_4 \cdot 6H_2O$). K-struvite has the similar structure of struvite; the only difference is the replacement of NH_4^+ into a smaller K ion. Optimum pH of 9–10 for K-struvite formation fall into the same range as the struvite has (Mathew et al. 1982). Therefore, the competition of NH_4^+ with K might be observed during the ammonium removal from potassium rich wastewaters like leachate, urine and dairy manure.

9.5 STRUVITE QUALITY

9.5.1 Composition of struvite

The composition of struvite, especially molar ratio of constituent ions in precipitate, is important for its potential use as a fertilizer. As mentioned above, many different compounds might contribute to the struvite precipitate due to the presence of competitive ions. Beside that, some organic and inorganic ions will be swept down by taking inside the precipitates during the crystallization and precipitation of struvite. Therefore, purity of struvite precipitate might change in a wide range depending on the composition of the wastewater and together with other related operational parameters. Consequently, the presence of impurities varies the theoretical composition of struvite. For the determination of struvite quality, percentage of struvite in precipitates as well as the composition and impurity content of precipitate have been determined by several researchers as given in Tables 9.5 and 9.6. As seen from Table 9.5, the struvite ratio of precipitate recovered from the digested sludge was lower than that of synthetic wastewater. It points out that some ions may contribute to the precipitates related to the composition of wastewater. Calcium, iron and potassium were the main impurities encountered in struvite. These ions mainly contributed to the precipitate by reacting with phosphorus and/or magnesium. As observed by Quintana et al. (2004), incorporation of ammonium into precipitate could be increased by elevating the Mg:P ratio, which decreased the formation of calcium phosphate formation.

Table 9.5 Composition of precipitate and percentage of struvite in precipitate

Wastewater Type	Constituent (%)				Struvite Recovery (%)	References
	Mg	N	P	H_2O		
Theoretical value for pure struvite	9.9	5.7	12.6	44	100	Calculation
Synthetic struvite	9.7	5.5	12.7	–	99.5	Bhuiyan et al. (2008)
Digested sludge	9.6	5.4	12.5	–	98	Bhuiyan et al. (2008)
Digested sludge	9.1	5.1	12.4	39	–	Munch and Barr (2001)
Leachate	11.2	5.4	19.9	46	–	Li and Zhao (2002)

Table 9.6 Impurity content of precipitate

	Constituent (%)						
Wastewater Type	Mg	N	P	Ca	K	Fe	References
Synthetic wastewater MgO (Mg: P = 2)	9.78	4.84	12.67	0.54	N.D.	–	
Synthetic wastewater MgO (Mg: P = 2.6)	8.35	4.99	10.95	0.34	N.D.	–	Quintana et al. (2004)
Digested sludge (Mg: P = 2.5)	14.83	3.28	9.76	2.52	N.D.	0.67	
Digested sludge (Mg: P = 3)	13.23	3.38	8.83	0.92	N.D.	–	
Digested sludge	9.9	5.0	11.9	1.34	–	0.03	Huang et al. (2006)
Digested sludge	9.9	5.6	12.8	0.49	0.04	0.03	Britton et al. (2005)
Denitrified Calf manure	10.73	0.57	12.54	1.28	11.37	N.D.	Schuiling and Andrade (1999)

N.D.: not determined

9.5.2 NH_4^+-N Incorporation into struvite

The fate of nitrogen during struvite precipitation can be determined by setting up mass balances on nitrogen. The mass balance includes conversion into struvite crystals, loss as a nitrogen gas and soluble part which is not incorporated into struvite. According to this definition, some reported mass balances were represented in Table 9.7. It was found that the loss of organic nitrogen can be reduced with the addition of Mg and P salts in excess amount than stoichiometric ratio (Demeester et al. 2001; Jeong and Hwang 2005). Although the ammonium removal efficiency could be increased by increasing the Mg:P molar ratio, this might bring to the problem about the purity of precipitate in terms of struvite. Application of pretreatment was also effective on the efficiency of ammonia removal with struvite precipitation (Altinbas et al. 2002a; Altinbas et al. 2002b). As can be seen from Table 9.7, besides obtaining higher ammonium removal efficiencies, ammonium recovery in struvite precipitates was also increased with the application of chemical pretreatment in addition to biological treatment.

Table 9.7 Ammonium incorporation into struvite

Wastewater Type	Experimental Conditions	Initial NH$_4^+$ Concentration (mg N/L)	N loss (%)	N not incorporated into struvite (%)	N in struvite precipitate (%)	NH$_4$-N removal (%)	Struvite Recovery (%)	References
Synthetic struvite	M:N:P = 1.0:1.0:1.0	600	5	15	80	85	86.2	Demeester et al. (2001)
	M:N:P = 1.5:1.0:1.5	600	1	2	97	97	69.3	
	M:N:P = 2.0:1.0:2.0	600	2	3	95	97	50	
Aerobic composting reactor	M:N:P = 0.5:1.0:0.5	–	20	43	37	57	–	Jeong and Hwang (2005)
	M:N:P = 1:1:1	–	15	31	55	70	–	
	M:N:P = 1.5:1:1.5	–	0	0	100	100	–	
Opium alkaloid industry (M:N:P = 1:1:1)	pretreatment: anaerobic+aerobic +Fenton's oxidation	84	4	26	70	74	93	Altinbas et al. (2002a)
	pretreatment: anaerobic+aerobic	208	5	21	74	77	86	Altinbas et al. (2002a)
Baker's yeast industry (M:N:P = 1:1:1)	pretreatment: anaerobic+aerobic +Fenton's oxidation	184	3	17	80	83	94	Altinbas et al. (2002a)
	pretreatment: anaerobic+aerobic +Ozonation	180	4	18	78	81	93	
Domestic wastewater +3% leachate (M:N:P = 1:1:1)	pretreatment: anaerobic	109	4	32	64	68	86	Altinbas et al. (2002b)
	pretreatment: anaerobic+ Fenton's oxidation	115	4	28	68	72	89	

9.5.3 Struvite crystal size

The particle size of crystals has an effect on settling properties and solubility of precipitate. Most of the reactor design aimed to increase the struvite crystal size, and consequently, the amount of recovered struvite. Therefore, for the optimization of struvite precipitation process regarding crystal formation and growth, struvite crystal size should be taken into consideration. The crystal retention time (CRT) has been determined as the main factor in determining the crystal size (Huang *et al.* 2006). The reported maximum struvite crystal sizes for different reactor configurations are given in Table 9.8.

In the study of Le Corre *et al.* (2007) a significant decrease in struvite crystal size has been observed as a result of pH increase from 8 to 10. The more

Table 9.8 Struvite crystal size obtained in different reactor configurations (adapted from Le Corre *et al.* (2007))

Wastewater Type	Reactor Configuration	CRT (days)	Crystal size (μm)	References
	Stirred Reactor V = 0.25 L	25 to 90 min	10–30	Kofina and Koutsoukos (2004)
	Stirred Reactor V = 0.6 L	15 min	5.6–45.6	Matynia *et al.* (2006)
Synthetic wastewater	Stirred Reactor V = 2 L	5	425–2360	Stratful *et al.* (2004)
	Fluidized Bed Reactor, V = 10 L	1 hour	84	Le corre *et al.* (2007)
	Fluidized Bed Reactor, V = 24.4 L	7 to 17	2600–3700	Adnan *et al.* (2003)
	Air agitated column Reactor, V = 143 L	1	25–215	Münch and Barr (2001)
Anaerobically digested Sludge liquors	Fluidized bed Reactor, V = 19 L	60	1300	Britton *et al.* (2005)
	Fluidized bed Reactor, V = 38.1 L	20	3400	Huang *et al.* (2006)
	Fluidized Bed Reactor, V > 25 m^3	10	500–100	Ueno and Fujii (2001)

increases in pH ($\sim$10.5) led to increase in size of precipitates explained with the presence of other impurities. In addition to that, Jones (2002) reported that the presence of impurities in solutions could inhibit the crystal size by blocking the active growth sites.

The crystal size was also affected by the seeding materials and mixing strength. The studies have shown that the mixing energy was primarily responsible for controlling struvite growth rate and crystal habit (Ohlinger et al. 1999). The decrease in crystal size as a result of increase in specific power input to the reactor was reported as well (Franke and Mersmann 1995). Additionally, Durrant et al. (1999) indicated that excessive mixing could break crystals resulting in reduced particle size. Furthermore, Wang et al. (2006) observed that the crystal size and settleability could be increased by using struvite powder as the seed material and applying proper mixing strength.

9.6 ECONOMICAL ASPECTS

The utilization of struvite as an excellent slow-release and non-burning fertilizer has been established for improvement of plant growth compared to conventional fertilizers (Bridger et al. 1962). However, the cost of constituent ions necessary to balance the molar ratio of ammonium has been considered as a major obstacle for struvite production (Schuiling and Andrade 1999).

In most cases, it is required the addition of magnesium, phosphorus and also the bases for pH adjustment for the removal of nitrogen via struvite precipitation from nitrogen rich wastewaters. The high solubility of magnesium chloride ($MgCl_2$) is effective in completion of reaction in short reaction times. Owing to this advantage of $MgCl_2$, it is possible to design small volume full-scale reactors by reducing the hydraulic retention time (Jaffer et al. 2002). On the other hand, due to the higher price of $MgCl_2$ and its high requirement of base dosages for pH adjustment (Stratful et al. 2001), some alternative magnesium sources have been also suggested. Because of its cheaper price and less caustic requirements for pH adjustment, magnesium oxide (MgO) has been accepted as a potential Mg source, although a greater molar ratio and vigorous agitation of the reaction solution should be supplemented (Zeng and Li 2006). In addition, magnesit ($MgCO_3$), which is 18% cheaper than $MgCl_2$, has been suggested as a magnesium source for ammonium removal (Gunay et al. 2008). However, the main obstacle was the requirement of HCl addition for dissolving $MgCO_3$. Also, the application of some low-cost magnesium minerals, such as MgO-containing by-products (Ouintana et al. 2005), bittern and seawater (Lee et al. 2003) have been given worthwhile results in terms of efficient ammonium removal.

The effect of addition of different phosphate sources has been studied by Li *et al.* (2007). In this study, maximum ammonium removal efficiency of 95% was achieved by using Na_2HPO_4 with a molar ratio of 1.3:1.0 (P:N), while the 90% and 55% ammonium removal was obtained using NaH_2PO_4 (1.5:1.0) and Na_3PO_4 (1.3:1.0), respectively. Although Na_2HPO_4 was found as an effective phosphate source, addition of this salt caused pH decrease and subsequently the cost of chemical addition for pH adjustment increased.

9.7 TECHNOLOGIES

9.7.1 General considerations

The most commonly applied processes for struvite precipitation include completely stirred reactors, air agitated reactors and fluidized bed reactors. These reactors are mainly distinguished by the type of the crystal formation. While the struvite crystals are formed by interaction between struvite nuclei in completely stirred reactors, crystallization occurs on seed materials or on matured struvite crystals in fluidized bed reactors. The fluidized bed reactor does not only help the matured crystal formation, but also separates solids–liquid phases, effectively (Graveland *et al.* 1983; Van Dijk and Wilms 1991).

9.7.2 Case study – REM-NUT®

The REM-NUT® process (short for REMoval of NUTrients) was developed in 1970s at the Water Research Institute of Italy's National Research Council (IRSA-CNR) for the recovery of high grade struvite from municipal sewage by selective ion exchange and chemical precipitation (Eur. Pat. No.114, 038).

Ion exchange of ammonium and phosphorus in municipal sewage occurs stoichiometrically and selectively according to the following reactions:

$$Z - Na + NH_4^+ \leftrightarrow Z - NH_4^+ + Na^+$$
$$R - Cl + H_2PO_4 \leftrightarrow R - H_2PO_4 + Cl^-$$
$$2R - Cl + HPO_4 \leftrightarrow R_2 - (HPO_4) + 2Cl^-$$

where Z is a natural zeolite (Clinoptilolite, Phillipsite or Chabasite) and R is a strong base scavenger anion exchange resin, both regenerated in a *closed-loop* fashion by neutral 0.6 M NaCl solution (i.e. seawater).

After that step, the concentrated NH_4^+ and PO_4^{-2} ions are quantitatively recovered by chemical precipitation with Mg^{2+} in a second step. After the precipitation, the wastewater is recycled back to ion exchange basin allowing *zero-discharge* as shown in Figure 9.5.

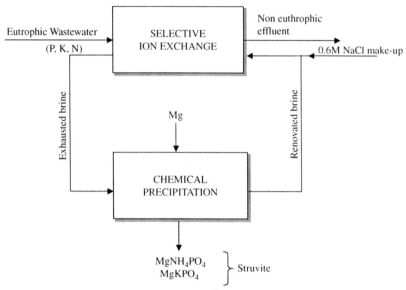

Figure 9.5 Flow diagram of the REM-NUT® process

Based on the results from 0.36 m³/d laboratory pilot plant, a skid-mounted fully automatic 240 m³/d REM-NUT® bench-scale plant was built (Figure 9.6) and used for demonstration campaigns at other municipal wastewater treatment plants in Italy (Manfredonia) and in the U.S.A. (South Lyon, MI), briefly described hereinafter.

With the financing by U.S. EPA, a 240 m³/d REM-NUT® plant was taken into operation in 1986 in South Lyon (MI), USA municipal wastewater treatment plant. In the 6 continuous months of test operation, production of slow-release premium-quality struvite-based fertilizer was observed.

Figure 9.6 Outside/inside views of 240 m3/d REM NUT mobile plant

The plant having the capacity of 6,624 m^3/d consisted of grit removal, primary sedimentation, rotary biological contactors (RBC), secondary sedimentation, filtration through rotary micro-screens and chlorination before discharge. FeCl$_3$ was used for P removal to meet discharge standard limitation of 1 mg P$_{Tot}$/L, while severe restrictions on ammonium discharge were under consideration by the State Authorities.

The REM-NUT$^\circledR$ plant consists of four stages as cationic resin, anionic basin, reservoir and settling thickeners. For ensuring treatment continuity, all sections consist of two basins and while one in service; the other one is being regenerated. Two cationic columns and two anionic columns, which particle size ranges between 20 and 50 mesh, contain 0.45 m^3 of natural zeolite, Clinoptilolite and 0.41 m^3 of Amberilte, respectively. While ammonium ions were removed selectively in cationic resin, the phosphorus ions were removed in anionic resin. These resins were connected to 1.2 m^3 of two reservoirs and 5 m^3 of two settling thickeners for regeneration. To regenerate cationic resin, 6 Bed Volumes (i.e. 6 × 0.45 = 270 m^3) were recycled more than four times to the system. The system was in operation for 8–10 hours and then both resins were regenerated in approximately 4 hours with closed loop procedure using neutral 0.6M NaCl solution (i.e. seawater). In reservoirs, wastewater softened by addition of Na$_2$CO$_3$ to pH 9–9.5. In settling thickeners, the struvite precipitation was realized with the addition of chemicals MgCl$_2$, H$_3$PO$_4$, Na$_2$CO$_3$ and NaOH for pH control around 8.5 after ammonium analysis. After precipitation, NaCl present in the supernatant was recycled. The anionic resin was regenerated using neutral 0.6M NaCl solution. ⩾90% of exchanged phosphates from reservoirs was used as a phosphate source for struvite precipitation. Produced struvite was stored in filter bags to remove the excess water. The drained water was recycled to the cationic resin column. In this way the regeneration of both resins occurred with *zero-discharge*. Results of the 6-months continuous operation of the system showed technically efficient system. Typical effluent concentrations of ⩽3 mg N/L (often ⩽1 mg N/L) and ⩽0.2 mg P/L were achieved after the REM-NUT$^\circledR$ treatment of both primary and secondary sedimentation effluents. Approximately 94% and 100% of ammonium and phosphorous was removed. Recovered struvite were composed of 3% N, 25% P$_2$O$_5$, 9% Mg, very close to the theoretical yield.

The economic evaluation of the REM-NUT$^\circledR$ treatment at South Lyon estimated a capital cost ⩾2.2 M$ and O&M costs ⩽0.6 M$/yr. The revenue breakeven point was determined between $0.95 and 1.45 per pound, to be compared with the $1.00 per pound retail price of a similar enhanced fertilizer (MAG-AM-P by Grace Co.).

9.7.3 Acknowledgements

Information for the preparation of case study sections of this chapter was kindly provided by Prof. Liberti Lorenzo from Technical University of Bari.

REFERENCES

Aage, H.K., Anderson, B.L., Blom, A. and Jensen, I. (1997) The solubility of struvite. *J. Radio. Anal. Nuclear Chem.* **223**(1), 213–215.

Abdelrazig, B.E.I. and Sharp, J.H. (1998) Phase changes on heating ammonium magnesium phosphate hydrate. *Therm. Acta* **129**, 197–215.

Adnan, A., Mavinic, D.S. and Koch, F.A. (2003) Pilot-scale study of phosphorus recovery through struvite crystallization examining the process feasibility. *J. Environ. Eng. Sci.* **2**(5), 315–324.

Ahn, Y.H. and Speece, R.E. (2006) Elutriated acid fermentation of municipal primary sludge. *Water Res.* **40**(11), 2210–2220.

Altinbas, M., Ozturk, I. and Aydin, A.F. (2002a) Ammonia recovery from high strength agro-industry effluents. *Water Sci. Technol.* **45**(12), 189–196.

Altinbas, M., Yangin, C. and Ozturk, I. (2002b) Struvite precipitation from anaerobically treated municipal and landfill wastewaters. *Water Sci. Technol.* **46**(9), 271–278.

Austermann-Haun, U. and Seyfried, C.F. (1994) Experiences gained in the operation of anaerobic treatment plants in Germany. *Water Sci. Technol.* **30**(12), 415–424.

Ban, Z.S. and Dave, G. (2004) Laboratory studies on recovery of N and P from human urine through struvite crystalisation and zeolite adsorption. *Environ. Technol.* **25**(1), 111–121.

Barnes, D., Li, X. and Chen, J. (2007) Determination of suitable pretreatment method for old-intermediate landfill leachate. *Environ. Technol.* **28**(2), 195–203.

Battistoni, P., Pavan, P., Cecchi, F. and Mata-Halvarez, J. (1998) Phosphate removal in real anaerobic supernatants: modeling and performance of a fluidized bed reactor. *Water Sci. Technol.* **38**(1), 275–283.

Battistoni, P., De Angelis, A., Pavan, P., Prisciandaro, M. and Cecchi, F. (2001) Phosphorus removal from a real anaerobic supernatant. *Water Res.* **35**(9), 2167–2178.

Bhuiyan, M.I.H., Mavinic, D.S. and Beckie, R.D. (2007) A solubility and thermodynamic study of struvite. *Environ. Technol.* **28**(9), 1015–1026.

Bhuiyan, M.I.H., Mavinic, D.S. and Koch, F.A. (2008) Thermal decomposition of struvite and its phase transition. *Chemosphere* **70**(8), 1347–1356.

Borgerding, J. (1972) Phosphate deposits in digestion systems. *J. Water Pollut. Control Fed.* **44**(5), 813–819.

Bridger, G.L., Salutsky, M.L. and Starosta, R.W. (1962) Metal ammonium phosphates as fertilizers. *J. Agr. Food Chem.* **10**, 181–188.

Britton, A., Koch, F.A., Mavinic, D.S., Adnan, A., Oldham, W.K. and Udala B. (2005) Pilot-scale struvite recovery from anaerobic digester supernatant at an enhanced biological phosphorus removal wastewater treatment plant. *J Environ. Eng. Sci.* **4**(4): 265–277.

Buchanan, J.R., Mote, C.R. and Robinson, R.B. (1994) Thermodynamics of struvite formation. *Trans. ASAE* **37**(2), 617–621.

Burns, R.T., Moody, L.B., Walker, F.R. and Raman, D.R. (2001) Laboratory and in-situ reductions of soluble phosphorus in swine waste slurries. *Environ. Technol.* **22**(11), 1273–1278.

Celen, I. and Türker, M. (2001) Recovery of ammonia from anaerobic digester effluents. *Environ. Technol.* **22**(11), 1263–1272.

Childs, C.W. (1970) A potentiometric study of equilibria in aqueous divalent metal orthophosphate solutions. *Inorg. Chem.* **9**, 2465–2469.

Chimenos, J.M., Fernandez, A.I., Villalba, G., Segarra, M., Urruticoechea, A., Artaza, B. and Espiell, F. (2003) Removal of ammonium and phosphates from wastewater resulting from the process of cochineal extraction using MgO-containing by-product. *Water Res.* **37**(7), 1601–1607.

Da Silva, S., Bernet, N., Delgenes, J.P. and Moletta, R. (2000) Effect of culture conditions on the formation of struvite by Myxococcus xanthus *Chemosphere* **40**(12), 1289–1296.

Demeestere, K., Smet, E., Van Langenhove, H. and Galbacs, Z. (2001) Optimalisation of magnesium ammonium phosphate precipitation and its applicability to the removal of ammonium. *Environ. Technol.* **22**(12), 1419–1428.

Dempsy, B.A. (1997) Removal and re-use of ammonia and phosphate by precipitation of struvite. In: *Proc. of the 52nd Industrial Waste Conference*, Purdue University, West Lafayette, Ind., Ann Arbor Press, Chelsea, Mich., pp. 32–45, 5–7 May.

Doyle, J.D., Oldring, K., Churchley, J. and Parsons, S.A. (2002) Struvite formation and the fouling propensity of different materials. *Water Res.*, **36**(16), 3971–3978.

Durrant, A.E., Scrimshaw, M.D., Stratful, I. and Lester, J.N. (1999) Review of the feasibility of recovering phosphate from wastewater for use as a raw material by the phosphate industry. *Environ. Technol.* **20**(7), 749–758.

Ganrot, Z., Slivka, A. and Dave, G. (2008) Nutrient recovery from human urine using pretreated zeolite and struvite precipitation in combination with freezing-thawing and plant availability tests on common wheat. *Clean* **36**(1), 45–52.

Graveland, A., van Dijk, J.C., de moel, P.J. and Oomen, J.H.C.M. (1983) Development in Water Softening by Means of Pellet Reactors. *J. Am. Water Works Assoc.* **75**(12), 619–625.

Griffith, D.P. and Osborne, C.A. (1987) Infection (urease) stones. *Miner. Electrolyte Metab.* **13**, 278–285.

Gunay, A., Karadag, D., Tosun, I. and Ozturk, M. (2008) Use of magnesit as a magnesium source for ammonium removal from leachate. *J. Hazard Mater.* **156**(1/3), 619–623.

Franke, J. and Mersmann, A. (1995) The influence of the operational conditions on the precipitation process. *Chem. Eng. Sci.* **50**(11), 1737–1753.

He, S., Zhang, Y., Yang, M., Du., W. and Harada, H. (2007) Repeated use of MAP decomposition residues for the removal of high ammonium concentration from landfill leachate. *Chemosphere* **66**(11), 2233–2238.

Hoffmann, J., Gluzinska, J. and Kwiecien, J. (2004) Struvite precipitation as the method of industrial wastewater treatment. In *International Conference for Struvite: its Role in Phosphorus Recovery and Reuse.* Cranfield University, UK.

Huang, H., Mavinic, D.S., Lo, K.V. and Koch, F.A. (2006) Production and basic morphology of struvite crystals from a pilot-scale crystallization process. *Environ. Technol.* **27**(3), 233–245.

Jaffer, Y., Clark, T.A., Pearce, P. and Parsons, S.A. (2002) Potential Phosphorus Recovery by Struvite Formation. *Water Res.* **36**(7), 1834–1842.

Jeong, Y.K. and Hwang, S.J. (2005) Optimum doses of Mg and P salts for precipitating ammonia into struvite crystals in aerobic composting. *Bioresource Technol.* **96**(1), 1–6.

Jeong, Y.K. and Kim, J.S. (2001) A new method for conservation of nitrogen in aerobic composting processes. *Bioresource Technol.* **79**(2), 129–133.

Jones, A.G. (2002) *Crystallization Process System*, Butterworth, Heinemann, UK.

Kabdasli, I., Gurel, M. and Tunay, M. (2000) Characterization and treatment of textile printing wastewaters. *Environ. Technol.* **21**(10), 1147–1155.

Kabdasli, I., Tunay, O., Islek, C., Erdinc, E., Huskalar, S. and Tatli, M.B. (2006) Nitrogen recovery by urea hydrolysis and struvite precipitation from anthropogenic urine. *Water Sci. Technol.* **53**(12), 305–312.

Kim, B.U., Lee, W.H., Lee, H.J. and Rim, J.M. (2004) Ammonium nitrogen removal from slurry-type swine wastewater by pretreatment using struvite crystallization for nitrogen control of anaerobic digestion. *Water Sci. Technol.* **49**(5/6), 215–222.

Kofina, A.N. and Koutsoukos, P.G. (2004) Spontaneous precipitation from synthetic wastewater solutions. *J. Cryst. Growth Des.* **5**, 489–496.

Le Corre, K.S., Valsami-Jones, E., Hobbs, P. and Parsons, A. (2005) Impact of calcium on struvite crystal size, shape and purity. *J. Cryst. Growth* **283**, 514–522.

Le Corre, K.S., Valsami-Jones, E., Hobbs, P. and Parsons, A. (2007) Impact of reactor operation on success of struvite precipitation from synthetic liquors. *Environ. Technol.* **28**(11), 1245–1256.

Lee, S.I., Weon, S.Y., Lee, C.W. and Koopman, B. (2003) Removal of nitrogen and phosphate from wastewater by addition of bittern. *Chemosphere* **51**(4), 265–271.

Li, X.Z. and Zhao, Q.L. (2002) MAP precipitation from landfill leachate and seawater bittern waste. *Environ. Technol.* **23**(9), 989–1000.

Li, Y., Yi, L., Ma, P. and Zhou, L. (2007) Industrial wastewater treatment by the combination of chemical precipitation and immobilized microorganism technologies. *Environ. Eng. Sci.* **24**(6), 736–744.

Maekawa, T., Liao, C.M. and Feng, X.D. (1995) Nitrogen and phosphorus removal for swine wastewater using intermittent aeration batch reactor followed by ammonium crystallization process. *Wat. Res.* **29**(12), 2643–2650.

Mamais, D., Pitt, P.A., Cheng, Y.W., Loiacono, J. and Jenkins, D. (1994) Determination of ferric chloride dose to control struvite precipitation in anaerobic sludge digesters. *Water Environ. Res.* **66**(7), 912–918.

Martell, A., Smith, R. and Motekaitis R. (1998) NIST Critically Selected Stability Constants of Metal Complexes, *Database Version 5.0 NIST Standard Reference Database 46*, Texas A&M University.

Mathew, B.Y.M., Kingsbury, P., Takagi, S. and Brown, W.E. (1982) A new struvite-type compound, sodium phosphate heptahydrate. *Acta Cryst.* **38**, 40–44.

Matynia, A., Koralewska, J., Wierzbowska, B. and Piotrowski, K. (2006) The influence of process parameters on struvite continuous crystallisation kinetics. *Chem. Eng. Commun.* **193**, 160–176.

Miles, A. and Ellis, T.G. (2001). Struvite precipitation potential for nutrient recovery from anaerobically treated wastes. *Water Sci. Technol.* **43**(11), 259–266.

Mohajit, X., Bhattarai, K.K., Taiganides, E.P. and Yap, B.C. (1989) Struvite deposits in pipes and aerators. *Biol. Wastes* **30**(2), 133–147.

Morel, F.M. and Hering, J.G. (1993) *Principles and Applications of Aquatic Chemistry*. John Wiley and Sons, Inc., New York.

Munch, E.V. and Barr, K. (2001) Controlled struvite crystallisation for removing phosphorus from anaerobic digester sidestreams. *Water Res.* **35**(1), 151–159.

Murray, J. (1858) Notices and Abstracts. British Association for the Advancement of Science, *Report of the 27th Meeting*, 54–55.

Mustovo, E.V., Ekama, G.A., Wentzel, M.C. and Loewenthal, R.E. (2000) Extension and application of the three-phase weak acid/base kinetic model to the aeration treatment of anaerobic digester liquors. *Water SA* **26**(4), 417–438.

Neethling, J.B. and Benisch, M. (2004) Struvite control through process and facility design as well as operation strategy. *Water Sci. Technol.* **49**(2), 191–199.

Nelson, N.O., Mikkelsen, R.L. and Hesterberg, D.L. (2003) Struvite precipitation in anaerobic swine lagoon liquid: effect of pH and Mg:P ratio and determination of rate constant. *Bioresource Technol.* **89**(3), 229–236.

Ohlinger, K.N., Young, T.M. and Schroeder, E.D. (1998) Predicting struvite formation in digestion. *Water Res.* **32**(12), 3607–3614.

Ohlinger, K.N., Young, T.M. and Schroeder, E.D. (1999) Kinetics effects on preferential struvite accumulation in wastewater *J. Environ. Eng. ASCE* **125**(8), 730–737.

Ozturk, I., Altinbas, M., Koyuncu, I., Arikan, O. and Gömec-Yangin, C. (2003) Advanced physico-chemical treatment experiences on young municipal landfill leachates. *Waste Manage.* **23**(5), 441–446.

Quintana, M., Colmenarejo, M.F., Barrera, J., Garcia, G., Garcia, E. and Bustos, A. (2004) Use of a byproduct of magnesium oxide production to precipitate phosphorus and nitrogen as struvite from wastewater treatment liquors. *J. Agric. Food Chem.* **52**(2), 294–299.

Quintana, M., Sanchez, E., Colmenarejo, M.F., Barrera, J., Garcia, G. and Borja, R. (2005) Kinetics of phosphorus removal and struvite formation by the utilization of by-product of magnesium oxide production. *Chem. Eng. J.* **111**(1), 45–52.

Rahaman, M.S., Mavinic, D.S., Bhuiyan, M.I.H. and Koch, A. (2006) Exploring the determination of struvite solubility product from analytical results. *Environ. Technol.* **27**(9), 951–961.

Rawn, A.M., Perry Banta, A. and Pomeroy, R. (1939) Multiple stage sewage digestion. *Trans Am Soc Agric Eng* **105**, 93–132.

Ryu, H.D., Kim, D. and Lee, S.I. (2008) Application of struvite precipitation in treating ammonium nitrogen from semiconductor wastewater. *J. Hazard. Mater.* **156**(1/3), 163–169.

Sarkar, A.K. (1991) Hydration/dehydration characteristics of struvite and dittmarite pertaining to magnesium ammonium phosphate cement systems. *J. Mater. Sci.* **26**(9), 2514–2518.

Schneider, H.J. (1985) Epidemiology of urolithiasis. In: *Schneider HJ, editor. Urolithiasis: Etiology Diagnosis*. Berlin, Springer Verlag, 137–184.

Schuiling, R.D. and Andrade, A. (1999) Recovery of struvite from calf manure. *Environ. Technol.* **20**(7), 765–768.

Smith, R.M. and Martell, A.E. (1989) *Critical Stability Constants*, Vol. 6, Plenum Press, New York.

Snoeyink, V. and Jenkins, D. (1980) *Water chemistry.* John Wiley & Sons, NewYork.

Stratful, I., Scrimshaw, M.D. and Lester, J.N. (2001) Conditions influencing the precipitation of magnesium ammonium phosphate. *Water Res.* **35**(17), 4191–4199.

Stratful, I., Scrimshaw, M.D. and Lester, J.N. (2004) Removal of struvite to prevent problems associated with its accumulation in wastewater treatment works. *Water Environ. Res.* **76**(5), 437–443.

Stumm, W. and Morgan, J.J. (1970) *Aquatic Chemistry.* John Wiley & Sons, New York.

Taylor, A.W., Frazier, A.W., Gurney, E.L. and Smith, J.P. (1963a) Solubility products of di-and trimagnesium ammonium and dissociation of magnesium phosphate solutions. *Trans. Farad. Soc.* **59**, 1585–1589.

Taylor, A.W., Frazier, A.W. and Gurney, E.L. (1963b) Solubility products of ammonium and magnesium potassium phosphates. *Trans. Farad. Soc.* **59**, 1580–1584.

Toshev, S. (1973) Homogeneous nucleation. *Crystal growth: An introduction,* P. Hartman, ed., North-Holland, Amsterdam, 1–49.

Tunay, O., Kabdasli, I., Orhon, D. and Kolcak, S. (1997) Ammonia removal by magnesium ammonium phosphate precipitation in industrial wastewaters. *Water Sci. Technol.* **36**(2–3), 225–228.

Ueno, Y. and Fujii, M. (2001) Three years experience of operating and selling recovered struvite from full-scale plant. *Environ. Technol.* **22**(11), 1373–1381.

Uludag-Demirer, S., Demirer, G.N. and Chen, S. (2005) Ammonia removal from anaerobically digested dairy manure by struvite precipitation. *Process Biochem.* **40**(12), 3667–3674.

Van Dijk, J.C. and Wilms, D.A. (1991) Water Treatment without Waste Material — Fundamentals and State of the Art of Pellet Softening. *J. Water SRT-Aqua* **40**(5), 263–280.

Wang, J., Burken, J.G. and Zhang, X.J. (2006) Effect of seeding materials and mixing strength on struvite precipitation. *Water Environ. Res.* **78**(2), 125–132.

Wollaston, W.H. (1797) On gouty and urinary concretions. *Phil. Trans.* **ii**, 386–401.

Zeng, L. and Li, X. (2006) Nutrient removal from anaerobically digested cattle manure by struvite precipitation. *J. Environ. Eng. Sci.* **5**(4), 285–294.

Zengin, G., Olmez, T., Kabdasli, I. and Tunay, O. (2005). Assessment of source-based nitrogen removal alternatives in leather tanning industry wastewater. *Water Sci. Technol.* **45**(12), 205–215.

Zhang, W. and Lau, A. (2007) Reducing ammonia emission from poultry manure composting via struvite formation. *J Chem Technol Biotechnol.* **82**(6), 598–602.

10

Ion Exchange processes for Ammonium Removal

A. Thornton and S.A. Parsons

10.1 INTRODUCTION

The adverse effects of ammonium (NH_4^+) present in effluents resulting from the treatment of municipal and industrial wastewaters are well documented (Du *et al.* 2005; Karadag *et al.* 2006; Wang *et al.* 2006) and the control of its release into the surrounding environment is critical. Excess amounts introduced into receiving waters can result in depleted dissolved oxygen concentration (Wen *et al.* 2006) and increased toxicity to fish and other aquatic organisms. Conventional methods of ammonium removal from municipal wastewaters are performed using biological processes (Jorgensen and Weatherley 2003), which in most cases prove sufficiently efficient in reducing the effluent concentration to a safe level. However, some treatment plants find themselves under increasing strain and as a result complementary methods of ammonium reduction have to be considered as a possible solution to maintaining an effective removal performance. A number of methods have been concerned with the treatment of high strength internal recycle streams within the wastewater treatment works (WwTW) in order to alleviate the strain placed on the mainstream biological processes. These include; air stripping (Siegrist 1996); struvite precipitation (Celen and Turker 2001; Kabdasli et al. 2000) ion exchange, and more recently, the combined Sharon/Anammox process (Fux et al. 2003; Schmidt et al. 2003; Van Dongen et al. 2001), which have all achieved some degree of success.

Over the last 30 years much experimental work has been performed to investigate the merits of ion exchange on to zeolite (mainly clinoptilolite) as an effective treatment for the removal of the ammonium ion from both high concentration and low concentration wastewater streams (Table 10.1). The success of such research does vary somewhat, but results indicate that the ion exchange process can successfully be utilised to remove ammonium from problematic wastewaters.

10.2 ION EXCHANGE MATERIALS

The first systematic studies of ion exchange were carried out using naturally inorganic ion exchangers (Bolto and Pawlowski 1987) but these soon became displaced by their much more predictable organic counterparts. In general, organic ion exchangers exhibit greater capacity and more stability than inorganic materials and as such are much more widely used in industrial applications. However, inorganic materials still have a role to play, largely due to their selectivity for specific cations. One such group of materials that display this capability are zeolites, often referred to as molecular sieves. A number of species, most commonly clinoptilolite, (although, mordenite, sepiolite and MesoLite have also been investigated) are reported to show a high affinity for the ammonium ion and as such are of great interest in the wastewater treatment sector. That is not to say that organic exchangers are excluded, indeed they are used for a wide range of applications within the wastewater industry. However, in the context of this book, the remainder of this chapter concentrates solely on the application of inorganic materials, specifically zeolites, for the removal of ammonium from various wastewater streams.

10.2.1 Zeolites as ion exchangers

Zeolites are crystalline, hydrated aluminosilicates of group I and group II elements, in particular, sodium, potassium, magnesium, calcium, strontium and barium (Breck 1974). Their structure is based upon an infinite network of SiO_4

Table 10.1. Operating conditions and material characteristics for materials used in various trials

Material	NH_4^+-N Capacity (g kg^{-1})	Grain Size (mm)	Bed Volume (ml)	Flow Rate (BV h^{-1})	Influent NH_4^+-N Concentration (mg l^{-1})	Effluent NH_4^+-N Concentration (mg l^{-1})	pH	Reference
Clinoptilolite	4.8	0.3–0.84	–	15	18.8	Breakthrough of 1 mg l^{-1} reached between 100 and 240 BV's	7.9–8.1	Koon and Kauffmann (1975)
Clinoptilolite - Tokaj region of Hungary	0.94–2.41	1.4–2.0 2.5–5.0 3.2–5.0	200	10–12	26–33	0.6–4.1	7.0	Jorgensen et al. (1976)
Clinoptilolite Buckhorn N.M. Double Eagle Mining Company	3.9	0.3–0.84	700	5.55–6.88	20	2 mg l^{-1} breakthrough reported after 16.5–20 hours	6.5–8.0	Semmens and Porter (1979)
Clinoptilolite - Tokaj region of Hungary	1.37–9.15	0.5–1.0	40–4000	5–7	12–45	4.5 at breakthrough	7.1–8.2	Hlavay et al. (1982)
Clinoptilolite - Beli Plast region of Bulgaria	2.0–3.6	2.0–2.3	96	6–20	15–17	–	–	Beler Baykal and Guven (1997)
Mordenite and Clinoptilolite	3.7–6.5	0.25–0.5 2.0–2.83	77–179	0.056–3.12	100–400	Breakthrough of 1–1.2 mg l^{-1}	7.5–7.7	Nguyen and Tanner (1998)
MesoLite	45–55	3–9	107000	1–2.7	500–1000	5–15	7.1–10.2	Mackinnon et al. (2003)
Croatian Clinoptilolite	6.0–13.65	0.5–2.0	710	2+	130–850 (400 average)	Breakthrough reached at 14 and 32 BV's	7.3–7.77	Farkas et al. (2005)
Chinese Clinoptilolite - Jinyun	5.89–7.74	0.45–0.9	–	6–24	25	Breakthrough set at 5 mg l^{-1}	–	Du et al. (2005)
Transcarpathian Clinoptilolite	3.56–21.52	0.125–2.0	14	8–40	100	Breakthrough set at 2 mg l^{-1}	–	Sprynskyy et al. (2005)
MesoLite	47–51	1.4–2.5	100000	4–5	500–700	Breakthrough set at 50 mg l^{-1}	7.8–8.1	Thornton et al. (2007a)

and AlO_4 tetrahedra, linked to each other by the sharing of all the oxygen atoms (Breck 1974; Mumpton 1984; Flanigen 1984).

The term zeolite was firstly used by Swedish scientist Cronstedt in 1756 (Dyer 1988) to describe naturally occurring siliceous minerals, which became dehydrated when heated and the word zeolite is derived from the Greek word meaning boiling stones (DeSilva 1999). This is due to the fact that their structure contains channels and pores that are filled with a certain amount of water, which can only evaporate when the sample is heated to about $250°C$ (Gottardi 1978).

The presence of these channels or pores, which uniformly penetrate the entire volume of the solid, give the material a high internal surface area available for adsorption (Breck 1974). The external surface of the particles only contributes a small amount of the total available surface area of the adsorbent (Breck 1974) and an important property of zeolites as adsorbents is that some of the cations within the pores and channels may be exchangeable without any major change to the zeolite structure (Mumpton 1984). Characteristics of naturally occurring zeolites depend on the origin, synthesis and variations in the naturally occurring processes during genesis and there currently in excess of 50 naturally occurring species accounted for (Kamarowski and Yu 1997).

10.2.2 Structure, properties and classification of zeolites

All zeolites contain primary building units (PBU's) and secondary building units (SBU's) (Breck 1974). The simplest of these is the primary building unit, which consists of a tetrahedron of four oxygen ions surrounding a central ion of either Si^{4+}, or Al^{3+} as illustrated in Figure 10.1. The primary building units are linked together by oxygen atoms to form a framework structure where all of the oxygen ions are shared by two tetrahedra (Gottardi 1978). Each AlO_4 tetrahedron in the framework results in a net negative charge, which is balanced by a cation, usually a group I or a group II element (Tosheva 1999), for example potassium (Figure 10.1). These cations can then be exchanged for other cations (Curkovic *et al.* 1997). The secondary building units consist of single and double ring tetrahedra (Breck 1974), the arrangement of which determines the nature of the framework and forms an important part of the zeolite classification system (Dyer 1988).

The Si/Al ratio of a zeolite can determine its fundamental property as an ion exchanger (Colella 1995). As the ratio increases, the amount of cations available for exchange in the framework decreases (Tosheva 1999). This results in pure silica (SiO_2) having no ion exchange properties at all. The Si/Al ratio therefore determines the cation exchange capacity of (CEC) of any given zeolite (Colella 1995). The CEC can be defined as the number of equivalents of fixed charges per amount of exchanger (Bolto and Pawlowski 1987) and represents the amount of

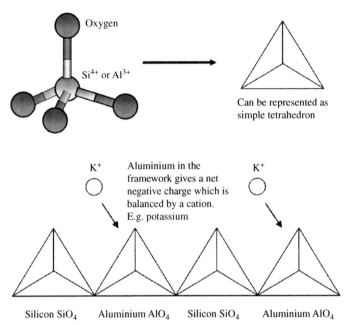

Figure 10.1. Primary building units of zeolites (Thornton 2007)

exchangeable cations (Semmens 1984). At lower Si/Al ratios there is a greater amount of aluminium in the framework. This leads to an increased deficiency in positive charge and means that a greater amount of cations are required for electrical neutrality (Mumpton 1984). The relationship between the Si/Al ratio and the CEC of some naturally occurring zeolites is presented in Table 10.2.

Table 10.2. Si/Al ratio versus CEC (Colella *et al.* 1995)

Zeolite Name	Structure Type	CEC Meq g^{-1}	Si/Al Ratio Ranges
Chabazite	CHA	3.84	1.43–4.18
Clinoptilolite	HEU	2.16	2.92–5.04
Erionite	ERI	3.12	3.05–3.99
Ferrierite	FER	2.33	3.79–6.14
Heulandite	HEU	2.91	2.85–4.31
Laumonite	LAU	4.25	1.95–2.25
Mordenite	MOR	2.29	4.19–5.79
Phillipsite	PHI	3.31	1.45–2.87
Faujasite	FAU	3.39	–

In addition, zeolite frameworks can be described as cage like structures, with micropores and channels (Tosheva 1999). The type of framework formed will determine the pore size and in naturally occurring zeolites these pores fall within the size range of 3–15 Å (Celik *et al.* 2001; Jung *et al.* 2004; Salerno and Mendioroz 2002). This allows zeolites to act as molecular sieves, which are selective for different cations on the basis of size. For this reason, certain species of zeolite are highly selective for ammonium. The resulting ammonium removal process relies predominantly on these two factors, although the type of exchanging cation in the framework does also have an influence (Booker *et al.* 1996). The ammonium ions are selectively adsorbed into the microstructure and ion exchange then takes place at the aluminium sites.

The classification of zeolites is complicated and as with all classification systems there are possible anomalies (Dyer 1988). It is the accepted convention that mineral species shall not be distinguished solely upon the framework Si/Al ratio, the exceptions to this being heulandite and clinoptilolite (Armbruster 2001). Dyer (1988) describes a classification based hierarchically on the following parameters; their SBU content; structure content; name (the prefix zeolite means that the material is only known as a synthetic material) and typical unit cell content. Each framework type is assigned a three letter structure code (Table 10.2) according to rules set up by IUPAC on zeolite nomenclature. This gives rise to many mineral types with each one possessing unique properties. This is by no means exhaustive and the International Zeolite Association (IZA) has a published handbook (Dyer 1988) which presents a more detailed view of zeolite frameworks.

10.3 ION EXCHANGE

In 1850, two agricultural chemists, Thompson and Way, noticed that when passing a liquid fertilizer containing ammonium through a soil sample, the ammonium was retained by the soil whilst the calcium was thrown off (Paterson 1970). They concluded that the aluminosilicates in the soil gave it ion exchange characteristics. The materials responsible for this phenomenon were identified chiefly by Lemberg and later by Wiegner as clays, glauconites, zeolites and humic acids (Helfferic 1962). In 1905, German scientist Gans, applied zeolites to the use of water softening by exchanging magnesium and calcium ions for sodium ions (DeSilva 1999). Since this early use, a greater understanding of the ion exchange process has been realised and there are now many practical applications of zeolite as the base for ion exchange processes (Varshney 2002).

10.3.1 Principles

Ion exchangers come in two forms, namely, cation exchangers and anion exchangers, although there are certain materials capable of both called amphoteric ion exchangers (Helfferic 1962). Zeolites are predominantly cation exchangers because the excess positive charge generated by the aluminium sites needs to be balanced by the presence of cations in the materials framework. Ion exchange is similar to adsorption, where in both cases, a dissolved species is taken up by an insoluble solid (Helfferic 1962). The difference between the two is the fact that during the ion exchange process, electroneutrality has to be preserved at all times, making the process a stoichiometric one (Paterson 1970). Therefore, every ion removed from the framework must be replaced by another ion of the same charge from solution. These ions are known as counter ions and are free to move within the framework as long as electroneutrality is observed.

When a cation exchanger, such as zeolite, is placed in a solution containing cations of those differing from the counter ions in the materials framework, a driving force is created, where ions will migrate from solution, into the zeolite framework and vice versa. That is to say ion exchange takes place. This process will continue until equilibrium is reached, where both the solution and the ion exchanger contain ions of both species. At this point no net exchange will take place unless further ions are introduced into the system.

For the case of zeolites, the ion exchange process between cation A^{ZA}, initially in solution, and cation B^{ZB}, initially in the zeolite framework, can be characterised as follows (Dyer 1988):

$$Z_B A^{ZA} + Z_A \underline{B}^{ZB} \leftrightarrow Z_B \underline{A}^{ZA} + Z_A B^{ZB} \qquad (10.1)$$

Where Z_A and Z_B are the valences of the ions and the underlined characters relate to a cation inside the zeolite framework. For the case of ammonium exchange, this is typically a uni-univalent exchange process, where the ammonium ion in solution is exchanged for a univalent ion, for example sodium, of the same charge from the zeolite framework (Equation 10.2). However, it is also possible for ions of differing valences to be exchanged as long as electroneutrality is observed (Equation 10.3). Here, a uni-divalent process is observed, where, every calcium ion removed from solution has to be balanced by two sodium counter ions leaving the zeolite framework.

$$NH_4^+ + \underline{Na}^+ \leftrightarrow \underline{NH_4^+} + Na^+ \qquad (10.2)$$

$$Ca^{2+} + 2\underline{Na}^+ \leftrightarrow \underline{Ca}^{2+} + 2Na^+ \qquad (10.3)$$

10.4 EQUIILIBRIUM STUDIES

Various models exist for predicting the uptake of impurities by ion exchange media with two of the most notable being the Langmuir and Freundlich isotherm models (Weatherly and Milandovic 2004). Although, many other models exist, these models are the most widely reported concerning the removal of ammonium from wastewaters using zeolite. The Langmuir isotherm relates Q_e (mg of impurity removed per gram of exchange media) and C_e (the equilibrium ammonium concentration in solution) as shown in Equation 10.4.

$$Q_e = \frac{KbC_e}{1 + KC_e} \qquad (10.4)$$

Where b is the maximum exchange capacity of 1 kg of exchanger (mg NH_4^+-N g^{-1}) and K is the Langmuir energy constant or binding index (1 mg^{-1}) (Nguyen and Tanner 1997). These constants can be determined by rearranging to Equation 10.5 and plotting $1/C_e$ versus $1/Q_e$. The resulting plot forms a straight line with an intercept of 1/b and a gradient of 1/kb.

$$\frac{1}{Q_e} = \frac{1}{KbC_e} + \frac{1}{b} \qquad (10.5)$$

The Freundlich model relates Q_e and C_e according to the following equation (Weatherly and Milandovic 2004).

$$Q_e = kC_e^{1/n} \qquad (10.6)$$

Where k and n are constants. The constant k is related to the capacity and 1/n represents the intensity of adsorption (Demir et al. 2002).

This equation can also be represented as

$$\log Q_e = \log k + 1/n \log C_e \qquad (10.7)$$

When $\log Q_e$ is plotted against $\log C_e$, the constants k and n can be calculated. Knowing all coefficients the Langmuir and Freundlich, theoretical Q_e can be calculated for any given ion concentration in solution.

Data for plotting isotherms is obtained by treating fixed volumes of liquid of known concentration with a series of known weights of media (Demir et al. 2002). Before isotherms can be produced it is necessary to ensure ion exchange equilibrium has been reached (Dyer 1988). Once the equilibrium time is determined the isotherm can be constructed. A representative sample of solution is chosen and a known volume is added to a series of flasks (a minimum of six is required and each flask contains the same volume of solution). Increasing weights of media are then added to the flasks and the flasks are then stirred at a constant temperature until equilibrium is reached (Wang *et al.* 2006). It is advisable that the solution: solid volume ratio should not be less than 20 (Dyer 1988). The media is then filtered out and the remaining solutions are again tested for residual concentration (C_e). The ammonium concentration transferred to the solid phase is calculated using the following mass balance equation (Thornton *et al.* 2007b).

$$Q_e = \frac{(C_0 - C_e)V}{M} \qquad (10.8)$$

Where:
Q_e is the amount of ammonium in the solid phase (mg g^{-1})
C_0 is the initial ammonium concentration in solution (mg l^{-1})
C_e is the ammonium remaining in solution at equilibrium (mg l^{-1})
V is the solution volume (l)
M is the mass of zeolite introduced (g)
A comparison of experimental data with that of the models can be made by plotting liquid ammonium concentration at equilibrium (C_e) against solids concentration (Q_e) to produce isotherms (Thornton *et al.* 2007b). The Langmuir and Freundlich isotherms are generated using Equations 10.4 and 10.6, experimental data for C_e and the values for K, n and b generated from Equations 10.5 and 10.7 to calculate theoretical Q_e values. A number of authors have compared the uptake of ammonium onto zeolite (Demir *et al.* 2002; Weatherley and Miladinovic 2004; Wang *et al.* 2006; Thornton *et al.* 2007b), concluding that the Langmuir isotherm best describes the process (Figure 10.2).

However, both isotherm models do have their limitations in accurately describing equilibrium ion exchange. The Freundlich model imposes no limit on the capacity of the material, where it is known that this capacity is limited by the number of available ion exchange sites. The Langmuir model fails to account for the true nature of ion exchange as a binary, displacement adsorption process, which involves at least two species, whose relative proportions and hence concentrations alter (Hankins *et al.* 2004).

A comparison of equilibrium studies is somewhat complicated by the fact that a number of factors can significantly affect the performance of the chosen media. The majority of studies have been performed using the same material, namely, clinoptilolite, but even these studies return a wide range of results

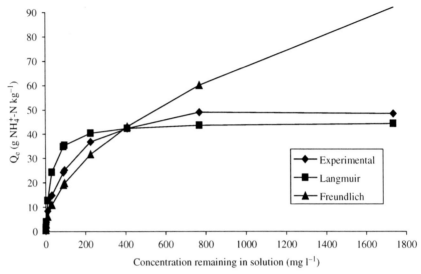

Figure 10.2. Ammonium equilibrium uptake on to MesoLite (Thornton *et al.* 2007b)

for capacity and K and n values. This can partially attributed to the way zeolites are classified. The classification of zeolites is complicated and as with all classification systems there are possible anomalies (Dyer 1988). This gives rise to many mineral types, each with each one possessing unique properties. Distribution of zeolites around the world is widespread and although classified into groups and types, each individual deposit will have its own individual characteristics. Various deposits have been used in ammonium removal experiments (Jorgensen *et al.* 1979; Beler Baykal 1998; Celik *et al.* 2001; Demir *et al.* 2002; Kovatcheva-Ninova *et al.* 2002; Tzvetcov and Stoytcheva 2003; Farkas *et al.* 2005; Du *et al.* 2005; Sprynskyy *et al.* 2005) and it is important to note that two separate deposits of different geographical location with the same classification will almost certainly possess differing characteristics.

A further complication is added by the level of purity of the deposit. Due to the nature of genesis of zeolites, it is highly unlikely that naturally occurring deposits will be 100% pure mineral ore. Structural imperfections and the presence of clay may lead to internal channel blockage, rendering much of the material ineffective as an adsorbent. Therefore, it is impossible to predict with complete authority, the finite properties of a deposit based solely on its classification (DeSilva 1999). This aside, a number of other factors also affect the performance of any given material.

The optimum pH range for ammonium removal is in the range 6–8. Above values of 8, an increasing proportion of ammoniacal nitrogen present in solution is in the unionised form (NH_3) and as such is unavailable for the exchange process. Therefore, as pH values further increase, removal efficiency rapidly decreases. At pH 9.25 only 50% of the ammoniacal nitrogen present is in the ionised form. Thus, at solution pH values above 8 it is unadvisable to consider the ion exchange process for ammonium removal. Conversely, below pH 8 the vast majority of ammonium is present in the ionic form and at pH 7 almost 100% is present as NH_4^+. Even so, removal efficiency rapidly diminishes below pH 6 and this can be attributed to the fact that the increased number of hydrogen ions in solution at lower pH provides added competition for exchange sites.

The solution strength has a 2-fold effect on the performance of any ion exchange media under batch conditions. The higher the solution strength, the higher the driving force behind the exchange process and as a result the initial uptake of ammonium ions is more rapid in the early stages (Figure 10.3). Consequently, at equilibrium, the total amount of ions removed from solution is much greater and as such a resulting increase in equilibrium capacity is observed until at some point a maximum is reached. This maximum value is indicative of

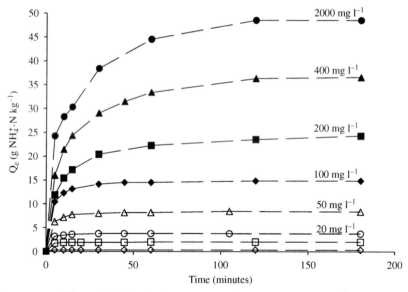

Figure 10.3. Effect of initial solution concentration on the rate of NH_4^+ uptake and equilibrium capacity of MesoLite (Thornton *et al.* 2007b)

the fact that the ion exchange surface becomes increasingly saturated with the ammonium ion (Karadag *et al.* 2006) until at some point, the ion exchange capacity of the material is reached. After this point, no further ions are removed from solution irrespective of further increases in concentration.

A number of authors (Booker *et al.* 1996; Hankins *et al.* 2004; Du *et al.* 2005; Wang *et al.* 2006; Thornton *et al.* 2007b) have investigated the effect of contact time on ammonium uptake. All conclude that the uptake of ammonium onto zeolite is initially a rapid process, which slows dramatically as equilibrium is approached. The majority of research indicates that a significant proportion of the process is complete within 10–15 minutes, even though it can take some hours to reach equilibrium. However, all conclude that after a four hour period, there is no further reduction of the ammonium ion in solution and as such the system is in a state of equilibrium. This is by no means an indication that the exchange process has ceased, it is just an indication that the rate of adsorption is equal to the rate of desorption and there is therefore no net alteration of the concentration of ions in the solution and solid phases (Thornton *et al.* 2007b).

Although the initial rate of uptake is less at lower concentrations, most of the exchange process is completed in a shorter period of time. This can be attributed to the fact that at lower concentrations, the vast majority of exchange sites remain unused, even at equilibrium. At high concentrations the rapid rate of uptake results in a greater saturation of the exchange sites in the early stages leaving the remaining ammonium in solution to "scavenge" for less accessible exchange sites. The opposite is true at lower concentrations where exchange sites are abundant throughout the process (Thornton *et al.* 2007b).

10.5 DYNAMIC ION EXCHANGE

For practical applications, the batch method of ion exchange, outlined above, does have its limitations. As mentioned previously, once equilibrium is reached, then no further net ion exchange takes place unless more counter ions are introduced into the system (i.e. only if the solution is replenished). Only then can further pollutants (in this case ammonium) be removed from the wastewater in question. A practical way of achieving this continuous replenishment of solution to the media is by the use of a continuous dynamic process.

In a continuous process, the media is suspended as a fixed bed in a column or series of columns and the solution to be treated is continually fed through the media bed (either upflow or downflow) to remove the polluting ion. In this application, flow rates are often quoted in bed volumes per hour (BV h^{-1}). Here, the driving force (Wang *et al.* 2006) behind the exchange process is continually being renewed and as a result a greater proportion of the materials capacity can be utilised. Therefore, most investigations into practical applications of ammonium removal using ion exchange have been concerned with the fixed bed column process (Koon and Kaufmann 1975; Semmens and Porter 1979; Beler Baykal et al. 1996; Kamarowski and Yu 1997; Lahav and Green 1998; Nguyen and Tanner 1998; Celik et al. 2001; Demir et al. 2002; Hankins et al. 2004; Du et al. 2005; Sprynskyy et al. 2005; Farkas et al. 2005; Thornton et al. 2007a).

During column operation, the bed is exposed to solution (concentration C_0) and the media becomes systematically saturated with contaminant from the inlet, progressing towards the outlet, until at some point a pre-determined breakthrough concentration (C_b) is observed in the column effluent (Cooney 1999) (Figure 10.4). At this point, the process is stopped and the media is either, discarded and replaced with fresh media, or regenerated so that the removal process can be recommenced. In practice, it is common to use two columns, alternating between the two, so that the removal process can operate uninterrupted by the regeneration process. As the solution moves through the bed, a series of equilibrations occurs between the ammonium ions and the ions in the solid phase on successive layers in the column (Paterson 1970). The top layers are continually bathed with fresh ammonium solution to the point of saturation and further exchange takes place lower down the column.

The zone where exchange occurs is referred to as the mass transfer zone (Barros et al. 2003) and the shorter its length, the longer the time to reach breakthrough (t_b) and as a result, more of the media's capacity is used. If the column were continually fed with solution until all zones were saturated, then the effluent concentration would equal that of the influent concentration (C_0). At this point, the whole bed (of length L) is in equilibrium and has reached its stoichiometric capacity (Barros et al. 2003). The time to reach this stoichiometric capacity is referred to as the stoichiometric time equivalent (t_t) and this can be used to determine the length of unused bed (LUB). The length of unused bed is dependant on C_b and at breakthrough can be determined from the following (Cooney 1999):

$$\text{LUB} = \left[1 - \frac{t_b}{t_t} \right] \text{L} \tag{10.9}$$

This LUB represents the mass transfer zone (Barros et al. 2003) and can be used as a design parameter, where the bed length can be designed to have the required capacity by making the column one LUB longer to account for unused capacity under actual conditions (Cooney 1999).

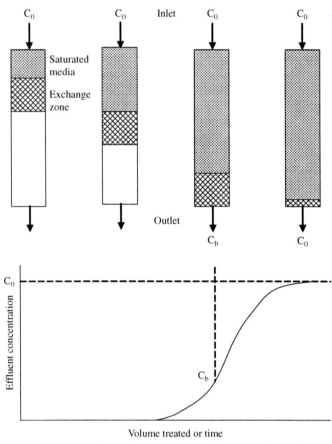

Figure 10.4. The progression of a mass transfer zone through a fixed bed (modified from Cooney 1999)

10.6 PROCESS PERFORMANCE AND MEDIA CAPACITY

Much research and investigation has gone into the understanding and establishment of the ammonium exchange capacity of various zeolites. However, under practical conditions for wastewater treatment, it is highly unlikely the full capacity of the ion exchanger will ever be realised. The process is in place to remove ammonium and as such the desired concentration in the effluent (the breakthrough concentration C_b) will be required to be considerably lower than that in the influent (C_0). It is theoretically possible to utilise the total capacity of the exchanger only where $C_b = C_0$. In this case, all the available exchange sites will be occupied by the ammonium ion (i.e. the material will be saturated), but there would be no further removal of the pollutant from solution.

10.6.1 Breakthrough concentration

Practically, this never occurs and the proportion of the materials capacity utilised depends on a number of factors. A significant parameter is the breakthrough concentration. Lower breakthrough concentrations are reached in shorter time periods and utilise less of the exchange sites resulting in lower operational capacities and more frequent regenerations. Therefore, the higher C_b is, the more efficiently the media can be used. Results using the modified bentonite, MesoLite, indicate that a reduction in C_b of 50% requires and extra media provision of 20% to treat the same amount of wastewater.

10.6.2 Contact time

Fortunately, the ion exchange process is a rapid one, resulting in a compact mass transfer zone within the column, thus restricting the amount of ammonium leaching into the effluent. However, elevated loading rates have the effect of elongating the mass transfer zone, resulting in breakthrough being reached much earlier.

The length of the mass transfer zone (Figure 10.4) is a major factor that impacts on the efficient use of the media and can be investigated by determining the length of unused bed (LUB) at breakthrough. The LUB represents the length of the mass transfer zone (Barros et al. 2003) and can be calculated from Equation 10.9. If mass transfer were instantaneous, then the breakthrough curve would be a vertical line at time $t_b = t_t$ (Cussler 1997) and the LUB would be zero. This would result in an ideal solution concentration profile across the bed as shown in Figure 10.5, where at any given time the all parts of the bed to the left of the mass transfer front will have reached stoichiometric capacity and all those to the right would still be available for exchange. However, in reality breakthrough profiles are not straight lines but gentle curves (as in Figure 10.4) and the shallower the gradient of the curve, the longer the mass transfer zone.

Results for trials of solution strength 30 mg l^{-1} show that the concentration profile across the bed at any given time (Figure 10.6) is not a step function but a gradual slope, which is indicative of an elongated mass transfer zone, rather than a mass transfer front. L represents the depth in the bed of media where solution samples were taken. The arrow indicates the direction of movement of the mass transfer zone through the column with time and the gradient of the curves is linked

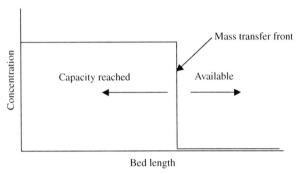

Figure 10.5. An idealised concentration profile (Adapted from Cussler 1997)

to the length of the mass transfer zone or LUB. The shallower the gradient is, the longer the mass transfer zone. For any given set of conditions (i.e. influent concentration and breakthrough set point), it is possible to increase the gradient of the curves by reducing the flow rate, thus resulting in a shorter length of unused bed.

The loading rate has a direct influence on the HRT and therefore influences the amount of media required for the treatment process. Ammonium removal efficiency is shown to decrease with increased flow rates (Aiyuk *et al.* 2003). As a result, the consensus of opinion is that HRT's of less than 10 minutes are not recommended.

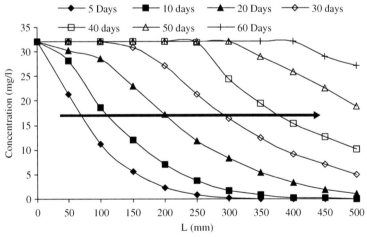

Figure 10.6. Solution concentration profiles across a 0.5 m bed of media at selected times throughout the exhaustion process (Thornton 2007)

10.6.3 Solution strength

There are two main possible applications within wastewater treatment for ammonium reduction using ion exchange. Firstly, columns of media maybe used to treat final effluent where there is a risk of consent breach (i.e. the process may be used to polish already treated wastewater streams). Secondly, recycle streams within the wastewater treatment process, can significantly increase the ammonium load going back to the head of the works. One such stream is digested sludge liquors, which can contribute in excess of 25% of the ammonium load to be treated. Again, ion exchange material can be used to treat these return liquors prior to their return to the main treatment process.

10.6.3.1 Final effluent treatment

The option of treating low strength solutions offers the possibility of reducing the risk of consent failure at sites which are struggling to achieve the consent limits imposed. A change in the nature of the incoming waste, due to population increase or industrial expansion can often lead to once compliant sites falling into this bracket. In addition to this, the ion exchange process also has the ability to remove any peaks, anomalous or otherwise, which may be present in the final effluent. Although biological treatments are very effective under constant loads, they are less capable under variable conditions, whereas ion exchange is extremely effective. It is worth noting that Beler Baykal and Guven (1997) successfully removed ammonium peaks of up to 20 mg l^{-1} from domestic wastewater effluent by ion exchange on to clinoptilolite, for a continual period of over 10 days.

During the treatment of low strength solutions, the duration of run cycles can be extremely long (months rather than days). This means that regeneration cycles are performed infrequently, resulting in the production of much less high concentration regenerant to dispose of. It can be argued that regeneration would not be required at all as it would be more cost effective to dispose of exhausted media, replacing it with fresh media each time it is fully loaded.

Trials on a final effluent of ammonium concentration 5 mg l^{-1} indicate that it is possible to operate a single column at the rate of 12 BV h^{-1} for 22 days when a breakthrough concentration of 1 mg l^{-1} is in place (Figure 10.7). This is a much higher than normal flowrate, but for this particular application the column doubled as a solids filter and this flow rate was chosen as more representative for such an application. At a more realistic flowrate of 5 BV h^{-1}, the run time would be increased to in excess of 50 days with the same breakthrough concentration in place. This is considerably longer than run times encountered during the treatment of high strength solutions, which last a number of hours.

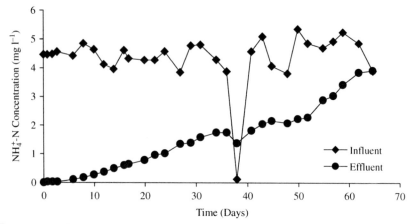

Figure 10.7. Breakthrough curves for trials performed on final effluent

However, the operational capacity returned when treating low concentration solutions is comparatively low. The operational capacity can be defined as the apparent capacity of the material when a breakthrough concentration is in place (i.e. the amount of ammonium removed per kg of media available for any given application). The trials outlined above were conducted using MesoLite media which has a reported capacity of 45–55 g NH_4^+-N kg^{-1} (Mackinnon *et al.* 2003). This would be the capacity of the material if the column was fully loaded. In reality the operational capacity proved to be 9.5 g NH_4^+-N kg^{-1}, approximately 20% of the total and even when the column was run to exhaustion after 65 days (Figure 10.7) a capacity of only 19 g NH_4^+-N kg^{-1} was observed.

This is partially due to the fact that at low influent concentrations, the driving force behind the process is significantly lower than at significantly higher concentrations and as such fewer exchange sites are utilised. That is to say equilibrium between the solid and solution phase is reached before exhaustion. However, another significant factor is that of competing cations, in particular calcium. Although, certain zeolites are selective for the ammonium ion, it is not necessarily true that they exclusively select the ammonium ion, especially if other ions are present in abundance. If competing ions are significantly more predominant, then the driving force behind their exchange will be high and as such some of the materials capacity will be rendered unusable for ammonium removal (i.e. the exchange sites become occupied by undesirable ions). Wastewater streams contain many cations in solution, most notable, calcium, magnesium, sodium and potassium. Of all these, calcium is reported to have the most detrimental effect (Hankins *et al.* 2004; Weatherley and Miladinovic 2004).

During the trials outlined above it was observed that a significant amount of calcium was removed from solution during the exhaustion process. MesoLite was supplied in the potassium form (i.e. the exchange sites on the media were predominantly occupied by potassium) and therefore, every potassium ion released by the media should see a corresponding ammonium ion being removed from solution to preserve electroneutrality. If this was the case, the number of milliequivalents (meq) of ammonium removed should be equal to the number of meq of potassium entering the solution, at any stage of the process. However, there was a significant shortfall between the amount of ammonium removed and the amount of potassium released (approximately 50%). On closer examination of the concentration of other cations in solution it was established that the removal of calcium accounted for this discrepancy (Figure 10.8). The detrimental effect of calcium is magnified by its valence. The exchange of potassium for calcium is a uni-divalent process (Equation 10.3) and as such, every calcium ion removed from solution results in two potassium ions being released from the media. That is to say each calcium ion occupies two exchange sites.

10.6.3.2 Digested sludge liquor treatment

During the treatment of sludge liquors, the high ammonium concentration (commonly 500–700 mg NH_4^+-N l^{-1} but can be up to 1500 mg NH_4^+-N l^{-1})

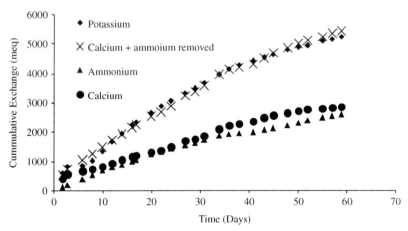

Figure 10.8. Cumulative exchange of ammonium, calcium and potassium (Thornton 2007)

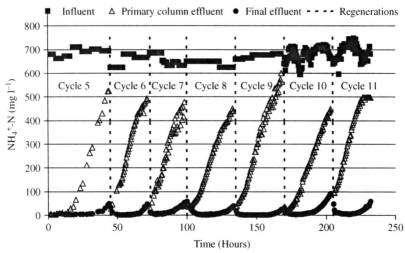

Figure 10.9. Breakthrough curves for trials performed on sludge liquors (Thornton *et al.* 2007a)

means that the ammonium ion is by far the most predominant ion in solution and as such the effect of competing cations is becomes less significant. This means that the removal of ammonium is much more predictable with repeatable results observed. It is common for such a process to consist of two columns in series, allowing regeneration to occur without interrupting the exhaustion cycle of the second column. The columns are alternated from one exhaustion cycle to the next with the primary column being the most recently regenerated.

Figure 10.9 shows a series of six breakthrough curves for the treatment of digested sludge liquors with an average concentration of 650 mg NH_4^+-N l^{-1} (Thornton *et al.* 2007a). In pilot scale trials, columns of MesoLite media were loaded at a rate of 5 BV h^{-1}. A total of 11 exhaustion cycles were performed with the first four taking the material to saturation, returning a total capacity of 47–51 g NH_4^+-N kg^{-1}. Subsequent trials were run with a breakthrough concentration of 50 mg NH_4^+-N l^{-1} in place. This resulted in average loading times of 33.5 hours and an operational capacity of 27–32 g NH_4^+-N kg^{-1}. This operational capacity is significantly higher than obtained when treating final effluent. This is due to the fact that the solute gradient and, therefore, the driving force behind the process is much greater with concentrated solutions and also the effect of competing cations is significantly reduced. However, the media becomes fully loaded in a much shorter amount of time and as such frequent regeneration cycles are required.

10.6.4 Regeneration

Regeneration frequency depends on a number of parameters including, the amount of exchanger used, the influent concentration, the effluent concentration, the loading rate and the breakthrough concentration selected (Hedstrom 2001). Regeneration is achieved by re-circulating a brine solution through the exhausted media (Booker *et al.* 1996; Celik *et al.* 2001; Witte and Keding 1992; Koon and Kaufmann 1975, Liberti *et al.* 1981; Semmens and Porter 1979; Hlvay *et al.* 1982). The cations in the brine exchange with the ammonium and a high concentration ammonium solution is formed. Once the regeneration cycle is complete, the media is then ready for use again in the exhaustion cycle. As mentioned previously, the most suitable cation for ammonium removal is sodium (Booker *et al.* 1996; Jorgensen *et al.* 1979). A number of different of different brines have been used to achieve this exchange of sodium for ammonium (Table 10.3).

Many authors have used sodium chloride as the chosen brine (Witte and Keding 1992; Koon and Kaufmann 1975; Semmens and Porter 1979; Liberti *et al.* 1981; Hlvay *et al.* 1982; Du *et al.* 2005). Two factors affect the duration of regeneration required. These are namely the concentration of the solution and the operating pH. This assumes that the amount of regenerant used is fixed for any given process. Koon and Kaufmann (1975) conducted experiments with pH conditions over a range of 11.5–12.5. At the lower pH, 20 BV of brine

Table 10.3. Regeneration characteristics

Regenerant Brine	Flow Rate (BV h^{-1})	Duration (h)	pH	Reference
NaCl	15	0.7–1.3	11.5–12.5	Koon and Kauffmann (1975)
NaOH	–	0.5	–	Jorgensen *et al.* (1976)
NaCl/NaNO$_3$	12–20	1.0	7.0–8.4	Semmens and Porter (1979)
NaCl	30	0.7	~7	Liberti *et al.* (1981)
NaCl	5–7.5	1.4–4.0	12.3	Hlvay *et al.* (1982)
NaCl	–	–	8.0	Beler Bykal and Guven (1997)
NaCl+NaOH	–	0.33–1.0	11.5	Celik *et al.* (2001)
Na$_2$CO$_3$	0.5–1.5	4	–	Mackinnon *et al.* (2003)
NaCl	5–10	2.5–5.0	11–12	Du *et al.* (2005)
NaOH	5	4	11–12	Thornton *et al.* (2007a)

(0.34 M NaCl) was required for regeneration. At an increased pH level of 12.5 only 10 Bed volumes of regenerant were needed at a lower concentration of 0.21 M NaCl. Other regenerants have been used as an alternative to sodium chloride, most notably sodium hydroxide (Celik *et al.* 2001) but the caustic nature of the brine can lead to attrition of the column media.

Several authors (Semmens *et al.* 1977; Semmens and Porter 1979; Green *et al.* 1996; Lahav and Green 1998; Chung *et al.* 2002) have used biological regeneration as an alternative to the use of passing brine solution. Biological regeneration is a combination of chemical regeneration and the introduction of nitrifying bacteria. After the exhaustion of the columns nitrifying sludge is introduced in such a manner as to fluidise the bed (Semmens *et al.* 1977). The system is dosed either with sodium nitrate (Semmens *et al.* 1977) or sodium bicarbonate (Lahav and Green 1996) and the media acts as a substrate for nitrifying bacteria. The addition of sodium compounds is to facilitate the exchange of sodium ions with the ammonium ions on the zeolite. The main driver for attempting biological regeneration was to reduce brine consumption. However, problems occur due to the re-adsorption of ammonium back onto the media and the regeneration rate is controlled by the nitrification rate as ammonium is displaced from the media column faster than the nitrifying bacteria were able to oxidise it (Semmens *et al.* 1977).

10.7 SUMMARY

Studies using ion exchange for this application indicate that, where conditions dictate, it can provide an effective solution for ammonium removal from problematic wastewaters. The continual centralisation of sludge treatment centres and increasingly stringent consent levels imposed, can only lead to the future requirement of more inventive and novel solutions. However, when considering this application, it is important to take into consideration the factors outlined in this chapter which affect the potential success of any installation.

10.8 ACKNOWLEDGEMENTS

The authors would like to thank Pete Pearce of Thames Water Utilities Ltd for his continued support and guidance throughout the projects undertaken to produce much of the data reported in this chapter. His contribution to their success proved invaluable.

REFERENCES

Aiyuk, S., Xu, H., Van Haandel, A. and Verstraete, W. (2003) Removal of ammonium from pretreated domestic sewage using a natural ion exchanger. (LabMET), Ghent University.

Armbruster, T. (2001) Clinoptilolite-heulandite: application and basic research. *Studies in Surface Science and Catalysis* **135**, 13–27.

Barros, M.A.S.D., Zola, A.S., Arroyo, P.A., Sousa-Aguiar, E.F. and Tavares, C.R.G. (2003) Binary ion exchange of metal ions in Y and X zeolites. *Brazilian Journal of Chemical Engineering* **20**(4), 413–421.

Beler Baykal, B., Oldenburg, M. and Sekoulov, I. (1996) The use of ion exchange in ammonia removal under constant and variable loads. *Environmental Technology* **17**, 717–726.

Beler Baykal, B. (1998) Clinoptilolite and multipurpose filters for upgrading effluent ammonia quality under peak loadings. *Water, Science and Technology* **37**(9), 235–242.

Beler Baykal, B. and Guven, D.A. (1997) Performance of clinoptilolite alone and in combination with sand filters for the removal of ammonia peaks from domestic wastewater. *Water, Science and Technology* **35**(7), 47–54.

Bolto, B. A. and Pawlowski, L. (1987) Wastewater treatment by ion exchange, E and FN Spon, New York.

Booker, N.A., Cooney, E.L. and Priestly, A.J. (1996) Ammonia removal from sewage using natural Australian zeolite. *Water, Science and Technology* **34**(9), 17–24.

Breck, D. (1974) Zeolite and molecular sieves structure and chemistry, John Wiley & Sons, New York.

Celen, I. and Turker, M. (2001) Recovery of ammonia as struvite from anaerobic digester effluents. *Environmental Technology* **22**, 1263–1272.

Celik, M.S., Ozdemir, B., Turan, M., Koyuncu, I., Atesok, G. and Sarikaya, H.Z. (2001) Removal of ammonia by natural clay minerals using fixed bed column reactors. *Water, Science and Technology: Water Supply* **1**(1), 81–88.

Colella, C., DE' Gennaro, M., Langella, A. and Pansini, M. (1995) In: Natural zeolites-cadmium removal from wastewaters using chabazite and phillipsite. Ming, D.W. and Mumpton, F.A. (Eds), pp 377–384, Int. Comm. Natural Zeolites, Brockport, New York.

Cooney, D.O. (1999), Adsorption design for wastewater treatment, Lewis Publishing, New York.

Curkovic, L., Cerjan-Stefanovic, S. and Filpan, T. (1997) Metal ion exchange by natural and modified zeolites. *Water Research* **31**(6), 1739–1382.

Cussler, E.L. (1997) Diffusion: Mass transfer in fluid systems, Cambridge University Press.

Demir, A., Gunay, A. and Debik, E. (2002) Ammonium removal from aqueous solution by ion exchange using packed bed natural zeolite. *Water SA* **28**(3), 329–336.

DeSilva, F.J. (1999) Essentials in ion exchange. Presented at the 25th Annual WQA Conference.

Du Qi, Shijun Liu, Q.D., Zhonghong, C. and Wang, Y. (2005) Ammonia removal from aqueous solution using natural Chinese Clinoptilolite. *Separation and Purification Technology* **44** 229–234.

Dyer, A. (1988) An introduction to zeolite molecular sieves, John Wiley and Sons Ltd, Chichester.

Farkas, A., Rozic, M. and Barbaric-Mikocevic, Z. (2005) Ammonium exchange in leakage waters of waste dumps using natural zeolite from the Krapina region, Croatia. *Journal of Hazardous Materials* B117 25–33.

Flanigen, E.M. (1984) In: Zeo-agriculture, use of natural zeolites in agriculture and aquaculture-adsorption properties of molecular sieve zeolites. Pond, W.G. and Mumpton, F.A. (Eds), pp 55–68, Westview Press, Boulder, Colorado.

Fux, C., Lange, K., Faessier, A., Huber, P., Grueniger, B. and Siegrist, H. (2003) Nitrogen removal from digester supernatant via nitrite – SBR or SHARON. *Water, Science and Technology* **48**(8), 9–18.

Gottardi, G. (1978) In: Natural zeolites occurrence and properties – mineralogy and chemistry of zeolites. Use Sand, L.B. (Ed) pp 31–44, Pergamon Press, Oxford.

Hankins, N.P., Pliankarom, S. and Hilal, N. (2004) An equilibrium ion exchange study on the removal of NH_4^+ ion from aqueous effluent using clinoptilolite. *Separation Science and Technology* **39**(15), 3639–3663.

Hedstrom, A. (2001) Ion exchange of ammonium in zeolites: A literature review. *Journal of Environmental Engineering* **August 2001**, 673–681.

Helfferich, F. (1962) Ion exchange, McGraw-Hill, New York.

Hlavay, J., Vigh, G., Olaszi, V. and Inczedy, J. (1982) Investigation on natural Hungarian zeolite for ammonia removal. *Water Research* **16**, 417–420.

Jorgensen, S.E., Libor, O., Graber, K.L. and Barkacs, K. (1976) ammonia removal by use of clinoptilolite. *Water Research* **10**, 213–224.

Jorgensen, S.E., Libor, O., Barkacs, K. and Kuna, L. (1979) Equilibrium and capacity data of clinoptilolite. *Water Research* **13**, 159–165.

Jorgensen, T.C. and Weatherley, L.R. (2003) Ammonia removal from wastewater by ion exchange in the presence of organic contaminants. *Water Research* **37**, 1273–1278.

Jung, J., Chung, Y., Shin, H. and Son, D. (2004) Enhanced ammonium nitrogen removal using consistent biological regeneration and ammonium exchange of zeolite in modified SBR process. *Water Research* **38**, 347–354.

Kabdasli, I., Tunay, O., Ozturk, I., Yilmaz, S. and Arikan, O. (2000) Ammonia removal from young landfill leachate by magnesium ammonium phosphate precipitation and air stripping. *Water, Science and Technology* **41**(1), 237–240.

Kamarowski, S. and Yu, Q. (1997) Ammonium ion removal from wastewater using Australian natural zeolite: Batch equilibrium and kinetics Studies. *Environmental Technology* **18**, 1085–1097.

Karadag, D., Koc, Y., Turan, M. and Armagan, A. (2006) Removal of ammonium ion from aqueous solution using natural Turkish clinoptilolite. *Journal of Hazardous Materials* B136 604–609.

Koon, H.J. and Kaufmann, W.J. (1975) Ammonia removal from municipal wastewaters by ion exchange. *Journal WPCF* **47**(3) 448–465.

Kovatcheva –Ninova, V., Nikolova, N. and Marinov, M. (2002) Investigation in the adsorption properties of natural adsorbents zeolite and bentonite towards copper Ions. *Annual of The University of Mining and Geology "St Ivan Rilski"* **45**(2), 93–97.

Lahav, O. and Green, M. (1998) Ammonium removal using ion exchange and biological regeneration. *Water Research* **32**(7), 2019–2028.

Liberti, L., Boari, G., Prtruzzelli, D. and Passino, R. (1981) Nutrient removal and recovery from wastewater by ion exchange. *Water Research* **15**, 337–342.

Mackinnon, I.D.R., Barr, K., Miller, E., Hunter, S. and Pinel, T. (2003) Nutrient removal from wastewaters using high performance materials. *Water, Science and Technology* **47**(11), 101–107.

Mumpton, F.A. (1984) In: Zeo-agriculture, use of natural zeolites in agriculture and aquaculture-natural zeolites. Pond, W.G. and Mumpton, F.A. (Eds), pp 33–43, Westview Press, Boulder, Colorado.

Nguyen, M.L. and Tanner C.C. (1998) Ammonium removal from wastewaters using natural New Zealand zeolites. *New Zealand Journal of Agricultural research* **41**, 427–466.

Paterson, R. (1970) An introduction to ion exchange, Heyden and Son Ltd London.

Salerno, P. and Mendioroz, S. (2002) Preparation of Al-pillared montmorillonite from concentrated dispersions. *Applied Clay Science* **22**, 115–123.

Schmidt, I., Sliekers, O., Schmid, M., Bock, E., Fuerst, J., Kuenen, J.G., Jetten, M.S.M. and Strous, M. (2003) New concepts of microbial treatment processes for the nitrogen removal in wastewater. *FEMS Microbiology Reviews* **27**, 481–492.

Semmens, M.J. (1984) In: Zeo-agriculture, use of natural zeolites in agriculture and aquaculture-cation exchange properties of natural zeolites. Pond, W.G. and Mumpton, F.A. (Eds), pp 45–53, Westview Press, Boulder, Colorado.

Semmens, M.J. and Porter, P.S. (1979) Ammonium Removal by Ion Exchange: Using Biologically Restored Regenerant. *Journal WPCF* **51**(12), 2928–2940.

Siegrist, H. (1996) Nitrogen removal from digester supernatant – comparison of chemical and biological methods. *Water, Science and Technology* **34**(1–2), 399–406.

Sprynskyy, M, Lebedynets, M, Terzyk, A.P, Kowalczyk, P, Namiesnik, J. and Buszewski, B. (2005) Ammonium sorption from aqueous solutions by the natural zeolite Transcarpathian clinoptilolite studied under dynamic conditions. *Colloid and Interface* Science **284**, 408–415.

Thornton, A., Pearce, P. and Parsons, S.A. (2007a) Ammonium removal from digested sludge liquors using ion exchange. *Water Research* **41**, 433–439.

Thornton, A., Pearce, P. and Parsons, S.A. (2007b) Ammonium removal from solution using ion exchange on to MesoLite: An equilibrium study. *Journal of Hazardous Materials* **147**, 883–889.

Thornton., A (2007). The application of MesoLite for ammonium removal from municipal wastewaters. Ph.D. Research Thesis, Centre for Water Science, Cranfield University.

Tosheva, L. (1999) Zeolite macrostructures. Licentiate Thesis, Lulea University of Technology, Sweden.

Tzvetcov, S. and Stoytcheva, M. (2003) The possibility of phosphates removal from waste waters and obtaining of mixed fertilizers with natural zeolites. *Annual of The University of Mining and Geology "St Ivan Rilski"* **46**(2), 123–125.

Van Dongen, L.G.J.M., Jetten, M.S.M. and van Loosdrecht, M.C.M. (2001) The combined Sharon/Anammox process: A sustainable method for N-removal from sludge water. STOWA report IWA Publishing, London.

Varshney, K.G. (2002) Synthetic ion exchange materials and their analytical applications. Aligarh Muslim University, Aligarh.

Wang, Y., Liu, S., Xu, Z., Han, T., Chaun, S. and Zhu, T. (2006) Ammonia removal from leachate solution using natural Chinese clinoptilolite. *Journal of Hazardous Materials* B136 735–740.

Weatherley, L.R. and Miladinovic, N.D. (2004) Comparison of the ion exchange uptake of ammonium ion onto New Zealand clinoptilolite and mordenite. *Water Research* **38**, 4305–4312.

Wen, D., Ho, Y.S. and Tang, X. (2006) Comparative sorption kinetic studies of ammonium onto zeolite. *Journal of Hazardous Materials* B133 252–256.

Witte, H. and Keding, M. (1992) Zeolite filters in wastewater treatment plants. In: Proceedings of the 5th Gothenburg Symposium Chemical and Wastewater Treatment II, 28–30 September, Nice, France. pp 467–484.

11

Denitrification of industrial flue gases

P. van der Maas

11.1 INTRODUCTION

The emission of nitrogen oxides (NO_x) to the atmosphere is a serious persistent environmental problem. The major nitrogen oxides comprise nitric oxide (NO) and nitrogen dioxide (NO_2), which is the reaction product of NO with oxygen (Jager 2001). NO is the primary NO_x form in combustion products, typically 95% of total NO_x content (Fritz and Pitchon 1997; Janssen 1999). Nitrogen oxides play an important role in the atmospheric ozone destruction and global warming (Stepanov and Korpela 1997). Furthermore, urban NO_x is regarded as one of the most important precursors to photochemical smog (Guicherit and Van den Hout 1982; Princiotta 1982). Smog products irritate eyes and throat, evoke asthmatic attacks, reduce visibility and damage plants and materials as well. Moreover, NO_x also contribute to acidification, because they dissolve in cloud and precipitation water to form the strong acids HNO_2 and HNO_3 (Hales 1982; Stepanov and Korpela 1997; Princiotta 1982). The presence of NO_x in ambient air is of great concern because of the toxicity of the individual compounds or the mixtures of nitrogen oxides. NO has cytotoxic properties and is implicated in the degeneration of neurons, associated with Parkinson's disease, dementia and stroke (Stepanov and Korpela 1997). NO_2 irritates the respiratory tract (Posthumus 1982).

11.1.1 Sources of NO_x

NO_x are naturally produced in the atmosphere due to lightning, volcanic activity and biological decay (Cole 1993; Stepanov and Korpela 1997). Naturally produced nitrogen oxides end up in the nitrogen cycle together with human-generated NO_x. For the last 20–30 years, the NO_x concentration in the atmosphere has increased due to human activities (Rasmussen and Khalil 1986), mainly due to the combustion of fossil fuels (Figure 11.1).

The presence of nitrogen oxides in flue gases can be sub-divided in two groups: thermal NO_x and fuel NO_x (Kremer 1982; Jager 2001). Thermal NO_x are formed by the combustion reactions where molecular nitrogen reacts with $O^{\cdot}$ and $OH^{\cdot}$ radicals and molecular O_2 in high temperature regions (>1800 K). The formation of this category of NO_x is highly influenced by the combustion temperature. Fuel NO_x is formed by oxidation of nitrogen compounds chemically bound in fossil fuel. Formation of fuel NO_x takes place at lower temperatures (1000–1400 K) than thermal NO_x.

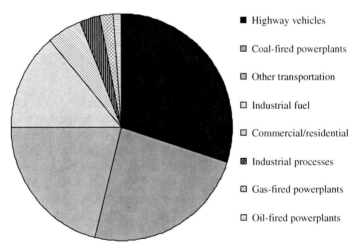

- Highway vehicles
- Coal-fired powerplants
- Other transportation
- Industrial fuel
- Commercial/residential
- Industrial processes
- Gas-fired powerplants
- Oil-fired powerplants

Figure 11.1 Relative contributions of anthropogenic NO_x emission sources (Schnelle and Brown 2002)

Table 11.1 shows typical concentrations in the flue gases from electric power generation plants as a function of the fuel utilised (Ramachandran *et al.* 2000). Natural gas, for example, is a relatively clean fuel with respect to NO_x formation when compared with heavy fuel or coal. Natural gas does not contain any fuel nitrogen, i.e. nitrogen atoms bound in the hydrocarbon molecules, and, therefore, the only NO_x production route is via oxidation of the molecular nitrogen contained in the combustion air.

Global NO_x emissions followed until recently an almost exponential increase over time, but over the last decade they seem to slow down. In Europe, the release of nitrogen oxides has even declined recently, but it still amounted to approximately 13 million tons in 2000 (European Environment Agency 2003). The NO_x emission reduction targets for the European Union are based on a 55% reduction of the total NO_x emission in 1990 by the year 2010 (Erisman *et al.* 2003). In 2001, a reduction of 28% compared to 1990 was achieved (Table 11.2). Approximately 24 million tons of NO was released to the atmosphere from US sources during 1998. Titles I and IV of the 1990 Clean Air Act Amendments regulate NO_x emissions from major stationary sources (Bradford *et al.* 2002). The overall goal of these programs was to achieve NO_x reductions of 2 million tons per year below 1980 levels by the year 2000. However, the growing economies, especially in Southeast Asia, likely overwhelm any reductions in NO_x emissions made in Europe and North America (Fawler *et al.* 1998). The rapid economic growth and the increasing consumption of fossil fuel there results in the emission of large amounts of NO_x into the atmosphere.

11.1.2 Removal of nitrogen oxides

In principle, NO_x emissions due to the combustion of fossil fuels can be abated via two different strategies (Figure 11.2): combustion modification (primary measures) and post-combustion techniques (end of pipe technologies).

Table 11.1 Typical concentrations of emissions in flue gases from power generation plants using different types of fuel (Ramachandran *et al.* 2000)

	Natural gas	Fuel oil	Coal
NO_x (ppm)	25–160	100–600	150–1000
SO_x (ppm)	<0.5–20	200–2000	200–2000
CO_2 (%)	5–12	12–14	10–15
O_2 (%)	3–18	2–5	3–5
H_2O (%)	8–19	9–12	7–10
N_2	balance	balance	balance

Table 11.2 NOₓ emissions in the European Union by sector (ktonnes) (European Environmental Agency 2003)

	1990	1991	1992	1993	1994	1995	1996	1997	1998	1999	2000	2001	Change 1990–2001 (%)
Energy Industries	4274	4019	3771	3433	3221	3191	3079	2858	2727	2612	2639	2640	-38%
Fugitive Emissions	40	34	34	35	35	34	36	35	34	42	38	35	-13%
Industry (Energy)	2282	2253	2109	1882	1814	1787	1739	1717	1646	1615	1533	1548	-32%
Industry (Processes)	287	246	208	179	182	157	158	158	160	158	152	148	-48%
Other (Energy)	805	835	795	861	781	748	840	780	743	742	706	732	-9%
Other (Non Energy)	0	0	0	1	1	1	0	0	0	0	0	0	-42%
Road Transport	7481	7401	7371	7133	7001	6685	6533	6255	6078	5838	5570	5331	-29%
Other Transport	2022	1967	1910	1882	1873	1900	1989	2021	2031	1997	1894	1874	-7%
Agriculture	144	142	137	129	129	134	146	148	147	143	153	146	1%
Waste	134	141	139	139	135	133	132	134	129	134	128	127	-5%
Total	17469	17038	16474	15674	15172	14770	14652	14106	13695	13281	12813	12581	-28%

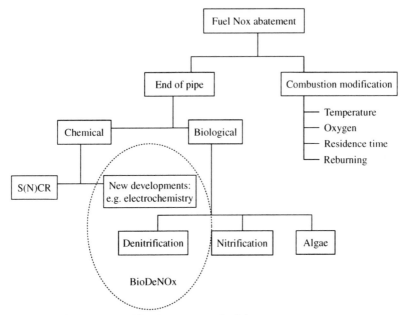

Figure 11.2 Overview of NO$_x$ abatement principles

Usually, post-combustion techniques are implemented in addition to primary measures when a high level of NO$_x$ abatement is required.

Combustion modification prevents or reduces the formation of NO$_x$ during combustion. These modifications can give moderate (20–50%) NO$_x$ reductions at relatively low costs (Princiotta 1982). The principles of combustion modifications are (Princiotta 1982; Jager 2001):
- decrease of the combustion temperature.
- decrease of the residence time in high temperature zones.
- decrease of the oxygen concentration in high temperature zones by lowering the excess air.
- Reburning/Infurnace NO$_x$ reduction: here a second combustion zone is created by injecting fuel into the furnace above the main combustion zone. In this second combustion zone, hydrocarbon radicals from the secondary fuel reduce the NO$_x$ to N$_2$.

The following paragraphs give an overview of the technology available for NOx removal from industrial gas steams. Special attention will be given to the BioDeNOx concept, a chemically enhanced technique for biological flue gas denitrification.

11.2 CHEMICAL NO$_X$ REMOVAL TECHNIQUES

11.2.1 Selective catalytic reduction

The most widely applied end of pipe technology for NO$_x$ removal from flue gases is Selective Catalytic Reduction (SCR). It is based on chemical NO reduction by ammonia or urea at high temperatures to form molecular nitrogen, according to (Bradford *et al.* 2002):

$$6NO + 4NH_3 \rightarrow 5N_2 + 6H_2O$$

$$2NO + 4NH_3 + 2O_2 \rightarrow 3N_2 + 6H_2O$$

$$6NO_2 + 8NH_3 \rightarrow 7N_2 + 12H_2O$$

Ammonia is sprayed into the exhaust gas, which passes then through a catalyst where NO$_x$ and NH$_3$ are converted into nitrogen and water at an optimum temperature between 300°C and 400°C. The catalyst is based on titanium oxide, vanadium oxide, zeolite, iron oxide, or activated carbon. The reactions are very efficient and therefore enable a very effective NH$_3$ injection control, based on feedback of the measured NO$_x$ concentration in the flue gas. Efficiencies up to 90% can be reached with this technique, mainly depending on the catalyst activity (Jager 2001). However, there are two significant problems associated with this process: a) the operating costs of the process are high due to catalyst deactivation and b) un-reacted ammonia either forms ammonium sulphate that can clog equipment or ends up in the effluent gas. Both destinations of ammonia are undesirable in the SCR process.

Selective non-catalytic reduction (SNCR) is similar to SCR, but without the use of a catalyst. The temperature is higher than in SCR: 850°C to 1100°C. The point of injection is located between the combustion chamber and an economiser in the area of the superheater. Adequate mixing of chemicals and temperature control are essential for maximum NO$_x$ reduction and minimum NH$_3$ slip (Jager 2001). For practical SNCR systems, NO$_x$ reduction efficiencies are only about 50%, but lower capital costs are necessary than for the SCR (Princiotta 1982).

11.2.2 Electrochemical NO$_x$ removal techniques

Kleifges *et al.* (1997) developed a novel NO$_x$ removal process, which is based on NO absorption into an aqueous Fe(II)EDTA^{2-} solution, followed by NO reduction using dithionite (S$_2$O$_4^{2-}$) as redox mediator. The latter is regenerated by cathodic reduction of SO$_3^{2-}$ in an electrochemical cell connected to the absorption column by a liquid loop (Figure 11.3).

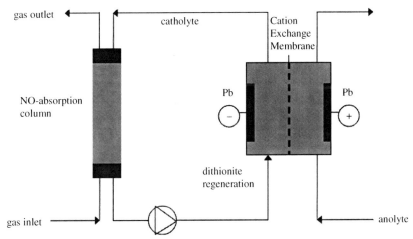

Figure 11.3. Schematic principle of electrochemical NO_x removal using Fe(II) $EDTA^{2-}$ as absorbent and dithionite as redox mediator (after Kleifges *et al.* 1997)

The use of Fe(II)EDTA^{2-} for NO absorption is essential to achieve satisfactory NO removal efficiencies. In the absence of Fe(II)EDTA^{2-}, only 17% of the NO was converted via $S_2O_4^{2-}$ oxidation. It was assumed that NO is reduced to N_2 via dithionite oxidation, but results of batch experiments showed that also $HON(SO_3)_2^{2-}$ and N_2O were formed as reaction products (Kleifges *et al.* 1997):

$$Fe(II)EDTA\text{-}NO^{2-} + S_2O_4^{2-} + H_2O \rightarrow Fe(II)EDTA^{2-} + 2SO_3^{2-} + \frac{1}{2}N_2 + 2H^+$$

$$4Fe(II)EDTA\text{-}NO^{2-} + 4HSO_3^- \rightarrow 4Fe(II)EDTA^{2-} + 2HON(SO_3)_2^{2-} + N_2O$$

11.3 BIOLOGICAL NO$_x$ REMOVAL TECHNIQUES

Biological treatment techniques offer advantages over chemical treatment techniques, especially when the concentration of the pollutant in the waste gas is low (Kirchner 1989). An important advantage over other processes is that biological treatment techniques operate at low pressures and temperatures (Cesario 1991). Basically, biological NO$_x$ removal technologies can be classified into three different groups: (i) nitrification; (ii) denitrification and (iii) algal removal, as reviewed by Jin *et al.* (2005).

11.3.1 Nitrification

Davidova *et al.* (1997) were the first who demonstrated NO removal from gas streams via nitrification. However, a 90% NO removal from a 100 ppm contaminated gas stream required relatively long residence times (~13.7 min). Insufficient biomass growth for effective nitrification was considered as the cause of the rather inefficient NO removal. The above work was further extended using nitrite as a substrate to develop and enhance growth of the nitrifying biofilm in biotrickling filters (Hudepohl *et al.* 2000) (Figure 11.4). The systems, operated at an empty bed retention time (EBRT) of 1 min, exhibited a limited removal efficiency (10–15%), attributed to mass transfer limitation of the poorly water soluble NO from the gas to the liquid phase. The estimated residence time for 90% removal was 6 min (Davidova *et al.* 1997). Further studies demonstrated that the relatively high EBRT of 6 min was required due to mass-transfer limitations of the poorly soluble nitric oxide and oxygen (Nascimento *et al.* 1999; 2000). Thus, the key to economical biological treatment of NO is to reduce mass transfer limitations. For biotrickling filters, a maximisation of the surface-to-volume ratio can support the required biofilm surface without clogging of the pores.

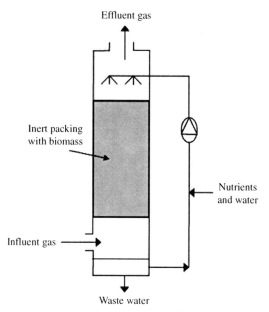

Figure 11.4 Schematic diagram of a biotrickling filter (after Van Groenestijn and Hesselink 1994)

NO removal using an aerobic biotrickling filter with a specific surface area of $120 \text{ m}^2 \cdot \text{m}^{-3}$ was investigated by Chou *et al.* (2000). Approximately 6 weeks were required to develop a biofilm for NO removal, using activated sludge as inoculum, while glucose, yeast powder, phosphate and $NaHCO_3$ were added as supplementary nutrients. At a gas residence time of two min, removal efficiencies of 80% were obtained, but it should be noted that the influent NO concentrations were relatively high (892–1237 ppmv NO). When glucose addition was ceased for two weeks, NO removal declined to 48%. This organic carbon deficiency resulted in the detachment of part of the biofilm from the packing surface. Also here, more than 90% of the eliminated NO was converted into nitrate, suggesting NO removal via nitrification. In other studies, soil was used as packing material for NO removing biofilters (Lukow and Diekmann 1997; Yani *et al.* 2000; Kim *et al.* 2000). The NO removal efficiency was 60% on an average at gas residence times of 1 min.

To maximize the surface-to-volume ratio, the application of a Hollow Fiber Membrane Bioreactor (HFMB) might be promising. NO_x removal from waste gas using a HFMB configuration was evaluated by Min *et al.* (2002). The NO gas diffuses through the membrane pores into a nitrifying biofilm (Figure 5), where it is oxidised to nitrate. The hollow fibre membranes serve as a support for the microbial populations and provide a large surface area for NO and oxygen mass transfer. Maximal NO removal rates of $27 \text{ g} \cdot \text{m}^{-3} \cdot \text{d}^{-1}$ were found for an overall gas residence time of 5 min, corresponding to a NO removal efficiency of 74%.

11.3.2 Denitrification

NO removal using heterotrophic denitrification has been observed in an aerobic toluene-treating biofilter (DuPlessis *et al.* 1998). When treating flue gas containing 17% O_2, the development of a thick biofilm is a pre-requisite for creating anaerobic underlayers where denitrification can take place. Removal efficiencies of 75% of 60 ppmv NO streams were observed at an EBRT of 6 min. Other researchers reported the occurrence of NO removal via denitrification in compost biofilters (Apel and Turick 1993; Samdami *et al.* 1993; Apel *et al.* 1995; Barnes *et al.* 1995; Lee *et al.* 1999). High removal efficiencies (up to 90%) were observed when treating a 500 ppmv NO gas stream at an EBRT of 1.3 min in the presence of a phosphate buffer containing either lactate or dextrose as electron donor and carbon source (Barnes *et al.* 1995, Klasson and Davison 2001; Lee *et al.* 2001; Lacey *et al.* 2002). NO removal efficiencies of over 85% NO were observed, but only at EBRT of 70–80 s. Also here, the effect of oxygen on the NO removal efficiency by these biofilters was significant. In steady anaerobic operation, both biofilters showed NO removal efficiencies exceeding 50%, but the NO removal dropped to 10–20% when 2% O_2 was present in the influent stream (Lee *et al.* 1999). Heterotrophic NO reduction in the presence of succinate, yeast extract and heat/alkali pre-treated municipal sewage sludge as carbon and energy source was investigated by Shanmugasundram *et al.* (1993) and by Dasu *et al.* (1993).

The oxygen-sensitivity of denitrification is a major complicating factor in the scope of applying denitrifying bioreactors for NO_x removal from oxygen containing gas streams. Since certain fungi exhibit denitrification under aerobic conditions, adoption of NO reducing fungi may be an interesting option. NO removal from oxygen containing flue gas streams by means of fungal denitrification was evaluated in a biofilter using toluene as a sole carbon and energy source (Woertz *et al.* 2001). The fungal bioreactor removed 93% of the inlet 250 ppmv NO at an EBRT of 1 min, with an inlet NO loading rate of $17.2 \ g \cdot m^{-3} \cdot h^{-1}$.

NO removal by *Thiobacillus denitrificans* via autotrophic denitrification using thiosulphate as electron donor was studied by Lee and Sublette (1990). They found up to 96% removal of nitric oxide when treating a gas stream containing 5000 ppmv NO. Autotrophic denitrification of gas streams via the oxidation of reduced sulphur compounds in principle represents a very interesting option since it may enable to combine NO and SO_2 removal from flue gases. Therefore, studies were performed to check the feasibility to simultaneously remove SO_x and NO_x from cooled flue gas by contact with cultures of the sulphate-reducing bacterium *Desulfovibrio desulfuricans*, converting SO_2 (via aqueous HSO_3^-/HSO_4^-) to H_2S, and *Thiobacillus denitrificans*, transforming H_2S to SO_4^{2-} using NO as the electron acceptor. However, a simultaneous process combining the SO_x/NO_x removal was technically not feasible due to NO inhibition of SO_4^{2-} reduction by *D. desulfuricans* (Lee and Sublette 1991). Chapter 12 presents an overview of autotrophic denitrifying processes to achieve the simultaneous removal of nitrogenous and sulphurous compounds from wastewaters.

Another system investigated is based on the pre-concentration of low concentrations of NO at a high volumetric flow onto activated carbon, followed by a thermal desorption of NO_x and biological denitrification treatment of the thermically desorbed gas (Chagnot *et al.* 1998). The NO_x flow was denitrified by a pure culture of *Thiobaccillus denitrificans* in a trickling biofilter with a gas superficial velocity of 0.5 m/h. The NO_x inlet concentration was 8000–16000 ppm and the EBRT was 22 min.

11.3.3 Nitric oxide removal by algae

Various studies demonstrate the capability of NO removal by algal cultures. A marine micro-algae was found to simultaneously eliminate nitric oxide and carbon dioxide from a model flue gas (Yoshihara *et al.* 1996). About 40 mg of NO and 3.5 g of CO_2 were eliminated per day in a 4 litre reactor column spiked with 300 ppm (v/v) NO and 15% (v/v) CO_2 in N_2 at a rate of 150 mL · min^{-1}. In order to understand the process of NO removal from flue gases by an algal culture more thoroughfully, a bioreactor system with the unicellular micro-alga *Dunaliella tertiolecta* was operated (Nagase *et al.* 1997). The authors found that the presence of both algal cells and O_2 is important for the reactor system and it was hypothesised that the principle of NO removal is based on NO oxidation, followed by assimilation of that oxidised nitrogen by the algal cells. The results of the study indicate that the dissolution of nitric oxide in the aqueous phase is the rate-limiting step in this reactor system. This was confirmed in another study made by Nagase *et al.* (1998) where the gas-liquid contact area was increased by reducing the bubble size, thereby ensuring a higher NO dissolution rate. NO removal efficiencies up to 96% were achieved with a counter-flow type airlift reactor aerated with small bubbles. The uptake pathway of nitric oxide by the algae *Dunaliella tertiolecta was* investigated in a bubble column type bioreactor (Nagase *et al.* 2001). It was found that most of the NO permeated directly into the cells by diffusion; only a small part of the NO-nitrogen was oxidised to nitrite and nitrate. Nitric oxide taken up in the algal cells was then preferentially utilised as a nitrogen source for cell growth rather than nitrate.

11.3.4 Evaluation of NO removal techniques

Major drawbacks of the commonly applied chemical NO_x abatement techniques are the consumption of catalysts (SCR), high energy consumption because of high operation temperatures (SCR and SNCR), and the slip ammonia or urea to the environment (SCR and SNCR). Since biological processes work at ambient temperatures, without chemical catalysts in the absence of ammonia or urea, biological NO_x removal techniques principally are more sustainable than SCR and SNCR. However, the poor water solubility of NO implies that the mass transfer of NO from the gas into the water phase is relatively slow. Consequently, relatively long gas retention times (Table 11.3) and big reactor volumes are needed for NO_x removal in biofilters or biotrickling filters. This is a major (economic) drawback when biological NO_x removal has to be applied at voluminous gas streams, e.g. at power plants with gas flows exceeding 1000000 Nm^3 · h^{-1}.

Table 11.3 Biological NO removal techniques: removal efficiencies (RE) and empty bed retention times (EBRT) required (after Jin et al. 2005)

	NO conc. (ppmv)	EBRT (min)	Load $(gNO \cdot m^{-3} \cdot h^{-1})$	Removal (%)
Nitrification				
HFMB	100	5	1.48	69–75
celite	100	1	7.4	okt-15
slag	982–1237	2	2.75–22.4	80
biosoil	1.6	1	0.12	60
Denitrification				
compost, heterotrophic	250	1	18	99
compost, heterotrophic	500	1	37	90
compost, heterotrophic	500	1.3	28	95
perlite, heterotrophic	500	1.18	31	94
biofoam, heterotrophic	500	1.18	31	85.5
silicate pellets, heterotrophic	60	3	0.75	75
sulphur+marel, autotrophic	8978	22	$4.15 \, mmol \cdot h^{-1}$	10
silicate pellets, heterotrophic	250	1	18.4	93
Microalgae	300	26.7	$22.3 \, mg \cdot d^{-1}$	50–60

11.4 THE BIODENOX CONCEPT

One promising technique for the NO_x removal from industrial flue gasses is BioDeNOx, an integrated physico-chemical and biological process (Buisman et al. 1999). Basically the concept consists of two parts: 1) wet absorption of nitric oxide (NO) into an iron chelate solution and 2) regeneration of this iron chelate solution by a biological denitrification process (Figure 11.5).

11.4.1 Flue gas denitrification by aqueous Fe(II)EDTA^{2-} solutions

The slow mass transfer of NO from the gas to the liquid phase can be overcome by the application of aqueous solutions of ferrous chelates, e.g. Fe(II)EDTA^{2-}.

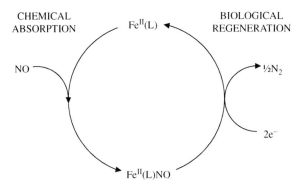

CHEMICAL
ABSORPTION

$Fe^{II}(L)$

BIOLOGICAL
REGENERATION

NO

$\frac{1}{2}N_2$

$2e^-$

$Fe^{II}(L)NO$

Figure 11.5 Schematic principle of BioDeNOx: 1) wet absorption of NO into a chelate solution and 2) regeneration of this chelate solution by a biological denitrification process

The absorption of NO into aqueous $Fe(II)EDTA^{2-}$ solutions proceeds according to Wubs and Beenackers (1993):

$$NO\ (g) \leftrightarrow NO\ (aq)$$

$$NO\ (aq) + Fe(II)EDTA^{2-} \leftrightarrow Fe(II)EDTA\text{-}NO^{2-}$$

The kinetics of NO absorption into aqueous $Fe(II)EDTA^{2-}$ solutions have been investigated thoroughly in the last two decades. Several researchers reported that this reaction is first order in both NO and $Fe(II)EDTA^{2-}$ (Demmink 2000; Schneppensieper *et al.* 2001). Recent studies have unequivocally proven that the reaction is reversible (Schneppensieper *et al.* 2001). The reaction is intrinsically very fast and mass transfer limitation of both NO and the $Fe(II)EDTA^{2-}$ complex plays an important role (Demmink 2000). As a consequence, the reaction often takes place in the instantaneous regime of reactive gas absorption.

When aqueous $Fe(II)EDTA^{2-}$ solutions are applied for NO removal from oxygen containing industrial flue gasses (O_2 content depends on fuel type), part of the $Fe(II)EDTA^{2-}$ will be oxidised to $Fe(III)EDTA^-$ according to:

$$4Fe(II)EDTA^{2-} + O_2(aq) \rightarrow 4Fe(III)EDTA^- + 2H_2O$$

The reaction of $Fe(II)EDTA^{2-}$ with oxygen has also been the subject of various studies and is known to be an irreversible reaction (Zang and Van Eldik

1990; Wubs and Beenackers 1993). The reaction is first order in oxygen, whereas the order in iron is a function of the iron chelate concentration (Zang and Van Eldik 1990; Wubs and Beenackers 1993). At low Fe(II)EDTA^{2-} concentrations (<10 mol $\cdot$ m^{-3}), the reaction is first order in iron whereas it becomes second order at higher concentrations.

BioDeNOx reactors normally operate under thermophilic conditions, at around 50–55°C, which is the adiabatic temperature of scrubber liquors. When ethanol is used as electron donor, the denitrification reaction occurs according to the overall reaction (Buisman et al. 1999):

$$6Fe(II)EDTA - NO^{2-} + C_2H_5OH \rightarrow 6Fe(II)EDTA^{2-} + 3N_2 + CO_2 + 3H_2O$$

To regenerate the absorption liquor, the Fe(III)EDTA$^-$ that is formed via oxidation by oxygen has to be reduced back to Fe(II)EDTA^{2-}. Thus, besides NO reduction, reduction of EDTA chelated Fe(III) is a core reaction within the regeneration pathway of the BioDeNOx process (Buisman et al. 1999):

$$12Fe(III)EDTA^- + C_2H_5OH + 3H_2O \rightarrow 12Fe(II)EDTA^{2-} + 2CO_2 + 12H^+$$

Figure 11.6 schematically presents the gas and liquor flows, as well as the conversions involved in the BioDeNOx concept.

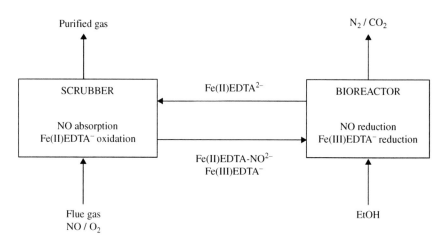

Figure 11.6 Flows and conversions in the BioDENOx concept

11.4.2 Iron dependent nitric oxide reduction

11.4.2.1 Fe(II)EDTA^{2-} as electron donor

Reduction of nitric oxide in aqueous Fe(II)EDTA^{2-} solutions is one of the core bioreactor processes (Figure 11.6). The scrubber liquor Fe(II)EDTA^{2-} is continuously regenerated from the nitrosylcomplex Fe(II)EDTA-NO^{2-} by means of NO reduction. Nitric oxide reduction to N_2 is biologically catalysed with nitrous oxide (N_2O) as an intermediate. Various sludge samples from full scale denitrifying and anaerobic reactors are capable to catalyse NO reduction under thermophilic conditions. Since NO reduction rate was not affected by the presence of any external electron donor (e.g. ethanol or acetate), it is very likely that Fe(II)EDTA^{2-} (i.e. the scrubber liquor itself) acts as electron donor for NO reduction (Van der Maas et al. 2003):

$$2Fe(II)EDTA\text{-}NO^{2-} + 2H^+ \rightarrow N_2O + H_2O + 2Fe(III)EDTA^-$$

$$N_2O + 2Fe(II)EDTA^{2-} + 2H^+ \rightarrow N_2 + H_2O + 2Fe(III)EDTA^-$$

The use of ferrous iron as electron donor for denitrification is described by Straub et al. (1996), who showed that chemolithotrophic bacteria can use Fe(II) as electron donor for the full reduction of nitrate to N_2. At neutral pH, the redox potential of NO reduction ($NO/N_2O = +1180$ mV and $N_2O/N_2 = +1350$ mV) is much more positive than that of iron (Fe(III)EDTA$^-$/Fe(II)EDTA^{2-} = +96 mV). Therefore, Fe(II)EDTA$^-$ is in principle a favourable electron donor for the reduction of both NO and N_2O, even though the oxidation of ethanol to carbon dioxide (CO_2) is energetically more favourable than the oxidation of Fe(II)EDTA^{2-} to Fe(III)EDTA$^-$ (Thaurer et al. 1977; Straub et al. 2001).

In BioDeNOx reactor liquors, the NO reduction rate showed to be independent on the presence of ethanol or acetate (Van der Maas et al. 2003). This strongly suggests that in BioDeNOx reactors, i.e. in aqueous Fe(II)EDTA^{2-} solutions, the chelated Fe(II) is the direct electron donor for NO reduction, and not ethanol. This is, at first sight, somewhat surprising, since ethanol is known to be an excellent electron donor for denitrification (Constantin and Fick 1997). However, Table 11.4 shows that the redox properties of FeEDTA and bacterial Nitric Oxide Reductase (NOR), the denitrification enzyme that catalyses NO reduction to N_2O, allow that Fe(II)EDTA^{2-} can be directly involved in NOR reduction, with the concomitant formation of N_2O. Figure 11.7 schematically depicts the possible roles of Fe(II)EDTA^{2-} as electron donor in NO reduction. Thus, Fe(II)EDTA^{2-} not only enhances the NO transfer from the gas into the liquid phase, i.e. the reason why it is applied in BioDeNOx reactors, but also plays a key role in the biological denitrification process.

Table 11.4 Standard potentials ($E^{0'}$) of various redox couples relevant for NO reduction.

Redox couple	$E^{0'}$ (mV)	Reference
NO/N_2O	+1177	Zumft, 1997
N_2O/N_2	+1352	Zumft, 1997
NOR_{ox}/NOR_{red}	between +280 and +320	Wasser et al. 2002
N_2OR_{ox}/N_2OR_{red}	+260	Coyle et al. 1985
Cyt c_{ox}/Cyt c_{red}	between +148 and +253	Gray et al. 1986
amicyanin	+294	Gray et al. 1986
Fe(III)EDTA$^-$/ Fe(II)EDTA^{2-}	+96	Kolthoff and Auerbach, 1952

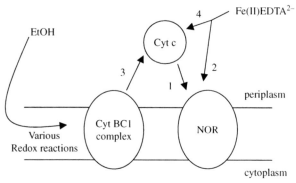

Figure 11.7 Possible electron transfer pathways for NOR reduction (Van der Maas et al. 2008a). 1, NOR reduction by reduced Cytochrome c; 2, Direct NOR reduction by Fe(II)EDTA^{2-}; 3, The conventional electron transfer chain: Cytochrome c reduction by the Cyt bc1 complex; 4, Cytochrome c reduction by Fe(II)EDTA^{2-}

11.4.2.2 Fe(II)EDTA^{2-} - enzyme interactions

The Fe(II)EDTA^{2-}/Fe(III)EDTA$^-$ system is redox reversible and has a midpoint redox potential of +96 mV (Kolthoff and Auerbach 1952). This implies that it easily accepts electrons from a system with a lower midpoint redox potential and that it donates electrons to any redox sensitive system with a higher standard redox potential (Straub et al. 2001). Based on the molecular weight (360 Da), both Fe(II)EDA^{2-} and Fe(III)EDTA$^-$ may penetrate into the periplasmic space, where these compounds can be involved in various biochemical redox reactions (Figure 11.7). Bacterial NOR is a membrane bound enzyme that receives both its electron donor and acceptor (NO) from the periplasmic side (Berks *et al.* 1995; Hendriks *et al.* 2002). The midpoint redox potential of bacterial NO reductases ranges between $+280$ and $+320$ mV (Wasser *et al.* 2002). This means that, from thermodynamic point of view, $Fe(II)EDTA^{2-}$ is a suitable electron donor for bacterial NOR reduction.

$$NOR_{ox} + 2Fe(II)EDTA^{2-} \leftrightarrow NOR_{red} + 2Fe(III)EDTA^{-}$$

Alternatively, $Fe(II)EDTA^{2-}$ is an excellent electron donor for Cytochrome c reduction, reported by Hodges *et al.* (1974):

$$CytC_{ox} + Fe(II)EDTA^{2-} \leftrightarrow CytC_{red} + Fe(III)EDTA^{-}$$

Cytochrome c is a physiological electron donor of cNOR, the most abundant NOR type reported so far (Hendriks *et al.* 2002; Wasser *et al.* 2002). NOR reduction by Cytochrome c, and the associated NO reduction to N_2O by reduced bacterial NOR, proceeds subsequently according to the following reactions:

$$NOR_{ox} + 2Cytc_{red} \leftrightarrow NOR_{red} + 2Cyt_{ox}$$

$$NOR_{red} + 2NO \leftrightarrow NOR_{ox} + N_2O$$

In case of NOR reduction by $Fe(II)EDTA^{2-}$, whether directly or indirectly via Cytochrome c, both the electron acceptor (NO) and donor ($Fe(II)EDTA^{2-}$) are coming from the periplasmic side. Therefore, the proposed electron transfer chain does not generate a proton motive force (PMF). Energy production and cell growth can, however, still be maintained by a PMF generated via a-specific $Fe(III)EDTA^{-}$ reduction (Rosen and Klebanoff 1981) by an enzyme-system that translocates protons. This alternative electron acceptor, $Fe(III)EDTA^{-}$, is produced by NO reduction and by oxidation of $Fe(II)EDTA^{2-}$ with O_2 present in the flue gas. Note that the properties of $Fe(II)EDTA^{2-}$ do not exclude that NO reduction occurs via the more conventional electron supply chain, i.e. via the Cytochrome $bc1$ complex and prior redox couples, following ethanol oxidation (Figure 11.7, pathway 3).

11.4.3 Regeneration capacity of BioDeNOx reactors: iron reduction

In principle, the regeneration capacity of the bioreactor is determined by the volumetric rate of two processes: reduction of NO to N_2 and reduction of

Fe(III)EDTA$^-$ to Fe(II)EDTA^{2-}. However, under normal conditions, NO reduction is not limiting the regeneration capacity of the bioreactor. In contrast, to NO reduction, the reduction of EDTA chelated Fe(III) is often the limiting process for the regeneration of scrubber liquor (Van der Maas et al. 2006). For successful (commercial) application of BioDeNOx, it is, therefore, important to optimise the iron reduction process, as well as to minimise the influence of oxygen in the flue gas on the iron oxidation rate. Minimisation of the iron oxidation rate can possibly be achieved by the application lower FeEDTA concentrations or by replacing EDTA by another ligand that is less sensitive for oxidation by oxygen, e.g. methyliminodiacetic acid (MIDA) or ethylene-diamine-N,N'-diacetic acid (EDDA). However, a strong correlation was shown between the NO absorption properties of a ligand and its sensitivity for oxidation by oxygen (Schneppensieper et al. 2001). Although that research was performed at a lower pH (pH = 5), replacing the ligand might lead to lower NO removal rates.

The principle of biological Fe(III) reduction when bound to strong chelating agents like EDTA is still a matter of debate. Chelating agents like NTA and EDTA might stimulate dissimilatory iron reduction because they keep Fe(III) soluble, and therefore they promote the bio-availability under neutral pH conditions (Lovley et al. 1990). In contrast, others concluded that EDTA chelated Fe(III) was not bio-available for Shewanella putrefaciens (Haas and DiChristina 2002). They presumed that the strong binding properties of EDTA inhibited the bacterial uptake of Fe(III). However, the set-up of that study (Haas and DiChristina 2002) suggests the presence of high concentrations of free EDTA, e.g. not complexed to iron or any other transition metal. This is important to notice, since free EDTA can strongly inhibit bacterial activity (Leive 1968).

Addition of small amounts of sulphide greatly accelerated Fe(III)EDTA$^-$ reduction in BioDeNOx mixed reactor liquors (Van der Maas et al. 2008). This indicates that biological Fe(III)EDTA$^-$ reduction in BioDeNOx reactor environments (i.e. in presence of sulphur and ethanol) is not a direct, enzymatic conversion, but an indirect reduction with the involvement of an electron mediating compound, presumably poly-sulphides, facilitating an electron shuttle between the microbes and Fe(III)EDTA$^-$ (Figure 11.8).

Further research is required to demonstrate that higher volumetric iron reduction rates can be obtained when reduced sulphur compounds are supplied to enhance the electron transfer to Fe(III)EDTA$^-$. Alternatively, ethanol might be replaced by carbohydrates like glucose or molasses, since these substrates showed higher volumetric Fe(III)EDTA$^-$ reduction rates (Van der Maas et al. 2005).

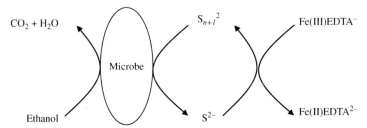

Figure 11.8 Proposed electron flow of biological driven Fe(III)EDTA⁻ reduction with ethanol as electron donor in presence of small amounts of sulphide (Van der Maas *et al.* 2008b)

REFERENCES

Apel W.A. and Turick C.E. (1993). The use of denitrifying bacteria for the removal of nitrogen-oxides from combustion gases. *Fuel* **72**, 1715–1718.

Apel W.A., Barnes J.M. and Barrett K.B. (1995). Biofiltration of nitric oxide from fuel combustion gas streams. Proceedings, 88th Annual Meeting, Air and Waste Management Association, San Antonio, TX, June 18–23, 1995.

Barnes J.M., Apel W.A. and Barrett K.B. (1995) Removal of nitrogen oxides from gas streams using biofiltration. *J. Hazard. Mater.* **41**, 315–326.

Berks B.C., Ferguson S.J., Moir J.W.B. and Richardson D.J. (1995). Enzymes and associated electron transport systems that catalyse the respiratory reduction of nitrogen en nitrogen oxides and oxyanions. *Biochim. Biophys. Acta*, **1232**, 97–173 and literature cited.

Bradford M., Grover R. and Paul P. (2002). Controlling NO$_x$ emissions: part 2. *Chem. Eng. Prog.*, **98**, 38–42.

Buisman C.J.N., Dijkman H., Verbraak P.L. and Den Hartog A.J. (1999). Process for purifying flue gas containing nitrogen oxides. United States Patent US5891408.

Cesario M.T., Beeftink H.H. and Tramper J. Biological treatment of waste gases containing poorly soluble pollutants. In *Proceedings of the International Symposium on Nitrogen Oxides*, Maastricht, The Netherlands. Elsevier, 1991.

Chagnot E., Taha S., Martin G. and Vicard J.F. (1998). Treatment of nitrogen oxides on a percolating biofilter after pre-concentration on activated carbon. *Process Biochem.* **33**, 617–624.

Chou M.S. and Lin J.H. (2000). Biotrickling filtration of nitric oxide. *J. AWMA.* **50**, 502–508.

Cole J. (1993). Controlling environmental nitrogen through microbial metabolism. *Trends Biotechnol.* **11**, 368–372.

Constantin H. and Fick M. (1997). Influence of C-sources on the denitrification rate of a high-nitrate concentrated industrial wastewater. *Water Res.* **31**, 583–589.

Coyle C.L., Zumft W.G., Kroneck P.M.H., Körner H. and Jakob W. (1985). Nitrous oxide reductase from denitrifying Pseudomonas perfectomarina: purification and properties of a novel multicopper enzyme. *Eur. J. Biochem.* **153**, 459–467.

Dasu B.N., Deshmane V., Shanmugasundram R., Lee C.M. and Sublette K.L. (1993). Microbial reduction of sulfur dioxide and nitric oxide. *Fuel* **72**, 1705–1714.

Davidova Y.B., Schroeder E.D. and Chang D.P.Y. (1997). Biofiltration of nitric oxide. Proceedings, 90th Annual Meeting, Air and Waste Management Association, Pittsburgh, PA, June 18–22, 1997.

Demmink J.F. (2000). Removal of Hydrogen Sulfide and Nitric Oxide with Iron Chelates. Ph.D. thesis, University of Groningen, The Netherlands.

Du Plessis C.A., Kinney K.A., Schroeder E.D., Chang D.P.Y. and Scow K.M. (1998). Denitrification and nitric oxide reduction in an aerobic, toluene-treating biofilter. *Biotechnol. Bioeng.* **58**, 408–415.

European Environment Agency (2003). Report AP2 – EEA31 NO_x emissions (http://themes.eea.eu.int/Environmental_issues/air_quality/indicators/AP2%2C2003/ap2_emiss_NOx_FnlDrft_2003.pdf)

Fawler D. (1998). Fertilising the atmosphere with fixed nitrogen, the roles of fossil combustion and agriculture. Institute of Terrestrial Ecology, Edinburgh Research Station. The 22nd Macaulay Lecture, May 14th 1998.

Fritz A. and Pitchon V. (1997). The current state of research on automotive lean NOx catalysis. *Appl. Cata.l, B Environ.* **13**, 1–25.

Gray K.A., Knaff D.B., Husain M. and Davidson V.L. (1986). Measurement of the oxidation-reduction potentials of amicyanin and c-type cytochromes from *Paracoccus denitrificans. FEBS Lett.* **207**, 239–242.

Guicherit R. and Van den Hout D. (1982). The global NO_x cycle. Proceedings of the US-Dutch International Symposium on Nitrogen Oxides, Maastricht, The Netherlands. Elsevier.

Haas J.R. and Dichristina T.J. (2002). Effects of Fe(III) chemical speciation on dissimilatory Fe(III) reduction by *Shewanella putrefaciens. Environ. Sci. Technol.* **36**, 373–380.

Hales, J.M. (1982). The role of NOx as a precursor of acidic deposition. Proceedings of the US-Dutch International Symposium on Nitrogen Oxides, Maastricht, The Netherlands. Elsevier.

Hendriks J.H., Jasaitis A., Saraste M. and Verkhovsky M.I. (2002). Proton and electron pathways in the bacterial nitric oxide reductase. *Biochemistry* **41**, 2331–2340.

Hodges H.L., Holwerda R.A. and Gray H.B. (1974). Kinetic studies of the Reduction of Ferricytochrome c by $Fe(EDTA)^{2-}$. *J. Am. Chem. Soc.* **96**, 3132–3137.

Hudepohl N.J., Schroeder E.D. and Chang D.P.Y. (1998). Oxidation of nitric oxide in a biofilter. In *Proceedings, 91st Annual Meeting, Air and Waste Management Association*, San Diego, CA, June 14–18, 1998.

Jager D. (2001). Emission reduction of non-CO_2 greenhouse gases. Dutch National Research Programme on Global Air Pollution and Climate Change: second phase. Theme III, Societal causes and solutions; 410 200 094.

Janssen F. (1999). Emission control from stationary sources. In: Jansen F.J.J., van Santen R.A., editors. Environmental catalysis, p 234–293. Imperial College Press.

Jin Y., Veiga M.C. and Kennes C. (2005). Bioprocesses for the removal of nitrogen oxides from polluted air. *J. Chem. Technol. Biotechnol.* **80**, 483–494.

Kim N.J., Hirai M. and Shoda M. (2000). Comparison of organic and inorganic packing materials in the removal of ammonia gas in bioflters. *J. Hazard. Mater.* **72**, 77–90.

Kirchner K., Schlachter U. and Rehm H.J. (1989). Biological purification of exhaust air using fixed bacterial monocultures. *Appl. Microbiol. Biotechnol.* **31**, 629–632.

Klasson K.T. and Davison B.H. (2001). Effect of temperature on biofiltration of nitric oxide. *Appl. Biochem. Biotechnol.* **91**, 205–211.

Kleifges K.H., Juzeliunas E. and Jüttner K. (1997). Electrochemical study of direct and indirect NO reduction with complexing agents and redox mediator. *Electrochim. Acta* **42**, 2947–2953.

Kolthoff I.M. and Auerbach C. (1952). Studies on the system iron-ethylenediamine tetraacetate. *J. Am. Chem. Soc.* **74**, 1452–1456.

Kremer H. (1982). Chemical and physical aspects of NOx formation. Proceedings of the US-Dutch International Symposium on Nitrogen Oxides, Maastricht, The Netherlands. Elsevier.

Lacey J.A., Lee B.D. and Apel W.A. (2002). PCR-DGGE as a method to monitor microbial populations in nitric oxide biofilters. In *Proceedings, 95th Annual Meeting, Air and Waste Management Association*, Baltimore, MD, June 23–27, 2002.

Lee B.D., Apel W.A. and Smith W.A. (1999). Effect of oxygen on thermophillic denitrifying populations in biofilters treating nitric oxide containing off-gas streams. Proceedings, 92nd Annual Meeting, Air and Waste Management Association, St. Louis, Missouri, June 20–24, 1999.

Lee B.D., Apel W.A. and Smith W.A. (2001). Oxygen effects on thermophilic microbial populations in biofilters treating nitric oxide containing off-gas streams. *Environ. Prog.* **20**, 157–166.

Lee K.H. and Sublette K.L. (1990). Reduction of nitric oxide to elemental nitrogen by *Thiobacillus denitrificans*. *Appl. Biochem. Biotechnol.* **24**, 441–445.

Lee K.H. and Sublette K.L. (1991). Simultaneous combined microbial removal of sulfur dioxide and nitric oxide from a gas stream. *Appl. Biochem. Biotechnol.* **28**, 623–634.

Lukow T. and Diekmann H. (1997). Aerobic denitrification by a newly isolated heterotrophic bacterium strain TL1. *Biotechnol. Lett.*, **19**, 1157–1159.

Min K.N., Ergas S.J. and Harrison J.M. (2002). Hollow-fiber membrane bioreactor for nitric oxide removal. *Environ. Eng. Sci.* **19**, 575–583.

Nagase H., Yoshihara K., Eguchi K., Yokota Y., Matsui R., Hirata K. and Miyamoto K. (1997). Characteristics of biological NOx removal from flue gas in a *Dunaliella tertiolecta* culture system. *J. Ferment. Bioeng.* **83**, 461–465.

Nagase H., Eguchi K., Yoshihara K., Hirata K. and Miyamoto K. (1998). Improvement of microalgal NOx removal in bubble column and airlift reactors. *J. Ferment. Bioeng.* **86**, 421–423.

Nagase H., Yoshihara K., Eguchi K., Okamoto Y., Murasaki S., Yamashita R., Hirata K. and Miyamoto K. (2001). Uptake pathway and continuous removal of nitric oxide from flue gas using microalgae. *Biochem. Eng. J.* **7**, 241–246.

Nascimento D.M., Davidova Y., Du Plessis C.A., Schroeder E.D. and Chang D.P.Y. (1999). Biofilter technology for NOx control. Final Report, Project 96–304, California Air Resources Board Research Division, Sacramento, CA.

Nascimento D.M., Hudepohl N.J., Schroeder E.D. and Chang D.P.Y. (2000). Bio-oxidation of nitric oxide in a nitrifying aerobic filter. In *Proceedings, 93rd Annual Meeting, Air and Waste Management Association*, Salt Lake City, Utah, June 18–22, 2000.

Princiotta F.T. (1982). Stationary source NOx control technology overview. Proceedings of the US-Dutch International Symposium on Nitrogen Oxides, Maastricht, The Netherlands. Elsevier.

Posthumus A.C. (1982). Ecological effects associated with NOx, especially on plants and vegetation. Proceedings of the US-Dutch International Symposium on Nitrogen Oxides, Maastricht, The Netherlands. Elsevier.

Ramachandran B., Hermann R.G., Choi S., Stenger H.G., Lyman C.E. and Sale J.W. (2000). Testing zeolite SCR catalyst under protocol conditions for NOx abatement from stationary emission sources. *Catal. Today.* **55**, 281–290.

Rasmussen R.A. and Khalil M.A.K. (1986). Atmospheric trace gases: trends and distributions over the last decade. *Science* **232**, 1623–1624.

Rosen H. and Klebanoff S.J. (1981). Role of iron and EDTA in the bactericidal activity of a superoxide anion-generating system. *Arch. Biochem. Biophys.* **208**, 512–519.

Schnelle K.B. and Brown C.A. (2002). Air pollution control technology handbook. CRC Press.

Schneppensieper T., Finkler S., Czap A., Van Eldik R., Heus M., Nieuwenhuizen P., Wreesman C. and Abma W. (2001). Tuning the reversible of NO to iron(II) amino-carboxylate and related complexes in aqueous solution. *Eur. J. Inorg. Chem.* 491–501.

Shanmugasundram R., Lee C.M. and Sublette K.L. (1993). Reduction of nitric oxide by denitrifying bacteria. *Appl. Biochem. Biotechnol.* 39–40, 727–737.

Stepanov A.L. and Korpela T.K. (1997). Microbial basis for the biotechnological removal of nitrogen oxides from flue gases. *Biotechnol. Appl. Biochem.* **25**, 97–104.

Straub K.L., Benz M., Schink B. and Widdel F. (1996). Anaerobic, nitrate depending microbial oxidation of ferrous iron. *Appl. Environ. Microbiol.* **62**, 1458–1460.

Straub K.L., Benz M. and Schink B. (2001). Iron metabolism in anoxic environments at near neutral pH. *FEMS Microbiol. Ecol.* **34**, 181–186 and literature cited.

Thaurer R.K., Jungermann K. and Decker K. (1977). Energy conservation in chemotrophic anaerobic bacteria. *Bacteriol. Rev.* **41**, 100–180.

Van der Maas P.M.F., Van de Sandt T., Klapwijk A. and Lens P.N.L. (2003). Reduction of nitrogen oxides in aquous Fe-EDTA solutions. *Biotechnol. Progress* **19**, 1323–1328.

Van der Maas P.M.F., Peng S., Klapwijk A. and Lens P.N.L. (2005). Enzymatic versus nonenzymatic conversions during the reduction of EDTA-Chelated Fe(III) in BioDeNOx reactors. *Environ. Sci. Technol.* **39**, 2616–2623.

Van der Maas P.M.F., Van den Brink P., Utomo S., Klapwijk A. and Lens P.N.L. (2006). NO removal in continuous BioDeNOx reactors: Fe(II)EDTA^{2-} regeneration, biomass growth, and EDTA degradation. *Biotechnol.Bioeng.* **94**, 575-584.

Van der Maas P.M.F., Manconi I., Klapwijk A. and Lens P.N.L. (2008a). Nitric [My paper]oxide reduction in BioDeNOx reactors: kinetics and mechanism. *Biotechnol. Bioeng.* **100**, 1099–1107.

Van der Maas P.M.F., Van den Brink P., Klapwijk A. and Lens P.N.L. (2008b). Acceleration of the Fe(III)EDTA$^-$ reduction rate in BioDeNO$_x$ reactors by dosing electron mediating compounds. *Chemosphere* doi:10.1016/j.chemosphere. 2008.04.043.

Van Groenestijn J.W. and Hesselink, P.G.M. (1994). Biotechniques for Air Pollution Control. *Biodegradation* **4**, 282–301.

Wasser I.M., De Vries S., Moënne-Loccos P., Schröder I. and Karlin D. (2002). Nitric oxide in biological denitrification: Fe/Cu metalloenzyme and metal complex NO$_x$ redox chemistry. *Chem. Res.* **102**, 1201–1234 and litarature cited.

Woertz J.R., Kinney K.A. and Szaniszlo P.J. (2001). A fungal vapour-phase bioreactor for the removal of nitric oxide from waste gas streams. *J. AWMA* **51**, 895–902.

Wubs H.J. and Beenackers A.A.C.M. (1993). Kinetics of the oxidation of ferrous chelates of EDTA and HEDTA in aqueous solution. *Ind. Eng. Chem. Res.* **32**, 2580–2594.

Yani M., Hirai M. and Shoda M. (2000). Enhancement of ammonia removal in peat biofilter seeded with enriched nitrifying bacteria. *Environ. Technol.* **21**, 1199–1204.

Yoshihara K.I., Nagase H., Eguchi K., Hirata K. and Miyamoto K. (1996). Biological elimination of nitric oxide and carbon dioxide from flue gas by marine microalga NOA-113 cultivated in a long tubular photobioreactor. *J. Ferment. Bioeng.* **82**, 351–354.

Zang V. and Van Eldik R. (1990). Kinetics and mechanism of the autoxidation of iron (II) induced through chelation by ethylenediaminetetraacetate and related ligands. *Inorganic Chemistry* **29**, 1705–1711.

Zumft W.G. (1997). Cell biology and molecular basis of denitrification. *Microbiol. Molec. Biol. Rev.* **61**, 533–61.

12

Autotrophic denitrification for the removal of nitrogenous and sulphurous contaminants from wastewaters

M. Tandukar, S.G. Pavlostathis and F.J. Cervantes

12.1 INTRODUCTION

12.1.1 Basic concept of autotrophic denitrification

The classic paradigm of biological denitrification defines the process as a microbial reduction of nitrate or nitrite under anoxic conditions to nitrogen gas in the presence of an organic electron donor and carbon source. However, in nature there are a number of microorganisms that can carry out chemolithotrophic denitrification utilising various reduced inorganic compounds or hydrogen as electron donor, while they utilise an inorganic carbon source, such as CO_2 or HCO_3^-, for growth. Until now, autotrophic denitrifies are known to use three main inorganic electron donors: reduced sulphurous compounds, reduced iron and hydrogen. However, there are sporadic reports on concomitant manganese(II) or U(IV) oxidation and nitrate reduction.

Bacteria like *Thiobacillus denitrificans, Thiomicrospira denitrificans, Thiobacillus versutus, Thiosphaera pantatropha,* and *Thiobacillus thioparus* (Gayle *et al.* 1989; Reijerse *et al.* 2007) can utilise reduced sulphurous compounds, like S^{2-}, S^0, $S_2O_3^{2-}$, and SO_3^{2-}, as electron donor to reduce nitrate to nitrogen gas. Whereas, bacteria like, *Paracoccus denitrificans, Hydrogenophaga sp., Alcaligens eutrophus, Micrococcus denitrificans,* etc. can reduce nitrate by utilising H_2 as electron donor or reduced metals like Fe^0 as an energy source, while assimilating inorganic carbon for growth (Kurt *et al.* 1987; Gayle *et al.* 1989; Biswas and Bose 2005). *Paracoccus denitrificans* can also oxidise sulphurous compounds (Meyer *et al.* 2007). The denitrifying activities of these bacteria play a significant role in marine as well as global nitrogen cycle. Researchers like Ingrid and Rheinheimer (1991) and Kuenen and Bos (1987) have already demonstrated a significant existence of autotrophic denitrification in marine environments. They reported that autotrophic denitrification driven by sulphide oxidation is a major pathway for nitrogen loss from Central Baltic Sea, especially at the oxic-anoxic interface. Due to its environmental significance, autotrophic denitrification has a wide application in water and wastewater treatment as well as bioremediation including groundwater remediation. The advantages of autotrophic denitrification are as follows:

(a) *No need of organic carbon*: Autotrophic denitrifiers utilize CO_2 as carbon source for growth, which is beneficial for practical applications. Addition of organic carbon for heterotrophic denitrification is one of the factors that raise the cost and complicates the process.

(b) *Low biomass production*: Presence of a singular electron donor will limit the type and population of microorganisms that grow and carry out denitrification. One of the causes of high biomass production in heterotrophic denitrification is the abundance of organic carbon, which is also utilised by heterotrophs other than the denitrifiers resulting in massive growth.

(c) *Simultaneous sulphur oxidation*: Use of reduced sulphurous compounds for denitrification or sulphur-based denitrification is advantageous as the presence and removal of sulphurous compounds from wastewater is an enormous challenge due to its pungent and corrosive properties. Combining sulphurous and nitrogenous wastewaters for autotrophic denitrification can be an attractive option. In many cases, this process is beneficial not just for odour suppression, but also for leaching of heavy metals, which had been previously entrapped in the solids in the form of sulphide precipitates.

(d) *Suitable for nitrate removal from drinking or groundwater:* Drinking and groundwater generally contain relatively low concentrations of nitrate compared to municipal or industrial wastewaters. Nitrate removal from drinking water or contaminated groundwater by heterotrophic denitrification is generally not feasible, because of the need of addition of an external carbon source. This could cause a massive growth of biomass and could also result in residual organic carbon in the treated water, which is very undesirable. In this case, autotrophic denitrification is a better option.

12.1.2 Sulphur based denitrification

As stated above, various reduced forms of sulphur compounds can donate electrons for the reduction of nitrate while producing sulphate. Stoichiometric half equations for various sulphur-based denitrification are given below (Beristain-Cardoso et al. 2006).

Sulphide as electron donor,
Complete oxidation of sulphide to sulphate:

$$S^{2-} + 1.6NO_3^- + 1.6H^+ \rightarrow SO_4^{2-} + 0.8N_2 + 0.8H_2O \qquad (12.1)$$
$$\Delta G^{\circ\prime} = -743.9 \text{ (kJ/reaction)}$$

Partial oxidation of sulphide to elemental sulphur:

$$S^{2-} + 0.4NO_3^- + 2.4H^+ \rightarrow S^0 + 0.2N_2 + 1.2H_2O \qquad (12.2)$$
$$\Delta G^{\circ\prime} = -191.0 \text{ (kJ/reaction)}$$

Partial denitrification of nitrate to nitrite:

$$S^{2-} + NO_3^- + 2H^+ \rightarrow S^0 + NO_2^- + H_2O \qquad (12.3)$$
$$\Delta G^{\circ\prime} = -130.4 \text{ (kJ/reaction)}$$

$$S^{2-} + 4NO_3^- \rightarrow SO_4^{2-} + 4NO_2^- \qquad (12.4)$$
$$\Delta G^{\circ\prime} = -501.4 \text{ (kJ/reaction)}$$

Elemental sulphur as electron donor:

$$S^0 + 1.2NO_3 + 0.4H_2O \rightarrow 0.6N_2 + SO_4^{2-} + 0.8H^+ \qquad (12.5)$$
$$\Delta G^{\circ\prime} = -547.6 \text{ (kJ/reaction)}$$

Thiosulphate as electron donor:

$$S_2O_3^{2-} + 1.6NO_3^- + 0.2H_2O \rightarrow 0.8N_2 + 2SO_4^{2-} + 0.4H^+ \qquad (12.6)$$
$$\Delta G^{\circ\prime} = -765.7 \text{ (kJ/reaction)}$$

These reactions take place in anoxic environments since the presence of oxygen changes the electron transfer pathway away from nitrate when the reaction is mediated by facultative autotrophs.

Unlike heterotrophic denitrification, sulphur-based denitrification consumes alkalinity, resulting in a low pH, which might necessitate a well buffered system. To overcome this problem, some researchers have proposed a combination of elemental sulphur and limestone granules as a packing material for the reactor, which prevents the pH from dropping below 6.5 (Flere and Zhang 1999; Zhang and Lampe 1999). This process is commonly called sulphur-limestone autotrophic denitrification (SLAD). Nevertheless, there are certain shortcomings of this process, which will be discussed later.

Sulphur oxidising denitrifiers can accept different forms of sulphur. However, the pathways and enzymes involved could be different, as discussed in detail in the following section. Sulphide is the most reduced form of sulphur (oxidation state, -2) and has therefore the highest oxygen demand (2 mol O_2/ mol of S^{2-}) and its oxidation to sulphate yields 8 electrons, whereas, the oxidation of elemental sulphur to sulphate contributes only 6 electrons to the respiratory chain.

In a common wastewater treatment system, autotrophic denitrifiers have to compete with heterotrophic denitrifiers. Thermodynamically, the change in free energy under standard conditions (ΔG°) for the sulphide-nitrate couple is -743.9 kJ/mol H_2S compared to -647.9 kJ/mol of substrate when methanol is used as energy and carbon source (Beristain-Cardoso et al. 2006). It suggests higher spontaneity of the autotrophic reaction compared to heterotrophic denitrification. However, in practice, this is reversed, possibly by sulphide limitation or inhibition of the growth of the autotrophs, abundance of organic carbon for heterotrophic denitrifiers, etc. Again, the heterotrophic growth is kinetically favoured by higher cell mass incorporation selectivity (cell yield coefficients) and shorter doubling times (Rittmann and McCarty 2001).

In practice, delivery of sulphur to a denitrifying system is also an issue, if the wastewater lacks any form of reduced sulphur. The most common source of reduced sulphur for autotrophic denitrification is elemental sulphur. Granular elemental sulphur is more stable, insoluble, cheap and easy to handle. However, solubility and the available surface area are the controlling factors. In the SLAD process, sulphur is embedded into a solid matrix of limestone ($CaCO_3$) to neutralize the acid produced during the process (Zhang and Lampe 1999). Because the autotrophic denitrifiers can also initiate the sulphur oxidation pathway from sulphide, sulphite or thiosulphate, these sulphur compounds can also be added in liquid forms. However, it is always economical to combine sulphur and nitrate rich wastewater whenever possible.

12.1.3 Applications and limitations

Recent developments on autotrophic denitrification have demonstrated that it is a suitable treatment option not only for drinking and groundwater or post-denitrification of wastewater that contain small concentrations of nitrate or nitrite, but also for industrial wastewaters and landfill leachates that have relatively high concentrations of these contaminants. Higher rates of sulphur based autotrophic denitrification, even in the presence of organic carbon or some oxygen, have been demonstrated by many researchers (Smith *et al.* 1994; Oh *et al.* 2000). These results are encouraging as simplified maintenance and operation can be achieved while simultaneously treating sulphide and nitrate in the wastewater. Wastewaters from petrochemical refineries or latex industries contain relatively high concentrations of sulphide as well as nitrogen in different forms. However, a considerable amount of nitrogen also exists as ammonium nitrogen, which should first be nitrified. Nitrification in the presence of sulphide may be problematic as it has been shown in many cases that hydrogen sulphide inhibits nitrification activity (Joye and Hollibaugh 1995). However, autotrophic denitrification can be more relevant in the case of combined treatment processes, for example, co-treatment of sulphur-rich and nitrate-rich wastewater. Other-wise, the addition of reduced sulphur compounds or diffusion of hydrogen into the nitrate rich wastewater could increase the treatment costs.

Autotrophic denitrification is also an attractive option for reclamation of groundwater or the treatment of drinking water (Soares 2002). According to a survey conducted by Thurman and Roberts (1995) and Barcelona (1984), most of the groundwater in the US has dissolved organic carbon (DOC) as low as 0.7 mg/L, whereas the total organic carbon was 2.95 mg/L, which is not enough to fulfil the demand for nitrate reduction at or above 10 mg N/L, if it were to be carried out by heterotrophic denitrification (Kraft and Stites 2003).

External addition of organic carbon, like methanol, ethanol or acetate, to the groundwater will be expensive and less feasible due to the formation of unnecessary by-products and biomass growth.

There are some limitations of autotrophic denitrification. In sulphur based denitrification, protons and sulphate are produced, which form strong sulphuric acid and lower the pH. This marked reduction in the pH is one of the main concerns in sulphur based denitrification (Rittmann and McCarty 2001). The SLAD process can help to overcome this problem. But again the disadvantage of the SLAD process is that the hardness and solid content of the effluent increases. Also, in the case of a high-rate treatment, the use of limestone might not be sufficient to maintain the alkalinity as the rate of dissolution of $CaCO_3$ may be limiting. On the other hand, a high concentration of sulphide also inhibits the nitrate reduction process. Dalsgaard and Bak (1994) reported significant inhibition of nitrate reduction at sulphide concentrations above 46 μM. A sulphide concentration above 127 μM resulted in more that 70% inhibition of the process and claimed that a high sulphide concentration also retards the microbial growth. In addition, sulphur compounds in water can also lead to the proliferation of filamentous bacteria and sludge bulking.

12.2 INDUSTRIAL SOURCES OF CONTAMINATION BY NITROGENOUS AND SULPHUROUS COMPOUNDS

12.2.1 Industrial wastewaters containing nitrogenous and sulphurous compounds

Many industries discharge nitrogen and sulphur-rich wastewaters. Wastewaters from industries like textile, tannery, food processing, pulp and paper, rubber-latex, explosive manufacturing, pesticides, etc., contain very high concentrations of both nitrogenous and sulphurous compounds (Lens *et al.* 1998; Tait *et al.* 2009; Altas and Buyukgungor 2008).

Nitrogen in wastewater can exist in various forms, such as organic nitrogen, ammonia, nitrite, or nitrate. For example, a large number of nitrogen-fertiliser industries produce mainly ammonia and urea. In these plants, urea solution is concentrated in vacuum concentrators. During this process, vapours condense to form contaminated condensate, which contains around 350 mg N/L of urea and 650 mg N/L of ammonium-nitrogen (Gupta and Sharma 1996). Nitrogen oxides in the wastewater can be a direct consequence of various industrial processes or anaerobic treatment of nitrogenous wastewater (Gupta and Sharma 1996;

Ahn and Kim 2004; Wang *et al.* 2005b). The composition of nitrogen-rich wastewaters is given in Table 12.1.

On the other hand, sulphur compounds in wastewaters can be in reduced or oxidised forms. Many industries like paper mill, potato starch factories, etc., use a significant amount of sulphuric acid or sulphite as a bleaching agent because they are cheap and effective. These industries produce wastewater containing a large amount of sulphate (Silva *et al.* 2002). Likewise, in petrochemical refineries, presence of sulphur in the product is very important. Actually, crude oil is classified based on the sulphur content. Sweet crude oil is the one with low sulphur content and sour crude oil is the one with high sulphur content.

Wastewaters from petrochemical refineries, galvanising industries, mining industries, gas scrubbing at power plants, etc. contain high concentrations of reduced sulphurous compounds. Likewise, wastewaters from industries, such as paper mill, textile, pesticides, fertiliser manufacturing, rubber manufacturing, etc. contain high concentrations of sulphate. Anaerobic treatment of such wastewaters generate high concentrations of reduced sulphur compounds, like sulphide or elemental sulphur by the bacterial reduction of sulphate as shown in the reaction below (Silva *et al.* 2002; Kanyarat and Chaiprapat 2008).

$$SO_4^{2-} + organic\ matter \xrightarrow{SRB} HCO_3^-/CO_2 + H_2S \qquad (12.7)$$

(SRB: Sulphate reducing bacteria)

Table 12.2 presents a list of industrial wastewaters containing sulphur in various forms.

12.2.2 Feasibility of co-treating wastewater containing nitrate and sulphur

As discussed in the previous section, some of the industrial wastewaters contain both sulphur and nitrogen in high concentrations, which makes autotrophic denitrification a good option for simultaneous removal of both contaminants. But sulphur and nitrogen species in the wastewater are not always reduced sulphur and nitrogen oxides, which is the ideal combination for autotrophic denitrification. In practice, wastewaters usually contain a mixture of reduced and oxidised forms of sulphur and nitrogen, which makes it difficult to design the denitrification process. For example, fish processing wastewater contains high concentrations of nitrogen (TN, 200–1,000 mg N/L) and sulphur (SO_4^{2-}, 300–1,000 mg S/L) (Kleerebezem and Mendez 2002). However, most of the nitrogen is in the form of NH_4^+-N , which needs nitrification (aerobic process)

Table 12.1. Industrial wastewaters containing high-nitrogen concentrations

Wastewaters	NH$_4^+$-N (mg/L)	NO$_2^-$-N (mg/L)	NO$_3^-$-N (mg/L)	pH	COD (mg/L)	Reference
Fertiliser condensate	38,000–45,000	–	–	–	3,520–4,850	Gupta and Sharma (1996)
Piggery (mixed)	1850–1960	<1	<1	5.8–6.1	13,800–14,650	Joo et al. (2006)
Piggery (Solid-free)	830–1250	<1	<1	8.3–8.5	4,150–5,300	Joo et al. (2006)
Piggery (slurry)	2,150	1,800–2,500	–	–	25,700	Ahn and Kim (2004)
Pharmaceutical	30–50	–	–	–	1,100–5,500	Ghafari et al. (2008)
Nuclear weapon plant	–	–	~50,000	–	–	Ghafari et al. (2008)
TNT production	–	–	30,000~50,000	–	–	Glass and Silverstein (1999)
Shrimp processing	84	250	31	–	1,593	Lyles et al. (2008)
Fertiliser	183	1.3	100	8.7	–	Mijatovic et al. (2000)
Poultry manure	990 (1,830 TKN)	–	–	7.3	12,100	Yetilmezsoy and Sakar (2008)
Food processing	300–450	–	–	7–8	1,500–2,000	Wang et al. (2005a)
Landfill leachate	1,100–2,20	–	–	3.8–5.5	~20,000 (TOC)	Tait et al. (2009)
Poultry processing	186 (TN)	–	–	6.5–7.0	3,102	Nery et al. (2007)

TN, Total nitrogen; TKN, Total Kjeldahl nitrogen; TOC, Total organic carbon; TNT, trinitrotoluene

Table 12.2. Industrial wastewaters containing high-sulphur concentrations

Wastewaters	Sulphur (mg-S/L)	pH	COD (mg/L)	Reference
Latex	100–4000 (H_2S)	8.52		Rattanapan et al. (2009)
Crude oil processing	~150 (S^{2-})		~1,500	Sekoulov and Brinke-Seiferth (1999)
Petroleum Refinery	~150 (S^{2-})	8–9	~216	Altas and Buyukgungor (2008)
Rubber latex	1819 (SO_4^{2-})	4.7	5430	Kanyarat and Chaiprapat (2008)
Landfill leachate	1,100–3,800 (SO_4^{2-})	3.8–5.5	15,000–20,000 (TOC)	Tait et al. (2009)
Chemical industry	12,000–35,000 (SO_4^{2-})	2.0–12.0	15,000–40,000	Silva et al. (2002)
Molasses	2,800 (SO_4^{2-})	7.8	20,000	Percheron et al. (1997)
Tannery	750–1250 (SO_4^{2-}); 140–280 (S^{2-})	–	2,990–8,200	Genschow et al. (1996)
Pulp & paper	50–200 (S^{2-}); 1,200–1,500 (SO_4^{2-})	7–9	7,500–10,400	Lens et al. (1998)
Beamhouse	500–1,500 (S^{2-})	11.2	2,800–4,500 (BOD)	Lens et al. (1998)
Photographic	45,500 ($S_2O_3^{2-}$); 8,000 (SO_4^{2-}); 3,500 (SO_3^{2-})	–	67,800	Lens et al. (1998)
TNT manufacturing	51,400 (SO_4^{2-}); 5,500 (SO_3^{2-})	7.6	68,500	Lens et al. (1998)

COD, Chemical oxygen demand; BOD, Biochemical oxygen demand; TOC, Total organic carbon; TNT, trinitrotoluene and most of the sulphur is sulphate, which needs to be reduced (anaerobic process) for autotrophic denitrification. These difficult combinations could jeopardize the feasibility of treating this type of wastewater via autotrophic denitrification. Likewise, for sulphur based denitrification, high alkalinity is also preferred as pH is lowered by the production of a large amount of sulphate. One of the most suitable wastewaters for autotrophic denitrification is that originated from petroleum refineries.

In petroleum refineries, caustic solutions (NaOH) are generally used to remove H_2S and organic sulphur by adsorption. These spent solutions, named spent sulphidic caustic, typically have sulphide concentrations over 2–3% by weight and pH higher than 12 (Byun et al. 2005). Spent sulphidic caustics are classified as hazardous waste by U.S. Resource Conservation and Recovery Act (RCRA, 1976) and are generally incinerated for treatment and disposal, which is actually a waste of resource from the wastewater engineering point of view. Not much attempt has been made to reuse spent sulphidic caustic as a resource. However, high sulphur content and high alkalinity of the spent solution is an ideal condition for autotrophic denitrification, which can substitute commercial sulphur particles and limestone. Again in most of the industrial wastewaters, sulphide is stripped by sparking of air or CO_2. Possibility of utilising the same wastewater, without getting rid of the sulphide, for autotrophic denitrification can save cost and is environmentally more friendly (Vaiopoulou et al. 2005). Another possibility is mixing different wastewaters containing nitrate and reduced sulphurous compounds. However, the mixing should be based on the stoichiometry of sulphur and nitrogen to avoid production of any intermediate sulphurous compounds or partial denitrification (Table 12.3). However, transportation of wastewater could be the limiting factor.

Most wastewaters containing nitrogenous or sulphurous compounds also contain substantial concentrations of organic matter (Tables 12.1–12.2), which makes autotrophic denitrification somewhat difficult. However, several researchers have successfully demonstrated simultaneous heterotrophic and autotrophic denitrification (Kim and Bae 2000; Zhao and Liu 2009).

Table 12.3. Stoichiometric N/S ratios with different sulphurous compounds as electron donors (based on reactions 1, 5 & 6 in Section 12.1.1.1)

Combinations	Ratio (g N:g S)
Nitrate (NO_3^--N) : Sulphide (S^{2-}-S)	1 : 1.43
Nitrate (NO_3^--N) : Elemental Sulphur (S^0-S)	1 : 1.90
Nitrate (NO_3^--N) : Thiosulphate ($S_2O_3^{2-}$-S)	1 : 2.80

The above considerations suggest that there is a need for an advanced engineering design and operation strategy for the successful establishment of autotrophic denitrification for the simultaneous elimination of sulphur and nitrogen from industrial wastewaters.

12.3 MICROBIAL ASPECTS OF AUTOTROPHIC DENITRIFICATION

12.3.1 Denitrifying microorganisms

Denitrifying bacteria belong to a broad variety of groups and encompass a wide range of physiological traits. At least 30 genera of bacteria are known to reduce nitrate. Most of them randomly fall in the *alpha* and *beta* subdivisions of the phylum *Proteobacteria*. However, the microbial diversity of autotrophic denitrifiers is still not fully known and current knowledge is largely based on laboratory-scale denitrification reactors. Some important bacteria associated with sulphur based denitrification are discussed in this section.

12.3.2 Autotrophic denitrifiers

One of the best studied obligate chemolithotrophic species that enormously contributes to the autotrophic denitrification is *Thiobacillus denitrificans*. It possesses an environmentally relevant metabolic repertoire, which includes combined nitrate reduction and sulphur oxidation and nitrate-dependent oxidation of Fe(II) and U(IV). It is a non-filamentous β-proteobacterium, which was firstly isolated by Beijerinck (1904). *Thiobacillus denitrificans* is clearly described in the Systematic Bacteriology of Bergey's Manual (Sneath 1986) as colourless sulphur bacterium, which can be found in any place where reduced sulphur compounds are present. It is a chemolithotrophic microorganism growing below 55°C, gram-negative, rod-shaped, motile with polar flagella, or non-motile. No intracellular sulphur was visible under light microscope, but it may contain finely dispersed sulphur, visible under the electron microscope after staining with silver salts. It deposits sulphur in the colonies after prolonged incubation in the presence of sulphide, making the colonies yellowish.

The genome sequence analysis of *Thiobacillus denitrificans* by Letain *et al.* (2007) has revealed many interesting facts and possibilities about the metabolic activities of this microorganism. Beller *et al.* (2006) found that in its single circular chromosome of 2.9 Mb in length, there are all necessary genes that encode for four essential enzymes that catalyse denitrification: nitrate reductase (N*ar*), nitrite reductase (N*ir*), nitric oxide reductase (*Nor*) and nitrous oxide reductase (*N2or*), some of which occur in multiple copies. Hole *et al.* (1996) purified and characterized cytochrome *cd1* and Cu-containing nitrite reductase and nitrous oxide reductase from *Thiobacillus denitrificans*.

The genome of this bacterium also possesses 50 gene clusters associated with sulphur oxidation, including *sox* genes that encode thiosulphate-oxidising multi-enzyme complex, *dsr* genes that encode sulphite reductase and a gene associated with AMP-dependent oxidation of sulphite to sulphate (Beller *et al.* 2006). It is also capable of producing ribulose 1,5-bisphosphate carboxylase/oxygenase enzymes that fix CO_2, essential for autotrophs (Beller *et al.* 2006). Additionally, the genome of *Thiobacillus denitrificans* was also found to be able to encode for as many as 17 possible metal resistance systems, including those for Ni^{2+}, Pd^{2+}, Hg^{2+}, Cr^{6+}, Cu^{2+} and Ag^{2+}.

Because of the ability to synthesize various environmentally significant enzyme systems, this versatile species was found to have broader oxidative capabilities, such as nitrate-dependent oxidation of certain metals like iron or uranium. In fact, *Thiobacillus denitrificans* is the first and only chemolitho-trophic bacterium reported to carry out anaerobic nitrate-dependent oxidation of U(IV) oxide minerals as follows (Beller 2005):

$$UO_2(s) + 0.4NO_3^- + 2.4H^+ \rightarrow UO_2^{2+} + 0.2N_2 + 1.2H_2O \qquad (12.8)$$

This process is probably catalysed by *c-type* cytochromes and hydrogenase. Again, Beller *et al.* (2006) reported that genes encoding *c-type* cytochromes constitute 1–2% of the whole genome of this bacterium, which is a greater proportion than any other bacterial or archaeal species sequenced to date. Though the role of the hydrogenase in *Thiobacillus denitrificans* is still unclear, Beller (2005) speculated that it probably plays a role during the oxidation of U(IV) in the presence of nitrate. Hydrogenotrophic activity with only hydrogen as electron donor was not observed in this microorganism. The intriguing mechanism of dissimilatory oxidation of inorganic electron donors by this organism is not yet clearly understood.

Another major bacterial group is *Thiomicrospira denitrificans*. Ingrid and Rheinheimer (2006) reported that *Thiomicrospira denitrificans*-like ε-proteobac-terium is the main single taxon responsible for autotrophic denitrification in the Central Baltic Sea. Sulphur driven autotrophic denitrification by these species has been shown to be a major pathway for nitrogen loss from the Baltic Sea, which was reported to take place especially at the oxic-anoxic chemocline layer. These bacteria oxidise thiosulphate present in the seawater coupled to the reduction of nitrate to nitrogen gas. Hoor (1981) demonstrated that the cell yield of *Thiomicrospira denitrificans* grown in a chemostat was almost half that of

Thiobacillus denitrificans. Carbon fixation rates by *Thiomicrospira denitrificans* enriched from central Baltic Sea by Ingrid and Rheinheimer (2006) were between 49 and 57 μg C L^{-1} d^{-1}. The thiosulphate consumption was 0.72 mol per mol of nitrate reduced.

Thiobacillus denitrificans and *Thiomicrospira denitrificans* are virtually restricted to the autotrophic mode of growth, because they cannot obtain energy from organic compounds. They are thus classified as obligate chemolithotrophs. However, other groups of bacteria like, *Thiobacillus versutus*, *Thiobacillus thyasiris*, *Thiosphaera pantotropha*, and *Paracoccus denitrificans* can grow autotrophically as well as heterotrophically, making them facultative chemo-lithotrophs. These species play a bigger role in mixotrophic denitrification.

Another type of widely studied denitrifier, which is also known to autotrophically utilise hydrogen as well as sulphur, is *Paracoccus denitrificans*. It is a member of α-*proteobacteria*, also known as *Rhodobacter* group. It was first isolated by Beijerinck in 1908 and named *Micrococcus denitrificans*. It is a gram-negative facultative bacterium, which is found in a wide range of environments like, soil, sewage and sludge, wastewater treatment plants, etc., where the concentration of oxygen changes drastically. It has also been used as a model microorganism for aerobic electron transport and energy transduction (Goodhew *et al.* 1996). The genome of this microorganism (*strain Pd1222*) consists of three distinct DNA molecules with apparent molecular sizes of approximately 2.8, 1.7 and 0.64 Mb (Winterstein and Ludwig 1998). The smaller DNA is also called mini-chromosome or mega-plasmid, which harbours important genes for denitrification, hydrogen utilisation or heavy metal resistance. Presence of multiple chromosomes or megaplasmid has also been identified in other common denitrifiers like *Alcaligens eutrophus*, *Rhodobacter capsulatus*, *Rhodobacter sphaeroides* or *Pseudomonas denitrificans* (Diels *et al.* 1989; Fonstein and Haselkorn 1993; Suwanto and Kaplan 1992; Gerstenberg *et al.* 1982). The genome of *Paracoccus denitrificans* is known to possess all the necessary genes that encode for essential denitrifying enzymes, which are discussed in the latter section. It also has genes for electron transporters like *a*, *b* or *c-type* cytochrome and cytochrome oxidase. Besides, the genome also contains genes for hydrogenase, sulphur oxidation enzymes and enzymes for oxygen respiration (cytochrome aa_3). There are also several genes that allow this organism to utilize organic substrates (Winterstein and Ludwig 1998). This versatility allows it to have alternative electron transport chains and meta-bolic flexibility, being facultative, heterotroph or autotroph. The metabolic path-way switches from one to another depending upon the availability of substrate. This bacterium is also able to grow on C1 substrates such as methanol or methylamine (Baker *et al.* 1998).

12.4 BIOCHEMICAL ASPECTS OF AUTOTROPHIC DENITRIFICATION

12.4.1 Genes, enzymes and pathways

In this section, genes, enzymes and biochemical pathways involved in denitrification and sulphur oxidation will be discussed separately and collectively for autotrophic denitrification.

12.4.1.1 Nitrogen oxides reduction

The denitrification pathway is largely similar regardless of the type of microorganism involved. The pathway is replete with metalloenzymes, which are induced sequentially under anoxic conditions. The catalytic subunits of these denitrifying reductases encoded by a number of genes (e.g. *narG, napA, nirS, nirK, norB, and nosZ*) vary with the type of microorganism and are either plasmid or chromosomal (Tavares *et al.* 2006). The overall denitrification reaction can be expressed by a redox reaction as shown below:

$$2NO_3^- + 10e^- + 12H^+ \rightarrow N_2 + 6H_2O \quad \Delta G^{o\prime} = -1120.5 kJ/\text{reaction} \quad (12.9)$$

However, this process is accomplished in four individual enzymatic steps (Table 12.4).

The first step is the reduction of nitrate to nitrite, which is catalysed by nitrate reductase (*Nar*). The majority of nitrate reductases are mononuclear molybdenum-containing enzymes with additional redox cofactors like iron-sulphur and hemes (Moura and Moura 2001). The location of this enzyme in prokaryotes could be cytoplasm, membrane or periplasm depending upon the mode of nitrate reduction (assimilatory, dissimilatory or respiratory). It receives electrons from a quinone pool or *b-type* cytochrome to pass to nitrate forming nitrite and a water molecule. The second step is nitrite reduction, which is carried out by nitrite reductase (*Nir*). There are two different types of nitrite reductase, based on their

Table 12.4. Steps involved in the complete denitrification process, associated free energy and enzymes

Reaction step	$\Delta G^{o\prime}$ (kJ/reaction)	Enzymes
$NO_3^- + 2e^- + 2H^+ \rightarrow NO_2^- + H_2O$	−163.2	Nar
$NO_2^- + e^- + 2H^+ \rightarrow NO + H_2O$	−73.2	Nir
$2NO + 2e^- + 2H^+ \rightarrow N_2O + H_2O$	−306.3	Nor
$N_2O + 2e^- + 2H^+ \rightarrow N_2 + H_2O$	−341.4	N2or

prosthetic groups: cytochrome *cd1* containing and *copper* containing reductase. As stated earlier, the nitrite reductase of *Thiobacillus denitrificans* has both cytochrome *cd1* and copper containing prosthetic groups. The nitrite reductase receives electrons from cytochrome type *c*, pseudoazurin or azurin, to reduce nitrite to nitric oxide (Conrado *et al.* 1999). Nitric oxide is then reduced to nitrous oxide by nitrous oxide reductase (*Nos*), which is also a membrane bound protein and has two electron transfer sites. There are three different classes of nitric oxide reductases classified based on the type of subunits and the electron donor. They either accept electrons from cupredoxin, quinols, menaquinols or from cytochrome c_{551} (Conrado *et al.* 1999). The last step of the complete denitrification is the reduction of nitrous oxide to nitrogen gas (Figure 12.1). Nitrous oxide reductase (*N2or*) is a multi-copper protein and has greater environmental relevance as it is highly sensitive to oxygen. Even a small concentration of oxygen inhibits this enzyme, which breaks the denitrification pathway resulting in the accumulation and release of nitrous oxide, a strong greenhouse gas.

12.4.1.2 Sulphur oxidation

In sulphur based autotrophic denitrification, another half of the metabolic pathway is the oxidation of reduced sulphurous compounds to sulphate. The most common sulphurous compounds used as electron donors are sulphide (S^{2-}),

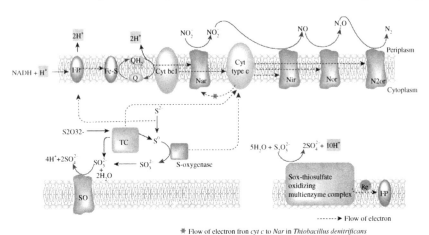

Figure 12.1. Proposed electron transfer pathways involved in simultaneous sulphur oxidation and denitrification in autotrophic denitrifiers. FP, Flavoprotein; Fe-S, Iron-sulphur protein; Q, Ubiquinone; Nar, Nitrate reductase; Nir, Nitrite reductase; Nor, Nitric oxide reductase; N2or, Nitrous oxide reductase; TC, Thiosulphate cleavage; SO, sulphite oxidase

elemental sulphur (S^0), and thiosulphate ($S_2O_3^{2-}$). The final product of this oxidation is sulphate (SO_4^{2-}), if there is sufficient electron acceptor. There are different postulations regarding the pathway, enzymes/enzyme complex and the genes involved in sulphur oxidation. According to the studies carried out by Kelly (1999), Bruser *et al.* (2000) and Friedrich *et al.* (2001; 2005), there are basically three pathways of thiosulphate oxidation:

(1) When the substrate is thiosulphate, it is degraded by thiosulphate dehydrogenase and tetrathionate hydrolase to sulphate. This pathway is common among chemolithotrophic sulphur oxidising bacteria living in extreme habitats.

(2) Thiosulphate is directly oxidised to sulphate by *Sox*-multienzyme complex and there is no formation of any elemental sulphur. The 8 electrons produced are then passed into the respiratory chain. The multi-enzyme complex is generally composed of four enzyme proteins, viz. *SoxXA, SoxYZ, SoxB* and *Sox(CD)*2, *which catalyses the thiosulphate oxidation at different steps. This enzyme system is found in most of the chemolithotrophic* α-proteobacteria *including some strains of* Para-coccus denitrificans. *Meyer* et al. *(2007) indicated that some strains of* Thiobacillus denitrificans *possess all the necessary genes for multi-enzyme complex except for* SoxCD.

(3) Thiosulphate is reduced to elemental sulphur and sulphate by a thio-sulphate cleaving enzyme. However, the free energy change during the formation of elemental sulphur from thiosulphate is approximately +26.8 kJ/mol S, which is still uncertain because of the lack of knowledge of how this step is achieved (Kelly 1999). Elemental sulphur is then oxidised to sulphite by *s*-oxygenase and then the resulting sulphite is finally oxidised to sulphate by sulphite oxidase. This pathway is common in *Thiobacillus denitrificans*. This bacterium is known to possess the ability to oxidise the stored elemental sulphur via the cytoplasmic reverse dissimilatory sulphite reductase, encoded by *DsrAB* gene.

In the case of sulphide, the first oxidation step results in the formation of elemental sulphur, which is either deposited in the cell as energy reserve or further oxidised to sulphite, which eventually converts to sulphate (Figure 12.1). Electrons from sulphide to elemental sulphur oxidation are passed to either *c-type* cytochrome (*cyt 550*) or quinone, which is mediated by the enzyme flavocytochrome *c*-sulphide dehydrogenase or the sulphide-quinone reductase (Friedrich *et al.* 2001).

Table 12.5. Different combinations of sulphur oxidation, associated free energy and enzymes involved

Reaction step	$\Delta G^{\circ\prime}$ (kJ/mol S)	Enzyme
$S_2O_3^{2-} \rightarrow SO_4^{2-}$	-750.1	Sox-MC
$S^{2-} \rightarrow SO_4^{2-}$	-701.8	Fcd/Sqr, S-ox, Sul-Ox
$S^{2-} \rightarrow S^0$	-168.6	Fcd/Sqr
$S^0 \rightarrow SO_4^{2-}$	-537.9	S-ox, Sul-Ox
$S^0 \rightarrow SO_3^{2-}$	-280.2	S-ox
$SO_3^{2-} \rightarrow SO_4^{2-}$	-230.1	Sul-Ox

Sox-MC: Sox-multi-enzyme complex; S-ox, S-oxigenase; Sul-ox, sulphite oxidase; Fcd, flavocytochrome c-sulphide dehydrogenase; Sqr, sulphide-quinone reductase.

$$S^{2-} + cyt\ c_{550(0x)} \rightarrow S^0 + cyt\ c_{550(red)}$$

Flavocytochrome c-sulphide dehydrogenase

$$S^{2-} + UQ_{0x} \rightarrow S^0 + UQ_{red}$$

Sulphide-quinone reductase (Friedrich *et al.* 2001)

Electrons are eventually passed down to nitrogen oxide reductase for denitrification. When elemental sulphur is used, the oxidation pathway is the same as that of intermediate elemental sulphur. Sulphur is oxidized to sulphite and then to sulphate. Different sulphur oxidation reactions and the associated free energy and enzymes are given in Table 12.5.

12.4.2 Autotrophic denitrification

Based on process observation and genetic studies of microorganisms responsible for autotrophic denitrification, the major pathway of nitrate reduction to nitrogen gas is the same as that in heterotrophic denitrification:

$$NO_3^- \rightarrow NO_2^- \rightarrow NO \rightarrow N_2O \rightarrow N_2 \qquad (12.10)$$

The electron transport pathway in autotrophic denitrifiers is also very similar to that of heterotrophic denitrifiers. As stated earlier, autotrophic denitrifiers are capable of synthesising all the necessary electron shuttles that are necessary for the electron transfer in denitrification. In heterotrophic denitrifiers, nitrate reductase is usually known to accept electrons only from *b*-type cytochrome, which are transferred through proton motive *Q*-cycle. This cytochrome, usually *b556*, has mid-point potential of $+120\ mV$ to $+149\ mV$ (Madigan and Martinko 2006). On the other hand, low potential *c*-type cytochromes shuttle electrons to nitrite reductase (*Nir*), nitric oxide reductase (*Nor*) or nitrous oxide reductase (*N2or*). Facultative microbes also possess *a* or *o*-type cytochrome (aa_3) that transport electrons to oxygen. However, c and d-type cytochromes are also capable of this electron transfer, e.g. in *E. coli* (Tavares *et al.* 2006). Figure 12.2 represents the electron transport chain and the corresponding reactions involved in microbial respiration in relation to standard reduction potential ($E^{0\prime}$).

In autotrophic denitrifiers, the electron transport pathway is slightly different depending on the type of substrate. *Thiobacillus denitrificans* has an unusual pathway of electron transfer during autotrophic denitrification. Instead of b-type cytochrome, c-type cytochrome shuttles electrons to nitrate reductase in this microorganism. During sulphur oxidation, electrons from the oxidation of thiosulphate or elemental sulphur are transferred to nitrate reductase via *c-type* cytochrome, whereas, electrons from the oxidation of sulphide are transferred to nitrite reductase via *FP*, *Fe-S*, *Q* and b-type cytochrome (Figure 12.1). Stouthamer (1988) suggested that different c-type cytochromes are involved in these two branches of respiratory chain: *cyt* c_{551} and *cyt cd* in case of sulphide oxidation and *cyt* c_{554} in case of sulphite oxidation.

Carbon dioxide fixation and cell growth in autotrophic denitrifiers take place by the reduction of NAD^+ to NADH, which enters the Calvin cycle along with ATP and corresponds to the reverse electron flow from cytochrome to NAD^+. This is common in the majority of autotrophic microorganisms too.

Many researchers also argued about the sidedness of the reduction of nitrogen oxides. Reduction of nitrite, nitric oxide and nitrous oxides are now known to take place in the periplasmic space of bacterial cell. However, the site of reduction of nitrate depends on the bacterial species. For example, in species like *Paracoccus denitrificans*, *Pseudomonas fluorescens* or *Rhodobacter capsulata*, nitrate reduction takes place in the cytoplasmic side of the membrane (Stouthamer 1988). The transport of nitrate into the cytoplasm through the membrane requires energy if it is driven by proton motive force. It is assumed that nitrate uptake is accompanied by two or more protons, and this mode is also known as PMF-dependent NO_3^-/H^+ symport. Once the nitrite is formed, the exchange between nitrite inside the membrane and nitrate outside the membrane can take place via nitrate-nitrite antiport system, which does not require any energy. There are also reports of passive nitrate uniport and ATP dependent uniport for nitrate uptake into the cytoplasm. Likewise, a membrane bound enzyme, *NarK* is now known as a nitrite exporter, which mediates electrogenic nitrite excretion (Conrado *et al.* 1999). A denitrifying *Staphylococcus carnosus* has been identified to possess the *narT* gene that encodes for a putative nitrite transporter. Additionally, the nitrite transporter gene, *nirC*, has been identified in *E. coli* and other denitrifying bacteria.

In other bacteria, like *Rhodobacter sphaeroides*, nitrate reduction is known to take place in the periplasmic space (Stouthamer 1988).

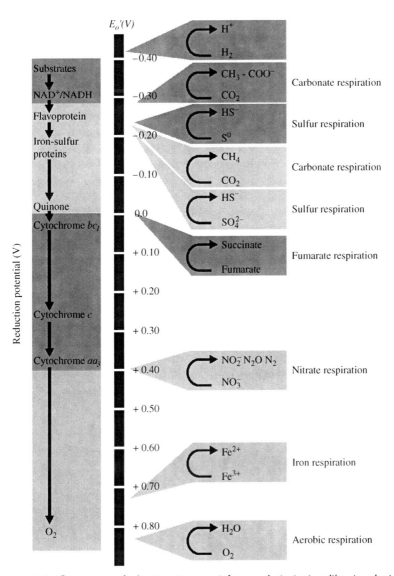

Figure 12.2. Sequence of electron transport from substrate to ultimate electron acceptor and the corresponding redox reaction essential for microbial metabolism in relation to the reduction potential ($E^{0'}$) (Reconstructed from Madigan and Martinko 2006)

On the other hand, regardless of the pathways, sulphur oxidation is suggested to take place in the periplasmic space (Kappler et al. 2001). Enzymes or a multi-enzyme complex associated with sulphur oxidations are membrane bound.

12.4.3 Energetics of autotrophic denitrification

In sulphur based denitrification, as stated earlier, the most common electron donors are sulphide (S^{2-}), elemental sulphur (S^0), and thiosulphate ($S_2O_3^{2-}$). In case of complete oxidation of these compounds to sulphate (SO_4^{2-}), the total number of electron yield is different. When one mole of sulphide (oxidation state, -2) is oxidised to sulphate (oxidation state, $+6$), 8 electrons are generated. However, when it is deposited in the bacterial cells as elemental sulphur (oxidation state, 0), only 2 electrons are generated. Oxidation of elemental sulphur to sulphate generates 6 electrons. Likewise, a mole of thiosulphate oxidised to sulphate (2 mol) generates 8 electrons, which takes place in different steps with the formation of elemental sulphur or sulphate (Figure 12.3). Madigan and Martinko (2006) discussed two ways of sulphite oxidation to sulphate. The most common one is the involvement of sulphite oxidase, which generates 2 electrons, which are transferred directly to c-type cytochrome. ATP is formed during this electron transfer and the formation of proton motive force. Oxidation of sulphite to sulphate via a reversal of the activity of adenosine phosphosulfate (APS) reductase, has also been discussed, which essentially is an enzyme for the metabolism of sulphur-reducing bacteria (SRB). This process yields one high-energy phosphate bond when AMP is converted to ADP (Madigan and Martinko

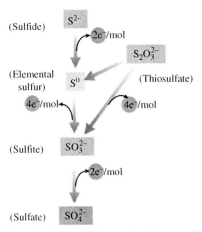

Figure 12.3. Different steps in the oxidation of sulphur and the yield of electrons

2006). On the other hand, in case of respiratory denitrification, the efficiency of oxidative phosphorylation with nitrate and oxygen is about the same (Stouthamer 1988).

12.5 KINETIC STUDIES OF AUTOTROPHIC DENITRIFICATION

12.5.1 General kinetic considerations for autotrophic denitrification

The kinetics of sulphur and hydrogenotrophic denitrification differs slightly depending on the availability of substrates and inhibition of the process by the substrates and by-products of the biochemical reaction.

In general, the rate of nitrate reduction can be defined by the following relationship (Metcalf and Eddy 2001).

$$r = -\left(\frac{kXS_{S/H}}{K_{s,(S/H)}}\right)\left(\frac{S_N}{K_{s,N} + S_N}\right)\left(\frac{K'_o}{K'_o + S_o}\right)\eta \qquad (12.11)$$

where,
r = nitrate reduction rate (mg/L-h),
k = maximum rate of substrate utilization (mg N/mg VSS/h),
X = autotrophic denitrifying biomass (mg/L),
$S_{S/H}$ = concentration of electron donor, (S) (mg/L),
$K_{s(S/H)}$ = half velocity constant for S or H_2 limited reaction (mg/L),
S_N = concentration of nitrate (mg/L),
$K_{s,N}$ = half velocity constant for nitrate limited reaction (mg/L),
K'_o = oxygen inhibition coefficient (mg/L),
S_o = concentration of dissolved oxygen (mg/L) &
η = ratio of substrate utilization rate with nitrate versus oxygen as the electron acceptor (dimensionless).

Equation (12.17) can be modified depending upon the limitation of one or more substrates or condition of the reaction and tolerance to dissolved oxygen.

Claus and Kutzner (1985) argued that the concentration of the end product of sulphur oxidation, sulphate, will be inhibitory to the denitrification by pure culture of *Thiobacillus denitrificans* rather than the concentration of thiosulphate or nitrate. They observed that the inhibition started at a sulphate concentration of 5 g/L and the activity stopped completely at a concentration of 20 g/L. In contrast, nitrate and thiosulphate showed inhibition at concentrations of 10 and 20 g/L, respectively. The inhibitory concentration of 5 g/L of sulphate corresponds to the oxidation of 2.9 g/L of thiosulphate and the reduction of 2 g/L of nitrate. These observations indicate that the concentration of sulphate in wastewater is a bigger limiting factor than the concentration of nitrate or thiosulphate. It also suggests the necessity of eliminating the sulphate from the reactor before it reaches an inhibitory concentration.

However, in mixed consortia of autotrophic denitrifiers, Oh *et al.* (2000) reported different values for substrate and by-product inhibition. Inhibition of denitrification was found when nitrate and sulphate concentrations reached about 0.7 and 2 g/L, respectively. Accumulation of nitrite was also detrimental to the process.

In practice, there are two major reactor configurations that utilize sulphur as electron donor: (1) packed bed, which forms biofilm, and (2) suspended growth. Packed bed type configuration generally uses elemental sulphur or elemental sulphur embedded in limestone as a substrate for the autotrophic biofilm that is developed on its surface. The solubility or the uptake of the elemental sulphur becomes limiting for the whole process. In such process it is important to stoichiometrically balance nitrogen loading based on the sulphur uptake rate to avoid partial denitrification.

General kinetic concepts of biofilm can also be applied to the biofilm of autotrophic denitrifiers with the consideration that the substrate is diffused into the biofilm from two different directions, sulphur from the inner part of the biofilm and nitrate from the outer bulk liquid. Two main phenomena that dominate the substrate utilization in the biofilm are the development and attachment of bacterial matrix on the surface and the diffusion and transport of the substrate and by-products in and out of the biofilm. In all biofilm studies the transport of substrates and reaction by-products are generally attributed to Fickian diffusion (Batchelor and Lawrence 1978).

12.5.2 Kinetic model for biofilm (Sulphur-lime packed-bed reactor)

In biofilm, it is generally assumed that the substrate consumption follow Monod kinetics. Depending on the values of the Monod saturation constant (K_s) the kinetics may be first-order or zero-order. From literature, it can be seen that, in most cases, the value of K_s (NO_3^--N) for autotrophic denitrification is very low (Batchelor and Lawrence 1978; Claus and Kutzner 1985; Zeng and Zhang 2003, Table 12.6) and thus the intrinsic reaction rate in the biofilm can be taken as zero-order kinetics. However, the zero-order intrinsic rate is modified by

Table 12.6. Kinetic parameters for autotrophic denitrifiers

Parameters	Claus and Kutzner (1985)	Oh et al. (2000)	Rittman and McCarty (2001)	Batchelor and Lawrence (1978)	Zeng and Zhang (2003)	Hasimoto et al. (1987)
Culture condition	Pure (S)	Mixed (S)	–	Mixed (S)	Mixed (B)	Mixed (S)
Electron donor	$S_2O_3^{2-}$	$S_2O_3^{2-}$	S^0	S^0	S^0	S^0
Max. specific growth rate (μ_{max}) (h^{-1})	0.11	0.12~0.2		–	0.006	–
Max. specific rate of N_2 evolution (mg N_2/g cells/h)	165 mg	–				
Max. specific NO_3^- reduction rate (g NO_3^-/g cells/h)	0.78	0.30 ~ 0.40		–	0.0062	–
Max. specific S oxidation rate (g S/g cells/h)	–	–	0.34	–	–	–
Saturation constant (K_s) (mg NO_3^--N/L)	0.045	3 ~ 10	–	0.03	0.39	–
Saturation constant (K_s) (mg S/L)	–	–	11.2	–	–	–
Yield Coef. (Y_{NO3-}) (g cells/g NO_3^-)	0.13	0.4 ~ 0.5	–	0.66	0.20–0.25	0.14
Yield Coef. (Y_s)(g cells/g S)	0.15	–	0.10	–	–	–

S: Suspended growth B: Biofilm

diffusional resistance to bulk zero-order or half-order equations with respect to the bulk concentration of the soluble substrate. Therefore, another term, the diffusion coefficient (D), which also depends on the solubility of the substrate, is introduced. Based on these assumptions, the substrate removal kinetics in a biofilm can be expressed by the following equations using a simplified pore diffusion model (Koenig and Liu 2001).

Zero-order bulk reaction:

$$R_a = k_{0a} \quad valid \ for \quad \beta = \sqrt{\frac{2DC}{k_{0a}\delta}} \geqslant 1 \tag{12.12}$$

Half-order bulk reaction:

$$R_a = k_{(1/2)a}C^{1/2} \quad valid \ for \quad \beta < 1 \tag{12.13}$$

where,
R_a = removal rate per unit biofilm area (mg/cm^2/h)
k_{oa} = zero-order reaction rate constant per unit biofilm area (mg/cm^2/h),
$k_{(1/2)a}$ = half-order reaction rate constant per unit biofilm area (mg$^{1/2}$/cm/h),
C = bulk concentration of substrate at the surface of the biofilm (mg/L),
β = dimensionless penetration ratio,
δ = thickness of the biofilm (cm)

In case of packed-bed reactor, we should also consider the volume of the reactor or the biofilm for the kinetic model. Thus, the spatial substrate removal rate can be given by:

$$\frac{dC}{dy} = -R_v \frac{A}{Q} \tag{12.14}$$

where,
A = cross-sectional area of the reactor (m^2);
Q = flow-rate (L/h),
R_v = removal rate per unit reactor volume, which is given by (mg/L/h);

R_v is given by:

$$R_v = \alpha R_a \tag{12.15}$$

where,

α = specific surface area of the reactor media per unit reactor volume (m^2/m^3),

R_a = substrate removal rate per unit biofilm area $(mg/m^2/h)$.

Likewise, the substrate concentration in the reactor at any height of the reactor in the effluent can be calculated by the following equation for two different reaction kinetics:

Zero-order reaction:

$$C_i = C_o - k_{0v}\theta_h \qquad (12.16)$$

Half-order reaction:

$$C_i^{1/2} = C_o^{1/2} - \frac{1}{2}k_{(1/2)v}\theta_h \qquad (12.17)$$

where,

C_i = substrate concentration at different time intervals or height of the reactor (mg/L);

C_0 = initial substrate concentration (mg/L);

k_{0v} = zero-order reaction rate constant per unit volume of the reactor $(mg/L/h)$;

$k_{(1/2)v}$ = half-order reaction rate constant per unit volume of the reactor $(mg^{1/2}/L^{1/2}/h)$;

θ_h = hydraulic retention time up to the height 'h' of the reactor.

For the substrate concentration in the effluent, $\theta_h = \theta$ which is the hydraulic retention time of the reactor.

Batchelor and Lawrence (1978) also attempted to develop a kinetic model for elemental sulphur based denitrification. They described the microbial uptake of sulphur as first-order, but nitrate consumption as zero-order suggesting that nitrate is not limiting. In the biofilm of autotrophic denitrifiers (Figure 12.4), it was assumed that the dissolution of sulphur takes place at a constant rate with its concentration represented by C_i. By assuming that mass transfer through the biofilm follows a relationship similar to Fick's law, they developed a second-order differential equation that can be applied to sulphur based system.

$$R_i = D_i\frac{d^2C_i}{dZ^2} \qquad (12.18)$$

where,

D_i = molecular diffusion coefficient (cm^2/h);

Z = depth of the biofilm (cm);

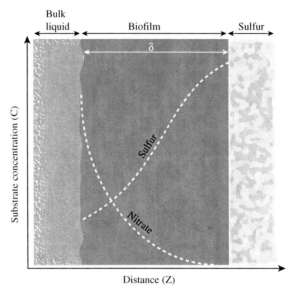

Figure 12.4. Cross-sectional schematic of sulphur-biofilm system and simulation of nitrate and sulphur concentration gradient in the biofilm (Batchelor and Lawrence 1978)

R_i = local rate of removal of component 'i' (sulphur or nitrate) per unit biofilm volume (mg/cm^3-h)

　　This kinetic model allows the computation of the removal rate of substrate in the biofilm.

12.5.3 Kinetic model for suspended growth (Thiosulphate or sulphide as electron donor)

The kinetic model for a suspended growth reactor differs slightly from that of the biofilm. However, the Monod model is largely proposed to describe the rate of substrate utilization and cell growth. When a continuous-flow system is operated under nitrate limiting conditions, the kinetics of growth of autotrophic denitrifiers and biodegradation rates can be described by Monod kinetics as follows:

For substrate utilization

$$\frac{dS}{dt} = -\frac{kSX}{(K_s + S)}$$

(12.19)

where,
S = concentration of substrate (mg/L);
k = maximum specific substrate utilisation rate, (mg N/mg-VSS/h);
K_s = half saturation constant of NO_3^--N (mg/L);
X = specific biomass concentration (mg VSS/L);
   For biomass growth

$$\frac{dX}{dt} = \left(\frac{\mu_{max}S}{K_s + S}\right)X - k_dX - \frac{1}{\theta}X_e \qquad (12.20)$$

where,
μ_{max} = maximum specific growth rate (h^{-1});
k_d = decay constant (h^{-1});
θ = hydraulic retention time (h);
X_e = concentration of autotrophic biomass exiting the reactor (mg VSS/L).
   The values of maximum specific growth rate (μ_{max}) and half saturation constant (K_s) can be calculated as proposed by Claus and Kutzner (1985).

$$\mu_{max} = D_c \frac{K_s + S_0}{S_0} \qquad (12.21)$$

$$K_S = S_D\left(\frac{\mu_{max} + D}{D}\right) \qquad (12.22)$$

where,
D_c = critical dilution rate (h^{-1});
S_0 = influent substrate concentration (mg/L);
S_D = substrate concentration in the reactor at dilution rate D (mg/L).
   Likewise, the concentration of substrates can be calculated using simple first-order kinetics. For a continuous-flow reactor using microporous membrane for nitrate diffusion into the bulk liquid, the nitrate diffusion coefficient and the mass transfer coefficient of the membrane should be taken into consideration. Then, the concentration of the nitrate can be calculated using the following model given by Mansell and Shroeder (1999; 2002).

$$C = C_0 \exp\left(\frac{-DA\varepsilon}{\delta_m Q}\right) \qquad (12.23)$$

where,

C = concentration of effluent NO_3^--N (mg/L);

C_0 = concentration of influent NO_3^--N (mg/L);

D = diffusion coefficient of NO_3^--N through microporous membrane pore area (cm^2/s);

ε = effective porosity of microporous membrane (dimensionless);

A = effective area of the microporous membrane (cm^2);

δ_m = thickness of the membrane (cm);

Q = volumetric flow rate of wastewater (cm^3/s)

12.6 PERFORMANCE OF AUTOTROPHIC DENITRIFYING SYSTEMS

During the last decade, several biological treatment systems have been explored to achieve the simultaneous removal of sulphurous and nitrogenous compounds from contaminated waters under different hydraulic conditions. The wide variety of treatment technologies evaluated includes suspended biomass processes as well as biofilm systems. Most treatment technologies have been tested under strict autotrophic conditions provided with different reduced sulphurous compounds as exclusive electron donors. Furthermore, a number of studies report on the performance of denitrifying reactors achieving the simultaneous removal of N, S and chemical oxygen demand (COD) under mixotrophic conditions (*e.g.* denitrifying cultures provided with both organic and inorganic electron donors). In the following sections, the main design criteria for autotrophic and mixotrophic denitrifying reactors will be discussed.

12.6.1 Performance of denitrifying systems under autotrophic conditions

Table 12.7 summarises the performance of autotrophic denitrifying systems during the simultaneous removal of sulphur and nitrogen from contaminated waters. Most denitrifying reactors were operated with sulphide as electron donor and with the goal of obtaining sulphate as the final sulphurous product. Sulphur loading rate (SLR) widely varies among the different research papers reported (from 0.186 to 13.82 kg S/m^3-d), which generally report high sulphur removal efficiencies (SRE, from 75 to 100%). Nevertheless, nitrogen removal efficiencies (NRE) observed in these denitrifying systems, operated at nitrogen loading rates (NLR) between 0.08 and 16.3 kg N/m^3-d, were much lower (up to 50.5%) than the sulphur removal efficiencies. Therefore, NLR seems to be the most critical

Table 12.7. Performance of different denitrifying systems for the simultaneous removal of S and N under autotrophic conditions

Reactor type (HRT, h)	S conversion	S loading (% removal)	N conversion	N loading (% removal)	Reference
DAF (6)	$S_2O_3^{-2} \rightarrow SO_4^{-2}$	2.7 (NA)	$NO_3^- \rightarrow N_2$	0.08 (high)	Furumai et al. (1996)
FBR (2–12)	$S^0 \rightarrow SO_4^{-2}$	NA	$NO_3^- \rightarrow N_2$	0.6 (64)	Flere and Zhang (1999)
CSTR (12)	$S^{-2} \rightarrow S^0$	0.6 (75)	$NO_3^- \rightarrow N_2$	0.11 (90)	Wang et al. (2005a)
CSTR (8.5–33.2)	$S^{-2} \rightarrow SO_4^{-2}$	0.28 (100)	$NO_3^- \rightarrow N_2$	0.11 (100)	Manconi et al. (2006)
SPBR (4.25)	$S^0 \rightarrow SO_4^{-2}$	NA	$NO_3^- \rightarrow N_2$	4.32 (> 95)	Kim et al. (2004)
UASB (13.4)	$S^{-2} \rightarrow SO_4^{-2} + S^0$	0.186 (> 99)	$NO_3^- \rightarrow N_2$	0.16 (89)	Sierra-Alvarez et al. (2005)
SBR (26–28)	$S^{-2} \rightarrow SO_4^{-2}$	NA	$NO_2^- \rightarrow N_2$	0.15 (76.9)	Pérez et al. (2007)
UASB (48)	$S^{-2} \rightarrow SO_4^{-2} + S^0$	1.46–3.29 (91.4–99.8)	$NO_3^- \rightarrow N_2$	0.42–0.81 (50.9–95.8)	Mahmood et al. (2007a)
UASB (48)	$S^{-2} \rightarrow SO_4^{-2} + S^0$	0.3–0.96 (88.9–99.7)	$NO_2^- \rightarrow N_2$	0.36–1.13 (50.5–78.6)	Mahmood et al. (2007a)
UASB (1.92–36)	$S^{-2} \rightarrow SO_4^{-2} + S^0$	0.77–13.82 (96.7–99.8)	$NO_2^- \rightarrow N_2$	0.9–16.3 (54.6–81.9)	Mahmood et al. (2007b)
UAF (1.1)	$S_2O_3^{-2} \rightarrow SO_4^{-2}$	NA	$NO_3^- \rightarrow N_2$	6.3 (~90)	Yamamoto-Ikemoto et al. (2000)

HRT, hydraulic retention time; DAF, down-flow anaerobic filter; FBR, fixed bed reactor; CSTR, completely stirred tank reactor; SPBR, sequential packed bed reactor; UASB, up-flow anaerobic sludge bed; SBR, sequential batch reactor; UAF, up-flow anoxic filter; NA, not available.

design criterion for autotrophic denitrifying reactors. Overloading of denitrifying reactors in terms of nitrogen feeding has been reflected in poor NRE even in treatment systems with high biomass content, such as upflow anaerobic sludge bed (UASB) and expanded granular sludge bed (EGSB) reactors. Several parameters, which will be discussed in section 12.7, have been underlined as responsible for the poor performance of denitrifying reactors.

12.6.2 Performance of denitrifying systems under mixotrophic conditions

Denitrifying reactors operated under mixotrophic conditions generally perform high SRE ($>90\%$), but the NRE could be as low as 54% depending on the NLR (Table 12.8). Most studies reported on the oxidation of sulphide to a mixture of elemental sulphur and sulphate. Moreover, it has been clear from this literature review that several key operational parameters determine the final sulphurous product obtained in denitrifying reactors (see section 12.7). Moreover, both the type and concentration of organic compounds present during the simultaneous removal of N, S and COD affect the efficiency of mixotrophic systems (see section 12.7).

12.6.3 Design criteria

As described above, the most important design criterion for all the autotrophic denitrifying systems reported is the NLR, which will be discussed in the following sections for the different reactor configurations. The effect of other operational parameters on both NRE and SRE will be discussed in section 12.7.

12.6.3.1 UASB and EGSB systems

Maximum denitrifying rates (MDR) observed in autotrophic UASB reactors, provided with reduced sulphurous compounds as electron donors, ranged between 0.14 and 1.73 kg NO_3^--N/m^3-d (Sierra-Alvarez et al. 2005; Mahmood et al. 2007a; 2007b) and up to 16.31 kg NO_2^--N/m^3-d (Mahmood et al. 2007b), when nitrate and nitrite served as terminal electron acceptors, respectively. However, it is important to emphasise that despite the much higher MDR achieved in nitrite-supplemented reactors compared to the results obtained in nitrate-supplemented systems, lower NRE were accomplished in the former (Table 12.7). Furthermore, MDR achieved in EGSB denitrifying reactors ranged between 0.78 and 0.96 kg NO_3^--N/m^3-d (Chen et al. 2008).

Table 12.8. Performance of different denitrifying systems for the simultaneous removal of COD, S and N under mixotrophic conditions

Reactor type (HRT, h)	S conversion	S loading (% removal)	N conversion	N loading (% removal)	Carbon source	COD-loading (% removal)	Reference
SPBR (14.1)	$S^0 \rightarrow SO_4^{-2}$	NA	$NO_3^- \rightarrow N_2$	1.1 (99)	LE	NA	Kim and Bae (2000)
SPBR (7.5)	$S^0 \rightarrow SO_4^{-2}$	NA	$NO_3^- \rightarrow N_2$	2.7 (97)	LE, ME	4.6 (90)	Oh et al. (2001)
CSTR (10.7)	$S^{-2} \rightarrow SO_4^{-2}$	0.46 (high)	$NO_3^- \rightarrow N_2$	0.28 (high)	ME	NA	Vaiopoulou et al. (2005)
CSTR (48)	$S^{-2} \rightarrow S^0$	0.29 (99)	$NO_3^- \rightarrow N_2$	0.21 (99)	AC	0.3 (65)	Reyes-Avila et al. (2004)
EGSB (6.4–11.2)	$S^{-2} \rightarrow S^0$	3.1 (>90)	$NO_3^- \rightarrow N_2$	1.45 (54–66)	AC	2.89 (>90)	Chen et al. (2008)
UASB (13.4)	$S^{-2} \rightarrow SO_4^{-2} + S^0$	0.18–0.39 (93.9–98.9)	$NO_3^- \rightarrow N_2$	0.159–0.28 (79.3–99.5)	p-cresol	0.39–0.85 (83.1–98.2)	Sierra-Alvarez et al. (2005)
FBR (NA)	$S^{-2} \rightarrow SO_4^{-2} + S^0$	3–4 (>99)	$NO_3^- \rightarrow N_2$	NA (>99)	NA	NA	Klerebezem and Mendez (2002)
IFBR (22)	$S^{-2} \rightarrow SO_4^{-2} + S^0$	0.08 (100)	$NO_3^- \rightarrow N_2$	0.23 (100)	AC	0.34 (100)	Beristain-Cardoso et al. (2008)
IFBR (45.6)	$S^{-2} \rightarrow SO_4^{-2}$	0.037 (100)	$NO_3^- \rightarrow N_2$	0.17 (100)	PHE	3.13 (90)	Beristain-Cardoso et al. (2009a)

HRT, hydraulic retention time; FBR, fixed bed reactor; IFBR, inversed fluidized bed reactor; CSTR, completely stirred tank reactor; SPBR, sequential packed bed reactor; UASB, up-flow anaerobic sludge bed; EGSB, expanded granular sludge bed; NA, not available. LE, leachate; ME, methanol; AC, acetate; PHE, phenol.

MDR should be carefully determined in lab-scale experiments for every denitrifying consortium before considering its application in full-scale installations. MDR could then be useful to estimate the reactor volume required for UASB and EGSB denitrifying systems, according to the following equation:

$$V = \frac{Q[NO_3^- - N]}{NLR} \tag{12.24}$$

where,
V = reactor volume (m³);
Q = flow rate (m³/d);
$NO_3 - N$ = influent nitrate concentration (kg/m³);
NLR = nitrate loading rate (kg N/m³-d).

Liquid up-flow velocity (V_{UP}) is another important design parameter for UASB and EGSB systems. A high V_{UP} increases the turbulence in the system enhancing the contact between the granular sludge and the incoming wastewater, avoiding the appearance of concentration gradients inside the system. However, an excessive V_{UP} could cause washout of active denitrifying biomass. Typical V_{UP} values are in the range of 0.5 to 2 m/h for UASB reactors (de Man et al. 1988); whereas EGSB reactors can be operated at V_{UP} as high as 10 m/h (Buitrón et al. 2006).

In the same way, the reactor height (H) has important implications for the performance and costs of the UASB reactor. V_{UP} is directly related to H and as it should not exceed a certain value to avoid biomass washout, H is also limited. Suggested H values for UASB reactors are between 4 to 6 m (Van Haandel and Lettinga 1994), whereas EGSB reactors have been constructed with much higher H. The relationship between V_{UP} and H for UASB and EGSB reactors is as follows:

$$V_{UP} = \frac{Q}{A} = \frac{V}{HRT\,A} = \frac{H}{HRT} \tag{12.25}$$

where,
A = reactor cross-section area (m²);
HRT = hydraulic retention time (h).

Further details for the design criteria of UASB and EGSB reactors, such as three-phase separators and influent distribution systems can be consulted elsewhere (Buitrón et al. 2006).

12.6.3.2 Fluidised bed reactors

Inversed fluidised bed (IFB) reactors have recently been explored for the simultaneous removal of sulphide and nitrate under mixotrophic conditions including acetate (Beristain-Cardoso *et al.* 2008) or phenol (Beristain-Cardoso *et al.* 2009a) as organic electron donor. MDR achieved in these treatment systems vary between 0.17 and 0.23 kg NO_3^--N /m^3-d achieving complete removal of N, S and high (>90%) COD removal.

During inverse fluidization, a down-flow liquid current is used as the continuous phase, since support media used are light with apparent densities varying from 75 to 930 kg/m^3, like polystyrene, polypropylene or polyethylene. Support particles are suspended or floating in the liquid phase and the liquid level control is made in the upper part of the reactor (Figure 12.5).

In this case, lower operating flow velocities used as a consequence of the support floatation capacity and low density, may also reduce the high cost associated to pumping equipment in the case of industrial reactors (Fan *et al.* 1982; Karamanev and Nikolov 1992; Krishnaia *et al.* 1993).

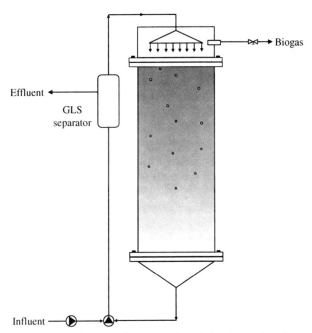

Figure 12.5. Schematic diagram of an experimental set- up of an inverse fluidized bed (Buitrón *et al.* 2006). GLS, gas-liquid-solid separator

Classical H/D ratios for IFB reactors may vary between 2 to 20, to avoid dragging of support particles from the fluidization zone. The reactor length includes a disengagement zone, consisting normally of a widening in column diameter, which allows the support particles to diminish its terminal velocities and go back to the bed.

Bed expansion is controlled by fluid vertical velocity and is considered to be 30 to 50% of the stagnant bed, but shall reach 100%, which will only be attained with high recirculation rates, using short HRT. For light support media used in IFB reactor, information about support particles terminal velocity is scarce and may be around 27 m/h for an spherical shaped mineral support with a density of 690 kg/m^3 (García-Calderón et al. 1998). Due to the floatation capacity of support particles used in this configuration, the down flow liquid velocity will be the necessary to exert the appropriate force contrary to floatation, to drag the particles down the bottom of the reactor. Although expansion velocities reported are found from 2 to12 m/h, the minimum velocities will be applied when gas velocity and hold up is the highest, as the case of the turbulent IFB systems (García-Calderón et al. 1998; Castilla et al. 2000).

The operational restriction for this prototype due to the down-flow pattern, during high strength wastewater treatment, is biomass accumulation onto the support that changes bio-particles density, the hydrodynamic regime and gas hold up. When the designed loading rate is surpassed, biogas and biomass accumulation increases, leading to an excessive bed expansion (Meraz et al. 1997; García-Calderón et al. 1998; Buffière et al. 1999). A particular concern with sulphide-supplemented denitrifying IFB reactors is that accumulation of S^0 particles may also occur inside the floating sludge particles, increasing its density and, consequently, causing the washout of the denitrifying sludge (Beristain-Cardoso et al. 2008). Therefore, calculation of the down-flow velocity in IFB systems should consider these changes in bio-particles density caused by accumulation of sludge and S^0.

Fluidization parameters such as liquid velocity (u) and bed voidage (ε) or bed expansion attained by liquid flow rate and gas hold up, are determinant in the design and operation of IFB reactors. Such variables related to support sphericity (ϕ_s) and density of liquid or gas flows ($\rho_{l,g}$), can be described with some accuracy through the following empirical equation:

$$\frac{1.75}{\varepsilon_{mf}^3 \phi_s} \mathrm{Re}_{p,mf}^2 + \frac{150(1 - \varepsilon_{mf})}{\varepsilon_{mf}^3 \phi_s^2} \mathrm{Re}_{p,mf} = Ar \qquad (12.26)$$

where the Reynolds number, $Re_{p,mf}$, which considers the particles diameters (d_p) at the onset of fluidization or minimum fluidization conditions (mf), is defined as,

$$Re_{p,mf} = \frac{d_p u_{mf} \rho_{l,g}}{\mu} \qquad (12.27)$$

and the Archimedes number, Ar, also at the onset of fluidization, is defined as,

$$Ar = \frac{d_p^3 \rho_{l,g}(\rho_s - \rho_{l,g})g}{\mu^2} \qquad (12.28)$$

where
g = acceleration of gravity;
μ = dispersion viscosity.

The equation states that when the onset of fluidization occurs, a balance between drag forces for upward or downward moving fluids, is equivalent to the support particles weight or floatation force (Muroyama and Fan 1985).

Gas hold-up, defined as the residence time of gas in the liquid phase, influences directly bed voidage and, consequently, the design volume as much as mass transfer between gas and liquid phase. The gas fraction to be present in the bed, ε_g, has to be taken also into account for IFB reactors design and numerous correlations have been proposed to relate gas fraction with bed expansion (Buitrón et al. 2006).

The most important design parameter in an IFB is bed height, which will depend directly on bed expansion. The solids residence time in the biological reactive zone will be determined by both. Also, bed expansion along with bed height and liquid velocity measurements, allows bio-film thickness and biomass concentration estimations (Nicollela et al. 1995; 1997).

12.6.3.3 Completely mixed systems

Completely mixed treatment systems, such as CSTR and SBR, have also been explored for the simultaneous removal of sulphurous and nitrogenous compounds through autotrophic denitrification. MDR achieved in these treatment systems range between 0.11 and 0.13 kg NO_3^--N/m^3-d, and up to 0.7 kg NO_3^--N /m^3-d, for CSTR (Vaiopoulou et al. 2005; Manconi et al. 2006) and SBR (Pérez et al. 2007), respectively. In all these cases reported in the literature high ($>95\%$) NRE and SRE have been achieved. Design criteria for these denitrifying reactors have been described in chapters 5 and 6.

12.6.4 Case study: Sulphide removal from oil-refining wastewater via autotrophic denitrification (Vaiopoulou *et al.* 2005)

12.6.4.1 Description

The DEA Wesseling Works in Germany is a refining and petrochemical complex with an annual crude-oil throughput of 6×10^6 tons, which directly discharges its wastewater into the Rhine River. A modern five-stage wastewater treatment plant (WWTP), which is based on the Bayer tower-biology principle (Zlokarnik 1985), treats an average of 5700 m^3/d corresponding to 4.8 tons of biochemical oxygen demand (BOD) per day (800 mg BOD/L). The wastewater consist mainly of two separate streams: (a) the so-called C/N-containing stream (130 m^3/h), which is characterised by a low concentration of organic content (400 mg BOD/L), while ammonia concentration averages 280 mg/L, and (b) the C-containing stream (40 m^3/h) that is characterised by a higher concentration of organic matter (2000 mg BOD/L). Separable oils are removed by phase separation in a coupled upstream physical separation stage. Stripping with CO_2 (2000 m^3/h) permits to lower the level of H_2S in wastewater to < 10 mg/L. This sulphide-containing water enters the WWTP along with the C/N-containing stream. The stripped H_2S is then converted to S^0 by the Claus process. Ammonia, with an initial concentration of 1500 mg/L, is removed by steam stripping, decreasing its concentration to 280 mg/L.

The WWTP (Figure 12.6) consists of five biological stages (A1, C, D, A2, E), which are operated at 28–30°C. In the denitrifying stage (A1) ~40% of the BOD from the raw wastewater is degraded using a nitrate-containing recycle stream from the nitrification stage (Nif-2). The remaining BOD removal, together with nitrification, is conducted in two steps (Nif-1 and Nif-2). Nitrate is

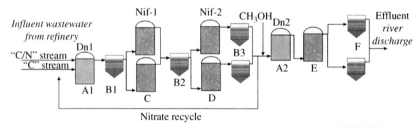

Figure 12.6. Schematic diagram of the DEA Wesseling WWTP: (a) denitrification (Dn1); intermediate clarification (B); (c) activated-sludge treatment with partial nitrification (Nif-1); (d) nitrification (Nif-2); (e) activated-sludge system (E); (f) final clarification (F)

then converted to N_2 in the second denitrification step (A2) with methanol as external electron donor (2700 kg/d). The activated sludge treatment phase (E) serves as protection against a possible BOD breakthrough. Since the energy requirement for the removal of H_2S from the oil-refining wastewater, using pressurised CO_2 stripping, is very high, a pilot-scale denitrifying CSTR system was built up to simultaneously remove sulphide with the recycled nitrate in the pre-denitrification stage.

12.6.4.2 Pilot plant set-up and performance

A CSTR reactor of 45 L was used for this study and operated at 30°C. The CSTR system was fed with the C/N-containing stream and the nitrification recycle, which were fixed at 3.0 and 1.2 L/h, respectively, giving a total influent flow of 4.2 L/h. The HRT was maintained at 10.7 h, which is exactly the applied HRT at the full-scale denitrification stage (A1, Fig. 12.6). The pH of the CSTR reactor was controlled at 7.3–7.4. During the start-up period (4 weeks), the SLR was gradually increased up to 0.165 g S/L-d, while the NLR remained constant at 0.127 g N/L-d. Afterwards, additional nitrate was incorporated to the storage tank in order to stoichiometrically balance the S/N ratio, which resulted in an inlet concentration of 130 mg NO_3^--N/L, corresponding to a NLR of 0.282 g NO_3^--N/L-d. After this, the SLR was further increased from 0.165 to 0.466 g S/L-d. The concentration of sulphide in the reactor effluent was <0.1 mg S/L during the whole experimental period of 100 days. Furthermore, the concentration of nitrate decreased gradually with increasing sulphide availability and was no detected after 80 days of operation. After this preliminary test, two main aspects were further investigated: (1) the influence of random variations of the sulphide concentration in the combined influent on the autotrophic denitrifying process; and (2) the influence of possibly inconvertible sulphide in the effluent of the denitrification stage on ammonia oxidation in the subsequent activated-sludge/nitrification process. The strategies implemented were very successful achieving high removal efficiencies of sulphide, ammonia and COD.

12.6.4.3 Full-scale implementation and benefits

Due to the promising results obtained during the operation of the pilot-scale CSTR denitrifying system, the operational mode was scaled-up at the DEA Wesseling WWTP. H_2S removal was then accomplished by a combination of autotrophic denitrification and sulphide stripping with CO_2, allowing a gradual decrease on the sulphide stripper up to 70%. Thus, a concomitant 70% decrease on the energy demand for CO_2 compression was also accomplished.

Additionally, the C-containing stream was introduced, bypassing the entrance denitrification stage, in the first aerated step in order to favour the selective enrichment of autotrophic denitrifying bacteria. An additional way to further optimise the flow pattern is to use the C-containing stream as an electron donor source, partly replacing the methanol dose in the second denitrification stage.

On the basis of a nitrate removal rate of 1150 kg/d in this denitrification stage, about 2870 kg methanol/d is required. This amount corresponds to a COD of 4180 kg or 2670 kg BOD. Since the average BOD load of the C-containing stream is 1900 kg/d, about 70% of the external C source supplementation could be saved.

12.7 KEY OPERATIONAL PARAMETERS IN AUTOTROPHIC DENITRIFYING REACTORS

12.7.1 Type and concentration of electron donor

Among the different reduced sulphurous compounds, which have been explored as electron donors for denitrifying processes, thiosulphate has yielded the highest denitrifying rates, followed by sulphide and elemental sulphur. For instance, the MDR observed in an autotrophic denitrifying culture with thiosulphate as electron donor was 4.6- and 9.5-fold higher than those obtained with sulphide and S^0, respectively. Moreover, during the same batch experiments, the rates of sulphate production in assays with thiosulphate were 4.8- and 25.3-fold higher compared to those achieved in sulphide- and S^0-supplemented cultures, respectively (Beristain-Cardoso et al. 2006).

Thermodynamics seems to be one of the factors determining the superior oxidation rate for thiosulphate compared with sulphide and S^0, in denitrifying processes. Certainly, the oxidation rate of thiosulphate coupled to nitrate reduction is thermodynamically more favourable than the reactions in which sulphide or S^0 serve as electron donors (Beristain-Cardoso et al. 2006). MDR reported with S^0 as electron donor are definitely lower than those observed in thiosulphate- and sulphide-amended denitrifying cultures, probably due to mass transfer limitations caused by the relatively low solubility of S^0 (0.16 μM, Steudel and Holdt 1998).

In a study documenting the effect of the initial S^0 concentration on the denitrifying activity in batch experiments, it was observed that the MDR increased linearly with S^0 concentration up to 7.6 mM, which was a large excess of S^0 with respect to the supplied nitrate. A continued increase on the MDR at concentrations of S^0 far exceeding the stoichiometric requirement was probably due to limited mass transfer from solid phase S^0 (Sierra-Alvarez et al. 2007).

Autotrophic denitrifying reactors have successfully ($>88\%$) remove sulphide at initial sulphide concentrations as high as 580 mg S/L and 1920 mg S/L for nitrate and nitrite-amended bio-reactors (Mahmood *et al.* 2007a; b). Nevertheless, high concentrations of sulphide have caused accumulation of nitrite in denitrifying process, which has been reflected in poor NRE (as low as ~50%, Mahmood *et al.* 2007a; b).

Inhibitory effects by sulphide, at concentrations as low as 20 mg S/L, have been documented causing a decrease on the rate of reduction of nitrate (Dalsgaard and Bak 1994), nitrite (Reyes-Avila *et al.* 2004; Meza-Escalante *et al.* 2008), and nitrous oxide (Sørensen *et al.* 1980), as well as on the rate of oxidation of *p*-cresol (Meza-Escalante *et al.* 2008) and phenol (Beristain-Cardoso *et al.* 2009a) in autotrophic and mixotrophic denitrifying processes. Increased sulphide concentrations have also caused accumulation of ammonia in denitrifying processes (Tugtas and Pavlostathis 2007; Sher *et al.* 2008).

12.7.2 Nitrogen and sulphur loading rates

The effect of NLR on autotrophic denitrifying processes varies largely depending on the type of reactor and operational conditions prevailing in the treatment systems. For example, NLR as high as 1.45 kg N/m^3-d have completely been removed in IFB (Beristain-Cardoso *et al.* 2009a) and EGSB (Chen *et al.* 2008) systems amended with sulphide as electron donor. In contrast, the NRE achieved in a fixed bed reactor, packed with S^0 as electron donor, decreased sharply from ~90% at NLR <0.3 kg N/m^3-d to ~40% at NLR >1 kg N/m^3-d (Flere and Zhang 1999).

A stronger effect of NLR on autotrophic denitrifying processes has been observed when using nitrite as terminal electron acceptor compared with nitrate. Two forms of evaluation of the effect of NLR were established in a nitrite-amended UASB reactor, by increasing the nitrite concentration (38–2265 mg N/L) at a fixed HRT of 2 days, and by decreasing the HRT (1.9–36 h) at a fixed nitrite concentration (1360 mg N/L). In both cases, the NRE decreased as low as ~50% by increasing the NLR (Mahmood *et al.* 2007a). During the same study, the SRE achieved was $>88\%$ at SLR as high as 0.96 kg S/m^3-d. However, the reactor operation was stopped after 135 days due to substrate toxicity caused by very high influent concentrations of nitrite and sulphide, i.e. 2265 mg N/L and 1920 mg S/L, respectively. During the second phase of these experiments, in which the HRT was decreased from 36 h to 1.9 h, the sulphide concentration was kept constant at 1152 mg S/L. Under these circumstances, the SRE declined sharply to 99.6% at 1.9 h HRT due to sulphide inhibition (Mahmood *et al.* 2007).

12.7.3 HRT

The performance of different autotrophic denitrifying reactors strongly depends on the hydraulic conditions prevailing during their operation. Moreover, it has been observed that changes in HRT on different reactor configurations cause distinct effects on the NRE depending on other factors, such as the type of sulphurous electron donor supplied and on whether the reactor is operated under strict autotrophic conditions or under mixotrophic conditions. For instance, packed bed denitrifying reactors, supplied with S^0 as unique electron donor, generally accomplish NRE $>90\%$ when operated at HRT $\geqslant 4$ h at nitrate concentration $\leqslant 100$ mg N/L (Flere and Zhang 1999; Oh $et\ al.$ 2001). However, when the HRT is lower than 4 h in these treatment systems, the NRE has dropped as low as 20–30% for the same concentration of nitrate (Flere and Zhang 1999; Oh $et\ al.$ 2001). One of the limiting factors, which has been underlined as the cause of the low NRE achieved in S^0-supplied denitrifying systems is the limited dissolution of this electron donor (Oh $et\ al.$ 2001). When an alternative electron donor, such as methanol or leachate, is available in S^0-amended systems (mixotrophic conditions), the NRE could remain $\geq 80\%$ even at HRT as low as 1 h (Oh $et\ al.$ 2001).

Autotrophic denitrifying reactors supplied with either sulphide or thiosulphate as electron donor generally accomplish high NRE ($\geqslant 90\%$) at relatively long HRT (Tables 12.1 and Table 12.2). Nevertheless, when short HRT are applied combined with high nitrate concentrations, poor NRE can occur even for high-rate denitrifying reactors, such as UASB (Mahmood $et\ al.$ 2007b) and EGSB (Chen $et\ al.$ 2008).

12.7.4 Temperature

Scarce information is available in the literature documenting the effect of temperature in autotrophic denitrifying processes. Most studies report on treatment systems operated under mesophilic conditions (25–35°C). Yavuz $et\ al.$ (2007) reported the sulphide oxidation rate (SOR) obtained under denitrifying conditions at different temperatures and pH values. SOR obtained at 25 and 30°C were very similar, although decreased from 0.12 to 0.09 g S/g VSS-h by raising the pH from 6.5 to 9. Furthermore, the highest SOR were accomplished at 35°C (up to 0.22 g S/g VSS-h), but were also strongly influenced by the pH. Indeed, the SOR achieved at this temperature decreased from 0.22 to 0.12 g S/g VSS-h by increasing the pH value from 6.5 to 9.

The effect of temperature has also been evident during the operation of denitrifying reactors installed at ambient temperature, which undergoes seasonal changes. For instance, during the removal of nitrate in an upflow anoxic filter the NRE decreased, as evidenced by an increase in the effluent nitrate concentration (up to ~60 mg/L), when the temperature went below 15°C (Yamamoto-Ikemoto *et al.* 2000).

Furthermore, higher NLR and better performance in terms of NRE and SRE were accomplished in a S^0-limestone denitrifying system operated at 30°C (Sierra-Alvarez *et al.* 2007) compared to similar treatment systems operated at 20–25°C (Flere and Zhang 1999; Koenig and Liu 2001; Derbi *et al.* 2003).

12.7.5 pH

Autotrophic sulphur denitrifying bacteria have an optimum pH range of 6–9 (Holt *et al.* 1994). Nevertheless, it has been observed a strong dependence of the denitrifying activities achieved in different reactor configurations on the pH prevailing in the treatment systems. Denitrifying activities generally decrease by raising the pH of the culture media. For instance, denitrifying activities linked to sulphide oxidation sharply decreased from 0.22 to 0.12 g S/g VSS-h when the pH was raised from 6.5 to 9. As an ionisable gas, hydrogen sulphide (H_2S) dissociates in water according to the following equations:

$$H_2S \Leftrightarrow H^+ + HS^-\ K_1 = 1.49 \times 10^{-7}\ (35°C)\ \text{(Speece 1996)}$$
$$HS^- \Leftrightarrow H^+ + S^{2-}\ K_2 = 0.8 \times 10^{-17}\ (20°C)\ \text{(Steudel 2000)}$$

The availability of sulphide is regarded as being pH-dependent because only the neutral un-dissociated H_2S molecule can pass through the cell membrane (Speece 1983). In the liquid phase, the total dissolved sulphide is present as the unionized form (H_2S) and as HS^-. As the pK_a value of this acid-base equilibrium is about 7, small pH variations in the pH range 6–8 will significantly affect the free (unionized) H_2S concentration. At neutral pH values, free H_2S accounts to 50% of total dissolved sulphide, whereas at pH 8 it is only around 10% (Figure 12.7). Therefore, availability of sulphide for autotrophic denitrifying bacteria is strongly affected by the pH.

The sulphide supplied to autotrophic denitrifying reactors will be distributed over the gas phase and the liquid phase according to the following expression:

$$[H_2S]_l = \alpha \cdot [H_2S]_g$$

in which $[H_2S]_l$ and $[H_2S]_g$ are, respectively, the concentrations of the H_2S in the liquid phase and the gas phase and α is a dimensionless distribution

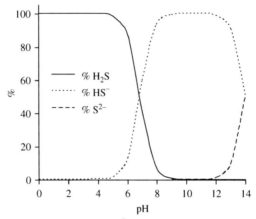

Figure 12.7. Equilibrium for $H_2S/HS^-/S^{2-}$ in aqueous solution as a function of pH (Fernández-Polanco and García-Encina 2006)

coefficient (Henry constant), whose value is 675 atm (mol fraction)$^{-1}$ at 35°C (Fernández-Polanco and García-Encina 2006).

According to the following equation, 1.28 mol of H^+ are produced when 1 mol nitrate is reduced (equivalent to 4.57 mg $CaCO_3$/mg NO_3^--N):

$$NO_3^- + 1.1S^0 + 0.4CO_2 + 0.76H_2O + 0.08NH_4^+$$
$$\rightarrow 0.5N_2 + 1.1SO_4^{-2} + 1.28H^+ + 0.08C_5H_7O_2N$$

Thus, pH control is essential in autotrophic denitrifying reactors, which could partly be controlled by alkalinity produced from heterotrophic denitrification in mixotrophic reactors. For instance, the alkalinity produced during denitrification with methanol as electron donor is equivalent to 3.57 mg $CaCO_3$/mg NO_3^--N reduced:

$$NO_3^- + 1.08CH_3OH + 0.24H_2CO_3 \rightarrow 0.056C_5H_7NO_2 + 0.47N_2 + 1.68H_2O$$
$$+ HCO_3^-$$

One of the strategies proposed to control pH variations in autotrophic packed reactors is the use of limestone, which besides serving as a buffer agent, also acts as a carbon source for autotrophic denitrifying bacteria (Flere and Zhang 1999; Oh et al. 2001).

12.7.6 Mass transfer limitations

A number of studies have reported mass transfer limitations in S^0-supplemented denitrifying processes. One of the limiting factors is the poor aqueous solubility of S^0 (0.16 µM, Steudel and Holdt 1998), which restricts the MDR achieve in denitrifying reactors amended with this electron donor (Sierra-Alvarez et al. 2007). However, increased surface area provided at high S^0 concentrations has resulted in improved mass transfer. In fact, several research studies confirm that the specific surface area of S^0 is a principal factor governing the kinetics of its biological oxidation (Germida and Janzen 1993; Watkinson and Blair 1993; Sholeh et al. 1997). The last observation was corroborated by Sierra-Alvarez et al. (2007), who reported that the surface normalised denitrifying rates (SNDR) achieved in batch incubations were comparable to those originated in continuous reactor studies. SNDR ranged 18.7–34.4 mmol/m^2-d during these experiments.

Contradictory values for the substrate saturation constant (K_s) have been reported in different denitrifying processes depending on the thickness of biofilm. Certainly, Zeng and Zhang (2005) determined a Ks of 28.4 µM nitrate for attached biomass with a thickness ranging 0.025–0.15 mm. On the other hand, increase in denitrification activities was observed in a granular sludge, which thickness was 0.5–3 mm, at concentrations as high as 4.8 mM (Sierra-Alvarez et al. 2007), which contrasts with the K_s value reported by Zeng and Zhang (2005). Limitations by diffusion caused by the thickness of biofilms may explain the different kinetic data reported in these studies. In fact, a positive relationship between the size of the biofilm granules and the apparent substrate saturation constants for activities has been demonstrated in several studies (Dolfing 1985; Gonzalez-Gil et al. 2001; Sierra-Alvarez et al. 2007).

12.7.7 S/N ratio

Oxidation of sulphide coupled to nitrate reduction proceeds according to the following stoichiometric equations:

$$12H^+ + 2NO_3^- + 5S^{2-} \rightarrow N_2 + 5S^0 + 6H_2O$$
$$5S^0 + 6NO_3^- + 8H_2O \rightarrow 5H_2SO_4 + 6OH^- + 3N_2$$

As can be seen, a molar S/N ratio of 5:2 is necessary to oxidise sulphide to S^0 via denitrification, whereas an excess of nitrate (S/N ratio of 5:6) promotes the oxidation of sulphide to sulphate.

Therefore, it is expected that the S/N ratio prevailing in autotrophic denitrifying reactors determines the final sulphurous product obtained. The last observation was verified in the study by Wang et al. (2005a), who tested the effects of different

S/N ratio in a denitrifying culture. Certainly, production of sulphate occurred in the denitrifying culture by decreasing the S/N from 5:1 to 5:4.

Furthermore, changes in the S/N ratio also affected the rate of both sulphide oxidation and nitrate reduction (Wang *et al.* 2005a). Regarding the oxidation of sulphide, it was observed higher rate and extent of conversion at S/N ratios of 5:3 and 5:4, compared to those observed at higher S/N ratios (5:1 and 5:2). Moreover, it was observed that the amount of nitrate consumed increased with the decrease of S/N ratio. Nevertheless, the amount of nitrate reduced was basically the same at the S/N ratios of 5:3 and 5:4, which suggested that the denitrifying capacity could hardly be improved when the S/N ratio is lower than 5:3 due to a lack of sufficient reducing power in the system (Wang *et al.* 2005a).

12.7.8 C/N ratio

Several studies have documented the feasibility to achieve the simultaneous removal of S, N and organic matter in mixotrophic denitrifying systems (Table 12.2). The C/N ratio has been pointed out as a key operational parameter in mixotrophic denitrifying processes as it drives the competition between autotrophic and organotrophic denitrifying bacteria in the treatment systems. The C/N ratio may determine the end sulphurous product obtained in mixotrophic denitrifying processes. For instance, Beristan-Cardoso *et al.* (2008) reported complete removal of sulphide, nitrate and acetate in an IFB denitrifying system operated at C/N ratios between 1.24 and 1.7. No accumulation of nitrogenous compounds occurred during these experiments; however, the sulphurous end product obtained varied depending on the C/N ratio. Indeed, 71% and 18% of the sulphide fed was recovered as S^0 and sulphate, respectively, at a C/N ratio of 1.7; whereas sulphate appeared as the unique sulphurous product at lower C/N ratios (1.24 and 1.44) due to an excess of electron accepting capacity.

Moreover, the C/N ratio may also affect the rate of sulphide oxidation as recently evidenced by Chen *et al.* (2008) during the simultaneous removal of sulphide, nitrate and acetate in a EGSB reactor. At molar C/N ratios of 0.85 and 1.05 complete removal of nitrate and acetate was achieved, but nitrite accumulation occurred owing to lower activity of organotrophic denitrifying bacteria compared to autotrophic denitrifiers. Furthermore, strong competition between autotrophic and organotrophic denitrifiers was observed at C/N ratios of 1.26 and 2, which was negatively reflected in a decrease on acetate oxidation efficiency from 95% to 60% and on a lower sulphide oxidation rate (from 0.46 to 0.17 kg S/m^3-d). S^0 was always obtained as the unique sulphurous product during these studies.

Another negative effect, which has been observed in autotrophic denitrifying processes when raising the C/N ratio, is that an increasing fraction of nitrate

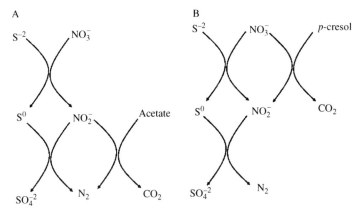

Figure 12.8. Mechanisms proposed during the simultaneous removal of nitrate, sulphide and organic substrates in mixotrophic denitrifying processes (A, from Reyes-Avila *et al.* 2004; B, from Meza-Escalante *et al.* 2008)

could end up as ammonia through dissimilatory nitrate reduction to ammonia (Tugtas and Pavlostathis 2007).

Different mechanisms have been reported during the simultaneous removal of nitrate, sulphide and organic compounds in denitrifying processes. Reyes-Avila *et al.* (2004) reported that the conversion of sulphide to S^0 was linked to the reduction of nitrate to nitrite, which was further reduced to N_2 coupled to the simultaneous oxidation of acetate and S^0 (to sulphate). On the other hand, Meza–Escalante *et al.* (2008) studied a denitrifying sludge, which achieve the simultaneous oxidation of sulphide (to S^0) and p-cresol (mainly to CO_2) coupled to the reduction of nitrate to nitrite; whereas S^0 was the sole electron donor sustaining the reduction of nitrite to N_2 in the denitrifying process (Figure 12.8).

Addition of organic electron donors to autotrophic denitrifying processes has improved the denitrification rates (Beristain-Cardoso *et al.* 2006). The last observation has important implications for the simultaneous removal of nitrogenous, sulphurous and organic compounds from wastewaters, such as those generated from the petrochemical sector (Kleerebezem and Mendez 2002; Sierra-Alvarez *et al.* 2005; Vaiopoulou *et al.* 2005).

12.8 SUMMARY

In the last two decades, a number of research reports have improved our understanding on the significance of autotrophic denitrification in natural systems. Autotrophic denitrification is one of the major modes of nitrate reduction leading to the production of molecular nitrogen. It has already been shown that autotrophic denitrification is significant in nature and a major contributor to the nitrogen and sulphur cycles in the ocean. From the environmental engineering standpoint, the process is very important as provides a good treatment option to simultaneously remove both sulphurous and nitrogenous contaminants from wastewaters. It is more economical as there is no need of external carbon source, which is generally added in the case of heterotrophic denitrification, consequently increasing the cost of the process.

Various obligate as well as facultative autotrophic microorganisms have been recognized, that possess all the necessary genes encoding the essential sulphur oxidising and denitrifying enzymes. Various electron flow pathways have been suggested and confirmed for some of the major autotrophs, such as *Thiobacillus denitrificans*, *Thiomicrospira denitrificans*, and *Paracoccus denitrificans*. Complete genome sequencing and analysis of some of these bacteria have already been performed, which suggest a high degree of metabolic versatility or responsiveness of these microorganisms to the environment. Regulatory mechanisms of the genes related to denitrification and sulphur oxidation have been sufficiently described and the requisite enzymes have also been described in terms of their structure and function. We now have sufficient information on the basic metabolic and genetic information of the autotrophic denitrifying microorganisms, which should facilitate their effective and successful application in the field of water and wastewater engineering. Nevertheless, a large group of microorganisms exists in the environment that can carry out autotrophic denitrification but are not properly identified or cultured. More research is needed for the identification, isolation and characterization of these microorganisms that could contribute even more in water and wastewater engineering applications.

In spite a number of research reports exist on the application of autotrophic denitrification, we still lack a well engineered system, pragmatic enough to establish autotrophic denitrification on a practical scale. Wastewater composition and pre-treatment are still a big issue, which limits the application of autotrophic denitrification. Different reactor configurations have been evaluated for the suitability of autotrophic denitrification. However, based on the type of electron donors, they are either suspended growth or biofilm processes. It has also been shown that autotrophic denitrification can be carried out under mixotrophic conditions, which gives more flexibility to the treatment process as real wastewaters contain a variety of pollutants, including organic carbon.

Finally, autotrophic denitrification is a cleaner and cheaper option not only for the treatment and remediation of drinking and groundwater, but also for industrial wastewaters that contain relatively high concentrations of nitrogen and sulphur. However, the current paucity of detailed knowledge on the microbial diversity and realistic configuration of denitrifying reactors should be addressed in future research efforts.

REFERENCES

Ahn Y.H. and Kim H.C. (2004) Nutrient removal and microbial granulation in an anaerobic process treating inorganic and organic nitrogenous wastewater *Water Sci. Technol.* **50**(6), 207–215.

Altas, L. and Buyukgungor, H. (2008) Sulfide removal in petroleum refinery waswater by chemical precipitation. *J. Hazard. Mater.* **153**, 462–469.

Beijerinck, M.W. (1904) Phénomènes de réduction produits par les microbes (Conférence avec demonstrations faite. *Arch. Neerl. Sci. Ser.* **9**(2), 131–157.

Baker, S.C., Stuart J.F., Ludwig, B., Page, M.D., Richter, O.H. and VanSpanning, R.J. M. (1998) Molecular Genetics of the Genus *Paracoccus*: Metabolically Versatile Bacteria with Bioenergetic Flexibility. *Microbiol. Mol. Biol. Rev.* **62**(4), 1046–1078.

Barcelona, M.J. (1985) TOC determination in groundwater. *Groundwater* **22**(1), 18–24.

Batchelor, B. and Lawrence, A.W. (1978) Autotrophic denitrification using elemental sulfur. *J. Wat. Pollut. Con. Fed.* **50**(8), 1986–2001.

Beller, H.R. (2005) Anaerobic, nitrate-dependent oxidation of U(IV) oxide minerals by the chemolitoautotrophic bacterium *Thiobacillus denitrificans*. *Appl. Environ. Microbiol.* **71**, 2170–2174.

Beller, H.R., Letain, T.E., Chakicherla, A., Kane, S.R., Legler, T.C. and Coleman, M.A. (2006) Whole-genome transcriptional analysis of chemolithoautotrophic thiosulfate oxidation by Thiobacillus denitrificans under aerobic versus denitrifying conditions. *J. Bacteriol.* **188**(19), 7005–7015.

Beristain-Cardoso, R., Sierra-Alvarez, R., Rowlette, P., Razo-Flores, E., Gómez, J. and Field, J.A. (2006) Sulfide oxidation under chemolithoautotrophic denitrifying conditions. *Biotechnol. Bioeng.* **95**, 1148–1157.

Beristain-Cardoso, R., Texier, A.C., Alpuche-Solís, A., Gómez, J. and Razo-Flores, E. (2009a) Phenol and sulfide oxidation in a denitrifying biofilm reactor and its microbial community analysis. *Process Biochem.* Doi:10.1016/j.procbio. 2008.09.002.

Beristain-Cardoso, R., Texier, A.C., Sierra-Alvarez, R., Field, J.A., Razo-Flores, E. and Gómez, J. (2008) Simultaneous sulfide and acetate oxidation under denitrifying conditions using an inverse fluidized bed reactor. *J. Chem. Technol. Biotechnol.* **83**, 1197–1203.

Beristain-Cardoso, R., Texier, A.C., Sierra-Alvarez, R., Razo-Flores, E., Field, J.A. and Gómez, J. (2009b) Effect of initial sulfide concentration on sulfide and phenol oxidation under denitrifying conditions. *Chemosphere* **74**, 200–205.

Bruser, T., Selmer, T. and Dahl, C. (2000) "ADP sulfurylase" from Thiobacillus denitrificans is an adenylylsulfate : phosphate adenylyltransferase and belongs to a new family of nucleotidyltransferases. *J. Biol. Chem.* **275**, 1691–1698.

Buitrón, G., Razo-Flores, E., Meraz, M. and Alatriste-Mondragon, F. (2006) Biological wastewater treatment systems. In *Advanced Biological Treatment Processes for*

Industrial Wastewaters: Principles and Applications. Cervantes F.J., Pavlostathis, S.G. and Van Haandel, A.C. (Eds.). London, IWA Publishing, pp. 141–185.

Byun, I-G., Ko, J-H., Jung, Y-R., Lee, T-H., Kim, C-W. and Park, T-J. (2005) The feasibility of using spent sulfidic caustic as alternative sulphur and alkalinity sources in autotrophic denitrification. *Kor. J. Chem. Eng.* **22**, 910–916.

Castilla, P., Meraz, M., O. Monroy and A. Noyola (2000) Anaerobic treatment of low concentration wastewater in an inverse fluidized bed reactor. *Water Sci. Technol.* **41**(4–5), 245–251.

Chen, C., Ren, N., Wang, A., Yu, Z. and Lee, D.J. (2008) Simultaneous biological removal of sulfur, nitrogen and carbon using EGSB reactor. *Appl. Microbiol. Biotechnol.* **78**, 1057–1063.

Claus, G. and Kutzner, H.J. (1985) Physiology and kinetics of autotrophic denitrification by *Thiobacillus denitrificans. Appl. Microbiol. Biotechnol.* **22**, 283–288.

Conrado, M-V., Cabello, P., Manuel, M-L., Blasco, R. and Castillo, F. (1999) Prokaryotic nitrate reduction: molecular properties and functional distinction among bacterial nitrate reductases. *J. Bacteriol.* **181**, 6573–6584.

Dalsgaard, T. and Bak, F. (1994) Nitrate reduction in a sulfate-reducing bacterium, *Desulfovibrio desulfuricans*, isolated from rice paddy soil: sulfide inhibition, kinetics, and regulation. *Appl. Environ. Microbiol.* **60**, 291–297.

Diels L., Sadouk A., Mergeay M. (1989) Large plasmids governing multiple resistance to heavy metals: a genetic approach. *Toxicol. Environ. Chem.* **23**, 79–89.

Dolfing J. (1985) Kinetics of methane formation by granular sludge at low substrate concentrations: the influence of mass transfer limitation. *Appl. Microbiol. Biotechnol.* **22**, 77–81.

Fan, L-S., Muroyama, K. and Chern, S-H. (1982) Hydrodynamic characteristics of inverse fluidization in liquid-solid and gas-liquid-solid systems. *Chem. Eng. J.* **24**, 143–150.

Fernández-Polanco, M. and García-Encina, P.A. (2006) Application of biological treatment for sulfate-rich wastewaters. In *Advanced Biological Treatment Processes for Industrial Wastewaters: Principles and Applications.* Cervantes, F.J., Pavlostathis, S.G. and Van Haandel, A.C. (Eds.). London, IWA Publishing. pp. 213–236.

Flere, J.M. and Zhang, T.C. (1999) Nitrate removal with sulfur-limestone autotrophic denitrification processes. *J. Environ. Eng.* **125**, 721–729.

Fonstein M., Haselkorn R. (1993) Chromosomal structure of *Rhodobacter capsulatus* strain SB 1003: cosmid encyclopedia and high-resolution physical and genetic map. *Proc Natl Acad Sci USA* **90**, 2522–2526.

Friedrich, C.G., Bardischewsky, F., Rother, D., Quentmeier, A. and Fischer, J. (2005) Prokaryotic sulfur oxidation. *Curr. Opin. Microbiol.* **8**, 253–259.

Friedrich, C.G., Rother, D., Bardischewsky, F., Quentmeier, A., and Fischer, J. (2001) Oxidation of reduced sulfur compounds by bacteria emergence of a common mechanism? *Appl. Environ. Microbiol.* **67**, 2873–2882.

Furumai, H., Tagui, H. and Fujita, K. (1996) Effects of pH and alkalinity on sulfur-denitrification in a biological granular filter. *Water Sci. Technol.* **34**, 355–362.

García-Calderón, D., Buffière, P., Moletta, R. and Elmaleh, S. (1998) Influence of biomass accumulation on bed expansion characteristics of a down-flow anaerobic fluidized bed. *Biotech. Bioeng.* **57**(2), 137–144.

Gayle, B.P., Boardman, G.D. and Sherrard, J.H. (1989) Biological denitrification of water. *J. Environ. Eng.-ASCE* **115**, 930–943.

Genschow, E., Hegemann, W. and Maschke, C. (1996) Biological sulfate removal from tannery wastewater in a two-stage anaerobic treatment. *Water Res.* **30**, 2072–2078.

Germida, J.J. and Janzen, H.H. (1993) Factors affecting the oxidation of elemental sulfur in soils. *Fertilizer Res.* **35**, 101–114.

Gerstenberg C., Friedrich B., Schlegel H.G. (1982) Physical evidence for plasmids in autotrophic, especially hydrogen-oxidizing bacteria. *Arch. Microbiol.* **133**, 90–96.

Ghafari, S., Hasan, M. and Aroua, M.K. (2008) Bio-electrochemical removal of nitrate from water and wastewater-A review. *Biores. Technol.* **99**, 3965–3974.

Glass, C. and Silverstein, J. (1999) Denitrification of high-nitrate, high-salinity wastewater. *Water Res.* **33**, 223–229.

Gonzalez-Gil, G., Seghezzo, L., Lettinga, G. and Kleerebezem, R. (2001) Kinetics and mass-transfer phenomena in anaerobic granular sludge. *Biotechnol. Bioeng.* **73**, 125–134.

Goodhew, C.F., Pittigrew, G.W., Devreese, B., Beeumen, J.V., Van Spanning, R.J.M., Baker, S.C., Saunders, N., Ferguson, S.J. and Thompson, I.P. (1996) The cytochromes c-550 of *Paracoccus denitrificans* and *Thiosphaera pantotropha*: a need for re-evaluation of the history of Paraccocus cultures. *FEMS Microbiol. Lett.* **137**, 95–101.

Gupta, S.K. and Sharma, R. (1996) Biological oxidation of high strength nitrogenous wastewater. *Water Res.* **30**, 593–600.

Hasimoto, S., Furukawa, K., Shioyoma, M.N. (1987) Autotrophic denitrification using elemental sulfur. *J. Ferment. Technol.* **65**, 683–692.

Hole, U.H., Vollack, K.U., Zumft, W.G., Eisenmann, E., Siddiqui, R.A., Friedrich, B. and Kroneck, P.M.H. (1996) Characterization of the membranous denitrification enzymes nitrite reductase (cytochrome cd1) and copper-containing nitrous oxide reductase from *Thiobacillus denitrificans*. *Arch. Microbiol.* **165**, 55–61.

Hoor, A.T. (1981) Cell yield and bioenergetics of *Thermomicrospira denitrificans* compared with *Thiobacillus denitrificans*. *Antonie Van Leeuwenhoek* **47**, 231–243.

Ingrid, B. and Rheinheimer, G. (1991) Denitrification in the Central Baltic: evidence for H$_2$S-oxidation as motor of denitrification at the oxic-anoxic interface. *Mar. Ecol. Prog. Ser.* **77**, 157–169.

Joo, H-S, Jirai, M. and Shoda, M. (2000) Piggery wastewater treatment using *Alcaligenes feacalis* strain No. 4 with heterotrophic nitrification and aerobic denitrification. *Water Res.* **40**, 3029–3036.

Joye, S.B. and Hollibaugh, J.T. (1995) Influence of sulfide inhibition of nitrification on nitrogen regeneration in sediments. *Science* **270**, 623–625.

Kanyarat, S. and Chaiprapat, S. (2008) Effects of pH adjustment by parawood ash and effluent recycle ratio on the performance of anaerobic baffled reactors treating high sulfate wastewater. *Biores. Technol.* **99**, 8987–8994.

Kappler, U., Friedrich, C.G. and Truper, H.G. (2001) Evidence for two pathways of thiosulfate oxidation in *Starkeya novella* (formerly *Thiobacillus novellus*). *Arch. Microbiol.* **175**, 102–111.

Karamanev, D.G. and Nikolov, L.N. (1996) Application of inverse fluidization in wastewater treatment: from laboratory to full-scale bioreactors. *Environ. Progress* **15**(3), 194–196.

Kelly, D.P. (1999) Thermodynamic aspects of energy conservation by chemolithotrophic sulfur bacteria in relation to the sulfur oxidation pathways. *Arch. Microbiol.* **171**, 219–229.

Kim, E.W. and Bae, J.H. (2000) Alkalinity requirements and possibility of simultaneous heterotrophic denitrification during sulfur utilizing autotrophic denitrification. *Water Sci. Technol.* **42**, 233–238.

Kim, S., Jung, H., Kim, K.-S. and Kim, I.S. (2004) Treatment of high nitrate-containing wastewater by sequential heterotrophic and autotrophic denitrification. *J. Environ. Eng.* **130**, 1475–1480.

Kleerebezem, R. and Mendez, R. (2002) Autotrophic denitrification for combined hydrogen sulfide removal from biogas and post-denitrification. *Water Sci. Technol.* **45**(10), 349–356.

Koenig, A. and Liu, L.H. (2001) Kinetic model of autotrophic denitrification in sulphur packed-bed reactors. *Water Res.* **35**, 1969–1978.

Kuenen, J.G., Bos, P. (1987) Habitats and ecological niches of chemolithoautotrophic bacteria. Brock/Springer Series in Contemporary Bioscience, *FEMS Symp.* **42**, Berlin: 53–80.

Kraft, G., J. and Stites, W. (2003) Nitrate impacts on groundwater from irrigated-vegetable systems in a humid north-central US sand plain. *Agricul. Ecosys. Environ.* **100**, 63–74

Kurt, M., Dunn, I.J. and Bourne, J.R. (1987) Biological denitrification of drinking-water using autotrophic organisms with H_2 in a fluidized-bed biofilm reactor. *Biotechnol. Bioeng.* **29**, 493–501.

Lens, P.N.L., Visser, A., Janssen, A.J.H., Hulshoff Pol, L.W. and Lettinga, G. (1998) Biotechnological treatment of sulfate-rich wastewaters. *Crit. Rev. Env. Sci. Tech.* **28**, 41–88.

Letain, T. E., Kane, S.R., Legler, T.C., Salazar, E.P., Agron, P.G., and Beller, H.R. (2007) Development of a genetic system for the chemolithoautotrophic bacterium *Thiobacillus denitrificans*. *Appl. Environ. Microbiol.* **73**, 3265–3271.

Lyles, C. Boopathy, R., Fontenot, Q. and Kilgen, M. (2008) Biological treatment of shrimp aquaculture wastewater using SBR. *Appl. Biochem. Biotechnol.* **151**, 474–479.

Madigan, M., T. and Martinko, J. M. (2006) Brock Biology of Microorganisms. 11 ed., Pearson Prentice Hall, Pearson Education Inc., NJ.

Mahmood, Q., Zheng, P., Cai, J., Wu, D., Hu, B., Islam, E. and Azim, M.R. (2007a) Comparison of anoxic sulfide biooxidation using nitrate/nitrite as electron acceptor. *Environ. Prog.* **26**, 169–177.

Mahmood, Q., Zheng, P., Cai, J., Wu, D., Hu, B., Islam, E. and Li, J. (2007b) Anoxic sulfide biooxidation using nitrite as electron acceptor. *J. Hazard. Mater.* **147**, 249–256.

Manconi, I., Carucci, A., Lens, P. and Rosetti, S. (2006) Simultaneous biological removal of sulphide and nitrate by autotrophic denitrification in an activated sludge system. *Water Sci. Tecnol.* **53**(12), 91–99.

Mansell, B.O. and Schroeder, E.D. (1999) Biological denitrification in a continuous flow membrane reactor. *Water Res.* **33**, 1845–1850.

Mansell, B.O. and Schroeder, E.D. (2002) Hydrogenotrophic denitrification in a microporous membrane bioreactor. *Water Res.* **36**, 4683–4690.

Meraz, M., Monroy, M. and Noyola, A. (1997) Start-up and operation of an inverse fluidized bed. In: *Proc. 8th Int. Conf. on Anaerobic Digestion* **3**, 248–251.

Metcalf and Eddy (2003) Wastewater Engineering Treatment and Reuse. Fourth Ed. McGraw-Hill Intl. Ed. New York, NY.

Meyer, B., Imhoff, J.F. and Kuever, J. (2007) Molecular analysis of the distribution and phylogeny of the soxB gene among sulfur-oxidizing bacteria-evolution of the Sox sulfur oxidation enzyme system. *Environ. Microbiol.* **9**, 2957–2977.

Meza-Escalante, E., Texier, A.C., Cuervo-López, F., Gómez, J. and Cervantes, F.J. (2008) Inhibition of sulfide on the simultaneous removal of nitrate and *p*-cresol by a denitrifying sludge. *J. Chem. Technol. Biotechnol.* **83**, 372–377.

Mijatovic, S.L.I., Cerjan-Stefanovic, S. and Hodzic, E. (2000) Nitrogen removal from fertilizer wastewater by ion exchange. *Water Res.* **34**, 185–190.

Moura, I. and Moura, J.J.G. (2001) Structural aspects of denitrifying enzymes. *Curr. Opin. Chem. Biol.* **5**, 168–175.

Muroyama, K. and Fan, L. S. (1985) Fundamentals of gas-liquid-solid fluidization. *AICHE J.* **31**, 1–34.

Nery, V.D., de Nardi, I.R., Damianovic, M.H.R.Z., Pozzi, E., Amorim, A.K.B. and Zaiat, M. (2007) Long-term operating performance of a poultry slaughter house wastewater treatment plant. *Res. Con. Recycl.* **50**, 102–114.

Oh, S.E., Kim, K.S., Choi, H.C. and Kim, I.S. (2000) Kinetics and physiological characteristics of autotrophic denitrification by denitrifying sulfur bacteria. *Water Sci. Technol.* **42**(3–4), 59–68.

Oh, S.E., Yoo, Y.B., Young, J.C. and Kim, I.S. (2001) Effects of organics on sulphur-utilizing autotrophic denitrification under mixotrophic conditions. *J. Biotechnol.* **92**, 1–8.

Percheron, G., Bernet, N. and Moletta, R. (1997) Start-up of anaerobic digestion of sulfate wastewater. *Biores. Technol.* **61**, 21–27.

Pérez, R., Galí, A., Dosta, J. and Mata-Alvarez, J. (2007) Biological nitrogen removal (BNR) using sulfides for autotrophic denitrification in a sequencing batch reactor (SBR) to treat reject water. *Ind. Eng. Chem. Res.* **46**, 6646–6649.

Rattanapan, C., Boonsawang, P. and Kantachote, D. (2009) Removal of H_2S in down-flow GAC biofiltration using sulfide oxidizing bacteria from concentrated latex wastewater. *Biores. Technol.* **100**, 125–130.

Reijerse, E.J., Sommerhalter, M., Hellwig, P., Quentmeier, A., Rother, D., Laurich, C., Bothe, E., Lubitz, W. and Friedrich, C.G. (2007) The unusual redox centers of SoxXA, a novel c-type heme-enzyme essential for chemolithotrophic sulfur-oxidation of *Paracoccus pantotrophus*. *Biochemistry* **46**, 7804–7810.

Reyes-Avila, J., Razo-Flores, E. and Gomez, J. (2004) Simultaneous biological removal of nitrogen, carbon and sulfur by denitrification. *Water Res.* **38**, 3313–3321.

Rittmann, B.E. and McCarty, P.L. (2001) Environmental Biotechnology: Principles and Applications. McGraw-Hill Intl. Ed. New York, NY.

Sekoulov, I. and Brinke-Seiferth, S. (1999) Application of biofiltration in the crude oil processing industry. *Water Sci. Technol.* **39**, 71–76.

Sher, Y., Schneider, K., Schwermer, C.U. and Van Rijn, J. (2008) Sulfide-induced nitrate reduction in the sludge of an anaerobic digester of a zero-discharge recirculating mariculture system. *Water Res.* **42**, 4386–4392.

Sholeh, Lefroy R.D.B., Blair G.J. (1997) Effect of nutrients and elemental sulfur particle size on elemental sulfur oxidation and the growth of Thiobacillus thiooxidans. *Aust. J. Agr. Resour. Ec.* **48**, 497–501.

Sierra-Alvarez, R., Beristain-Cardoso, R., Salazar, M. Gómez, J., Razo-Flores, E. and Field, J.A. (2007) Chemolithotrophic denitrification with elemental sulfur for groundwater treatment. *Water Res.* **41**, 1253–1262.

Sierra-Alvarez, R., Guerrero, F., Rowlette, P., Freeman, S. and Field, J.A. (2005) Comparison of chemo-, hetero- and mixotrophic denitrification in laboratory-scale UASBs. *Water Sci. Technol.* **52**(1–2), 337–342.

Silva, A.J., Varesche, M.B., Foresti, E. and Zaiat, M. (2002) Sulphate removal from industrial wastewater using a packed-bed anaerobic reactor. *Process Biochem.* **37**, 927–935.

Smith, R.L., Ceazan, M.L. and Brooks, M.H. (1994) Autotrophic, hydrogen-oxidizing, denitrifying bacteria in groundwater, potential agents for bioremediation of nitrate contamination. *Appl. Environ. Microbiol.* **60**, 1949–1955.

Sneath, P.H.A. (1986) Bergey's Manual of Systematic Bacteriology. 4 ed. 2, Williams & Wilkins, MD.

Soares, M.I.M. (2002) Denitrification of groundwater with elemental sulfur. *Water Res.* **36**, 1392–1395.

Sørensen, J., Tiedje, J.M. and Firestone, R.B. (1980) Inhibition by sulfide of nitric and nitrous oxide reduction by denitrifying *Pseudomonas fluorescens*. *Appl. Environ. Microbiol.* **39**, 105–108.

Speece R.E. (1983) Anaerobic biotechnology for industrial wastewater. *Env. Sci. Technol.* **17**, 416A-427A.

Speece R.E. (1996) Anaerobic Biotechnology for Industrial Wastewater. *Archae Press.* Nashville, Tennessee.

Steudel R. (2000) The chemical sulfur cycle. In *Environmental Technologies to Treat Sulfur Pollution. Principles and Engineering.* IWA Publishing.

Steudel, R. and Holdt, G. (1998) Solubilization of elemental sulfur in water by cationic and anionic surfactants. *Angew. Chem. Int. ed. Engl.* **27**, 1358–1359.

Stouthamer, A.H. (1988) Dissimilatory reduction of oxidized nitrogen compounds. Biology of Anaerobic Microorganisms. John Wiley & Sons, Inc., NY.

Suwanto, A., Kaplan, S. (1992) Chromosome transfer in *Rhodobacter spheroides*: Hfr formation and genetic evidence for two unique circular chromosomes. *J. Bacteriol.* **174**, 1135–1145.

Tait, S., Clarke, W.P., Keller, J. and Batstone, D.J. (2009) Removal of sulfate from high-strength wastewater by crystallization. *Water Res.* doi:10.1016/j.watres.2008.11.008.

Tavares, P., Pereira, A.S., Moura, J.J.G. and Moura, I. (2006) Metalloenzymes of the denitrification pathway. *J. Inorg. Biochem.* **100**, 2087–2100.

Thurman, J.L. and Roberts, R.T. (1995) New strategies for the water data center. *J. Soil Water Conserv.* **50**, 530–531.

Tugtas, A.E. and Pavlostathis, S.G. (2007) Effect of sulfide on nitrate reduction in mixed methanogenic cultures. *Biotechnol. Bioeng.* **97**, 1448–1459.

Vaiopoulou E., Melidis P., and Aivasidis A. (2005) Sulfide removal in wastewater from petrochemical industries by autotrophic denitrification. *Water Res.* **39**, 4101–4109.

Van Haandel, A.C. and Lettinga, G. (1994) Anaerobic Sewage Treatment. John Wiley & Sons. Singapore.

Wang, A.-J., Du, D.-Z., Ren, N.-Q. and Van Groenestijn, J.W. (2005a) An innovative process of simultaneous desulfurization and denitrification by *Thiobacillus denitrificans*. *J. Environ. Sci. Health* **40**, 1939–1949.

Wang., Y., Huang, X. and Yuan, Q. (2005b) Nitrogen and carbon removals from food processing wastewater by an anoxic/aerobic membrane bioreactor. *Process Biochem.* **40**, 1733–1739.

Watkinson, J.H. and Blair, G.J. (1993) Modelling the oxidation of elemental sulfur in soils. *Nutr. Cycl. Agroecosyst.* **35**, 115–126.

Winterstein, C. and Ludwig, B. (1998) Genes coding for respiratory complexes map on all three chromosomes of the *Paracoccus denitrificans* genome. *Arch. Microbial.* **169**, 275–281.

Yamamoto-Ikemoto, R., Komori, T., Nomura, M., Ide, Y. and Matsukami, T. (2000) Nitrogen removal from hydroponic culture wastewater by autotrophic denitrification using thiosulfate. *Water Sci. Technol.* **42**(3–4), 369–376.

Yavuz, B., Türker, M. and Engin, G.Ö. (2007) Autotrophic removal of sulphide from industrial wastewaters using oxygen and nitrate as electron acceptors. *Environ. Eng. Sci.* **24**, 457–470.

Yetilmezsoy, K. and Sakar, S. (2008) Improvement of COD and color removal from UASB treated poultry manure wastewater using Fenton's oxidation. *J. Hazard. Mater.* **151**, 547–558.

Zeng, H. and Zhang, T.C. (2005) Evaluation of kinetic parameters of a sulfur-limestone autotrophic denitrification biofilm process. *Water Res.* **39**, 4941–4952.

Zhang, T.C. and Lampe, D.G. (1999) Sulfur:limestone autotrophic denitrification processes for treatment of nitrate-contaminated water: batch experiments. *Water Res.* **33**, 599–608.

Zhao, W. L. Q-L. and Liu, H. (2009) Sulfide removal by simultaneous autotrophic and heterotrophic-denitrification process. *J. Hazard Mater.* **162**, 848–853.

13

Nitrogen removal in aerobic granular systems

M. Figueroa, A. Val del Río, N. Morales,
J.L. Campos, A. Mosquera-Corral and R. Méndez

13.1 INTRODUCTION

The biological treatment of wastewater is often accomplished by means of conventional activated sludge systems that generally require large surface area for implantation. This is caused by the relatively poor settling characteristics of activated sludge, resulting in low solids concentrations inside the aeration tanks and in low maximum hydraulic load of secondary sedimentation tanks. Therefore, the conventional wastewater treatment installations have some inherent disadvantages, like the low volumetric conversion capacities, focussed mainly on the removal of easily degradable organic compounds and the high amounts of sludge production.

© 2009 IWA Publishing. *Environmental Technologies to Treat Nitrogen Pollution: Principles and Engineering*, Edited by Francisco J. Cervantes. ISBN: 9781843392224. Published by IWA Publishing, London, UK.

Due to the sensitivity of nitrifying bacteria to environmental factors, as well as their lower growth rates, it is difficult to obtain and maintain sufficient nitrifying bacteria in conventional wastewater treatment plants. Thus, the nitrification process is the limiting step. This fact provokes the necessity of the use of high solid retention times (SRT) and of the large volume of the system. Therefore, the development of new processes and technologies is required in which the efficient nitrogen removal is obtained together with the organic matter removal.

To overcome the drawbacks associated with activated sludge systems, in the 1990s, the use of biofilm systems for aerobic processes was promoted. In such systems, biomass adheres to both fixed and mobile carriers, which allows that the biomass retention time does not depend on the settling characteristics of biomass (Tijhuis *et al.* 1994). The main feature of these technologies is their ability to treat high volumetric loads, occasionally without an independent sludge/effluent separation step. However, the main disadvantage of these systems is the relatively high investment costs (de Bruin *et al.* 2004) associated to the necessity of using support materials for the biomass attachment.

In order to improve the biofilm technologies and get the benefits of biofilm systems, but without the use of carrier materials, in recent years, research showed that it is possible to grow granular sludge in either continuous or batch-wise operated systems. Compared to conventional activated sludge flocs, aerobic granules present regular, dense, and strong physical structure, good settling ability, allow high biomass retention in the systems, and possess the ability to withstand shock load. Aerobic granules are usually produced to aerobically remove organic matter. Once aerobic granules are established inside the reactor, the processes of nitrification and denitrification can occur simultaneously inside the system.

The first studies in aerobic granular sludge were performed in the 1990s in continuous operated systems. The development of aerobic granules has been firstly reported by Mishima and Nakamura (1991). These authors used a continuous Aerobic Upflow Sludge Blanket (AUSB) reactor to treat a synthetic wastewater previously aerated with pure oxygen to oxidise organic matter and obtained well-settling aerobic granules with diameters from 2 to 8 mm. In recent years, an increasing number and diversity of studies reflecting research on aerobic granular sludge were developed and showed that it is possible to grow aerobic granules in Sequencing Batch Reactors (SBRs). This systems have been used for organic carbon removal (Beun *et al.* 1999; Dangcong *et al.* 1999; Etterer and Wilderer 2001; Tay *et al.* 2001a; Sun *et al.* 2006) and simultaneously for nutrient (nitrogen and phosphorus) removal (Beun *et al.* 2001; Lin *et al.* 2003; Yang *et al.* 2003; de Kreuk *et al.* 2005a; Qin *et al.* 2005; Wang *et al.* 2007).

Evidence shows that the problems encountered in the suspended growth systems for nitrogen removal, such as sludge bulking, large space requirement for the treatment plant installation, washout of nitrifying biomass and large production of waste sludge, would be overcome by developing and applying nitrogen removing granules (Yang *et al.* 2003).

13.2 FUNDAMENTALS OF AEROBIC GRANULATION

13.2.1 Type of reactors

The first works reported by van Benthum *et al.* (1996) about nitrifying granular sludge were developed in continuously operated biofilm airlift suspension (BAS) reactors with basalt as carrier material where continuously detached biofilm fragments led to the formation of nitrifying granules. Later, Campos *et al.* (2000) reported the formation of nitrifying granular sludge in an airlift system operated without biomass carrier and fed synthetic media. These authors oxidised an ammonia loading rate of 4.1 g NH_4^+-N/(L·d) to nitrate and accumulated 12 g VSS/L of nitrifying particles with a sludge volumetric index (SVI) of 17 mL/g VSS, a density of 100 g VSS/$L_{particle}$ and a particle average diameter from 0.22 to 0.36 mm. Tsuneda *et al.* (2003) also obtained nitrifying granules with a diameter of 0.35 mm in a continuous system (an aerobic upflow fluidized bed reactor, AUFB) for inorganic wastewater. The common feature of these works was that no control of the formation of the nitrifying granules was applied. The aerobic granulation of nitrifying organisms has been obtained in continuous culture only in scarce experimental works and no full scale continuous reactors have been reported to operate with aerobic granular sludge.

For this reason so far, most research has been focused in sequencing batch reactors (SBRs) and the aerobic granules have successfully been cultivated in discontinuous systems where simultaneous nitrogen and organic matter removals occurred (Tay *et al.* 2002a; Yang *et al.* 2003; Arrojo *et al.* 2004; de Kreuk *et al.* 2004). Compared to continuous operated reactor systems, a discontinuous operated system is advantageous for many applications and the good settling characteristics of the biomass allow in a SBR process a short standstill time for settling, allowing a more efficient use of reactor volume (Beun *et al.* 2002).

The SBR systems operate in cycles which comprise different phases: filling, aeration, settling, discharging, etc. In SBR systems, the settling phase substitutes the performance of the external settler in the continuously operated reactors with the consequent saving in surface requirements.

Regarding the reactor shape, the primary design criterion is based on the assumption that sludge granules will be formed if flocculent biomass is washed out. Sludge granules have a high settling velocity compared to sludge flocs, because granules are denser. Due to the fact that the settling velocity is an important selection criterion, a large H/D ratio (column height/column diameter) is advantageous. Values of the H/D ratio between 20 and 5 and the absence of an external settler result in a reactor with a small footprint (Beun *et al.* 1999; Arrojo *et al.* 2004). Furthermore, the good settling characteristics of the biomass allow a short stand-still time for settling, allowing more time for biological purification.

Furthermore, in a column-type upflow reactor, a large H/D ratio can ensure a longer circular flowing trajectory, creating a more effective hydraulic attrition to microbial aggregates (Liu and Tay 2002) and improving the mixture of the media.

13.2.2 Operational conditions

Evidence shows that aerobic granulation is a gradual process from seed sludge to compact aggregates, further to granular sludge and finally to mature granules (Tay *et al.* 2001b). Obviously, to obtain aerobic granulation, a number of conditions have to be fulfilled. Existing literature on aerobic granular sludge typically focuses on a few parameters that are identified to influence granule formation, although the major part of the works evaluate these factors by separate.

- Substrate composition
- Feast-Famine regime
- Hydrodynamic shear force
- Short settling times

13.2.2.1 Substrate composition

Aerobic granules have successfully been cultivated in SBRs with a wide variety of organic substrates including glucose, acetate, phenol, starch, ethanol, molasses, sucrose, peptone and other synthetic media components (Beun *et al.* 1999; Dangcong *et al.* 1999; Jiang *et al.* 2002; Moy *et al.* 2002; Tay *et al.* 2002b; Zheng *et al.* 2005; Sun *et al.* 2006; Adav *et al.* 2007a;b) and with wastewater from both domestic and industrial origin (Arrojo *et al.* 2004; Schwarzenbeck *et al.* 2004; Cassidy and Belia 2005; Inizan *et al.* 2005; Schwarzenbeck and Wilderer 2005; Su *et al.* 2005; Liu *et al.* 2007; Wang *et al.* 2007; Figueroa *et al.* 2008).

These researches indicate that the formation of aerobic granules is possible treating different substrates; however, evidence shows that the substrate influences the granule microstructure and species diversity (Tay *et al.* 2001b; Liu *et al.* 2003). As an example, the aerobic granules grown on glucose exhibited a filamentous structure, while those fed with acetate presented a non filamentous structure in which rod like species predominated (Tay *et al.* 2001b).

13.2.2.2 Feast-Famine regime

One condition to achieve aerobic granulation is the feeding pattern, which is established in the SBRs. The fact that the feeding is supplied in very short times allows the existence of two different phases during the reaction period: a phase in which the substrate is depleted to a minimum value, followed by an aerobic starvation phase in which the external substrate is no longer available. Thus, micro-organisms in SBR are subjected to a periodic feast and famine regime, called periodic starvation (Wilderer *et al.* 2001). Some authors state that under the periodic feast-famine conditions, bacteria become more hydrophobic and high cell hydrophobicity in turn facilitates microbial aggregation (Tay *et al.* 2001a; Liu *et al.* 2004). When bacteria are subjected to a periodic feast-famine regime, microbial aggregation could be an effective strategy for cells against starvation. In fact, the periodic feast-famine regime in SBR can be regarded as a kinetic microbial selection pressure that determines the physical properties of the formed aggregates (Chudoba *et al.* 1973). This theory hypothesises that the filamentous microorganisms (K-strategists) are slow-growing organisms that can be characterised by having maximum growth rates (μ_{max}) and affinity constant (K_s) lower than the floc-forming bacteria (r-strategists). In systems like continuously fed completely mixed reactors, where the substrate concentration in the liquid media is low and close to the Ks value, the operational conditions are such that filamentous bacteria exert a higher specific growth rate than floc-forming bacteria. In systems where the substrate concentration is high, like in SBR systems, the filamentous bacteria should be suppressed since their growth rate is expected to be lower than that for floc-forming bacteria.

13.2.2.3 Hydrodynamic shear force

Hydrodynamics are crucial in the aerobic granular systems in order to favour the substrates mass transfer and the physical properties of the biomass. The structure of aerobic granules is influenced by the hydrodynamic shear force present in a bioreactor. Evidence shows that a high shear force favours the formation of smooth, dense and stable aerobic granules (Beun *et al.* 1999).

Tay *et al.* (2001c) operated four column SBR reactors at different superficial upflow air velocities (from 0.3 to 3.6 cm/s) and found that aerobic granules could be formed only above a threshold shear force value in terms of superficial upflow air velocity above 1.2 cm/s. Adav *et al.* (2007c) compared the granulation processes in three reactors aerated at different intensities and observed that at low aeration intensity no granules were formed, at intermediate intensity granules with filaments appeared and at high aeration rate mature and stable granules were formed. These experiments linked the formation of aerobic granules not only to the shear force but also to the dissolved oxygen concentration, because the lower the aeration intensity the lower the oxygen concentration in the liquid phase. Deficiencies in dissolved oxygen (DO) or nutrients are favourable for filament growth (Gaval and Pernell 2003, Liu and Liu 2006; Mosquera-Corral *et al.* 2005a). Therefore, in these experiments a high hydrodynamic shear force would not only promote compact granules, but also would provide sufficient oxygen to suppress filament growth for a long-term operational stability (Adav *et al.* 2008).

Further research demonstrated that hydrodynamic shear force favours the production of extracellular polysaccharides (EPS) (Tay *et al.* 2001a; c). Microorganisms in high shear environments adhered by secreting EPS to resist damage of suspended cells by environmental forces (Trinet *et al.* 1991). It is well known that EPS can mediate both cohesion and adhesion of cells and play a crucial role in maintaining the structural integrity in a community of immobilised cells.

Chen *et al.* (2008) studied the effect of hydrodynamic shear force combined with the substrate loading and found that when high air superficial velocities are applied, the treated organic loading rates (OLRs) were better and granules with great settleability were obtained, while for low superficial air velocity values the granules deterioration occurred and the stable operation was limited to a narrow range of OLRs.

13.2.2.4 Short settling times

Strategies to limit the amount of flocs in an aerobic granule system include the use of short settling and discharging times. In a SBR, wastewater is treated in successive cycles each lasting a few hours. At the end of every cycle, the biomass is settled before the effluent is withdrawn. Sludge that cannot settle down within the given settling time is washed out of the reactor through a fixed discharge port. Basically, a short settling time preferentially selects for the growth of good settling bio-particles. Thus, the settling time exerts a major hydraulic selection pressure on the microbial community in the way that a short settling period will eventually select for biomass particles with a high settling velocity (Beun *et al.* 1999). As a consequence, during the start-up period due to the biomass wash-out, the quality of the effluent in terms of the suspended solid content is not appropriated for discharge. The performance of the system is not entirely predictable or controllable, so a post-treatment step will be required to fulfil the effluent standards for municipal wastewater treatment. An immediate discharge of effluents from aerobic granular sludge reactors directly into receiving waters, as proposed in the past, is hence not permissible.

In aerobic granulation research, a short settling time has been commonly used to enhance aerobic granulation in SBR (Jiang *et al.* 2002; Lin *et al.* 2003; Liu *et al.* 2003; Yang *et al.* 2003; Wang *et al.* 2004; Hu *et al.* 2005). In fact, at a long settling time, poorly settling sludge flocs cannot be withdrawn effectively; and they may in turn out-compete granule-forming bio-particles. As a result, aerobic granulation could fail in SBR run at long settling times.

Therefore, granular sludge requires less time to settle than flocs in the SBR cycle. The settling time is an important design parameter and is related to the shape of the reactor. The settling time has to be chosen in such a way that the settling velocity is higher than that corresponding to activated sludge flocs with a value of 9 m/h (Campos *et al.* 1999). In this way Arrojo *et al.* (2004) operated an aerobic granular SBR for organic matter and nitrogen removal designed to retain inside the system only aggregates of biomass with settling velocities higher than 9 m/h. Qin *et al.* (2004a, b) studied the effect of the settling time, by imposing settling velocities from 1.9 to 7.6 m/h, on aerobic granulation in an SBR and found that aerobic granules were successfully cultivated in the experiments performed with the larger value, while a mixture of aerobic granules and suspended sludge developed in the experiments run at lower settling velocities.

13.3 NITROGEN REMOVAL IN AEROBIC GRANULES

13.3.1 Biological processes in aerobic granules

Regarding their bioengineering functions, aerobic granules present in their structure different zones where different environmental conditions exist. These zones are located as concentric layers which enables simultaneous organic matter and nitrogen removal, the last via the combined nitrification-denitrification processes (Figure 13.1).

The way the different processes occur in the granules varies along the operational cycle since the reactor is sequentially fed and feast and famine periods exist. In the feast period (Figure 13.2a) the concentration of external organic carbon (for example, acetate) usually measured as chemical oxygen

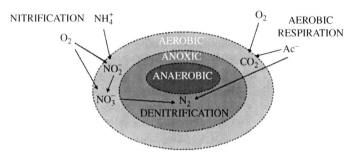

Figure 13.1. Biological processes occurring into the aerobic granule

demand (COD) is high. This substrate will therefore diffuse into the granules completely. Dissolved oxygen (DO) will have a much smaller penetration depth because it will be consumed very rapidly by autotrophic and heterotrophic organisms in the outer layers of the granules. In the feast period DO is used for nitrification, for aerobic conversion of acetate, and for aerobic biomass growth. Since the autotrophic micro-organism need oxygen, they tend to be located where this is available. In the case of granular sludge, this is the outermost layer of the granules. The autotrophic organisms convert ammonia (NH_4^+) into nitrogen oxides (NO_x^-). The formed nitrogen oxides (NO_x^-) will diffuse towards the centre of the granules, but also towards the liquid phase surrounding the granules. In the centre of the granules acetate can be stored anoxically (using NO_x^-) as poly-hydroxybutyrates (PHB) by the heterotrophic organisms. In the famine period (Figure 13.2b) the DO penetration depth is larger than in the feast period since the oxygen concentration in the liquid is higher and the oxygen

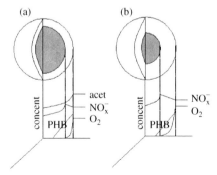

Figure 13.2. Schemes of the evolution of the concentration profiles of acetate, PHB, NO_x^- and O_2 during the feast (a) and the famine (b) period in a SBR

consumption inside the granule lower. In the centre of the granules NO_x^- is present. The stored PHB in the centre of the granules can be used as organic carbon source for the denitrification. Aerobic conversion of PHB occurs, and nitrification as long as there is NH_4^+ present. Nitrogen removal occurs though the combination of nitrification-denitrification processes inside the aerobic granules.

13.3.2 Characteristics of the granular biomass

The aerobic granules are characterised by presenting denser packed structures than those of conventional activated sludge (Figure 13.3), and are known to exhibit attributes of: regular, smooth and nearly round in shape; excellent settleability; dense and strong microbial structure; high biomass retention; ability to withstand at high organic loading and tolerance to toxicity (Adav et al. 2008).

A large number of physical parameters are conventionally measured to study the characteristics of aerobic granules: settling velocity, density, sludge volumetric index and particle diameter.

The settling velocity determines the efficiency of solid-liquid separation; thereby, high values of this parameter correspond to very large biomass retention inside of the system. Granular sludge demonstrates high settling velocities facilitating efficient solid-liquid separation. In previous researches, it has been established that the settling velocity of the granular biomass varied from 20 to 150 m/h (Tay et al. 2001b; Yang et al. 2003; Cassidy and Belia 2005; Wang et al. 2007) and this value was significantly higher than that of activated sludge flocs (lower than 9 m/h) (Campos et al. 1999).

The SVI is related to the settling velocity, low values of SVI indicate that the granules have a compact structure. Flocculent activated sludge used as inoculum

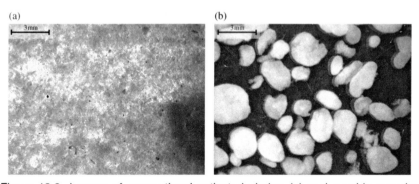

Figure 13.3. Images of conventional activated sludge (a) and aerobic granular biomass (b). The bar size is 3 mm

in many research works on aerobic granulation presented SVI values between 100 and 250 mL/g VSS. This biomass gradually evolved to granular biomass with SVI values between 20 and 60 mL/g VSS (Arrojo *et al.* 2004; Mosquera-Corral *et al.* 2005a; Tay *et al.* 2002a).

Granules have a larger density than flocs, allowing a bioreactor to maintain high biomass concentrations. The density of the granules can vary between 10 and 62 g VSS/L$_{granule}$ (Thanh *et al.* 2008).

The particle size of aerobically grown microbial granules depends on different parameters as the balance of biomass growth, production of cell binding EPS, cell detachment from the granule and the composition of the feeding media. Performed research works show that the granule size can vary in a wide range. Arrojo *et al.* (2004) obtained granular sludge in a SBR treating the effluent from a laboratory of analysis of dairy products for carbon and nitrogen removal with a size distribution comprehended between 0.25 and 4.0 mm. Figueroa *et al.* (2008) reported a size distribution for the aerobic granules from 0.6 to 6.5 mm treating a fish canning effluent to remove carbon and nitrogen. In view of biological viability and physical properties, the operational size-ranges for optimal performance and economic effective aerobic granular sludge, the average diameter recommended is around 1.0–3.0 mm (Toh *et al.* 2003). As the diameter increases the aerobic granules undergoes serial morphological and physical changes that could cause problems to the reactor operation. Furthermore, the bigger the aerobic granule the worse the mass transfer from the liquid and gas phases to the solid phase and the higher the amount of biomass which has no biological activity.

13.4 PARAMETERS INFLUENCING THE NITROGEN REMOVAL IN AEROBIC GRANULAR SYSTEMS

In aerobic granular biomass, simultaneous removal of organic matter and nitrogen compounds implies the coexistence of aerobic heterotrophic, nitrifying and denitrifying populations. Therefore, a series of parameters that influence the operation must be taken into account.

13.4.1 DO concentration

DO concentration is an important parameter in the operation of aerobic wastewater treatment systems. Performed experimental work shows that if aerobic conditions are maintained inside the reactor by sufficient aeration, the DO concentration is not a decisive parameter for aerobic granulation, but it has a pronounced effect on the efficiency of the denitrification process. Therefore, the amount of oxygen supply has to be optimised in order to minimise the energy consumption in full-scale aerobic systems and to promote nitrogen removal via denitrification.

Beun *et al.* (2001) predicted, through a simulation model, that the optimal DO value is expected to be around 40% air saturation for adequate nitrogen removal in aerobic granular systems, but this has not been experimentally verified. Later, Mosquera-Corral *et al.* (2005a) attempted to study experimentally this fact and revealed the significant role of oxygen on the simultaneous nitrification and denitrification processes. These authors studied the short- and long-term effects of decreased oxygen concentrations on the aerobic granular system and observed that a short-term oxygen reduction did not influence the organic substrate uptake rate and nitrogen removal was favoured by decreased DO concentrations, while long-term oxygen reduction (40% saturation level), promoted an increase of the nitrogen removal, but the beginning of the granules disintegration and the biomass washout.

13.4.2 Organic matter content

There is evidence suggesting that the presence of organic carbon can affect nitrification (Moreau *et al.* 1994; Ohashi *et al.* 1995; Ballinger *et al.* 2002), although raw wastewater often contains both organic matter and nitrogen. Therefore, aerobic granules capable of simultaneously removing organic carbon and nitrogen compounds are highly desired.

Mosquera-Corral *et al.* (2005b) operated an aerobic granular SBR to study the removal of nitrogen by assimilation and simultaneous nitrification-denitrification, the predominance of each mechanism being related to the organic matter to nitrogen (COD/N) ratio in the feeding media (Table 13.1). Granules exhibited different denitrifying activities depended on the COD/N ratio. As occurred in biofilms, DO was present only in the outer layers (Tijhuis *et al.* 1994) and the inner layers, which were maintained under anoxic conditions, received the carbon source and nitrate to support the denitrification process. If the carbon source was not high enough (low COD/N ratios), the denitrification process was not completed and nitrite was produced as intermediate. In spite of the changes in the feeding composition, the granules maintained their structures and a stable nitrifying granular sludge was obtained from heterotrophic granular sludge by decreasing the COD/N ratio.

The effect of the COD/N ratio (from 3.3 to 20 g/g) on the formation of aerobic granules was also studied by Yang *et al.* (2005). These authors observed that COD/N ratios in the range studied had no significant impact on the

Table 13.1. Percentages of nitrogen assimilated and denitrified referred to the inlet ammonia concentration (Mosquera-Corral et al. 2005b)

COD/N (g/g)	Nitrogen assimilated (%)	Nitrogen denitrified (%)
0	0	0
1.25	0	0
2.5	4	29–37
5.0	8	40–46
15.0	39	14–16

formation of aerobic granules; however, they had remarkable effects on the activity distribution of heterotrophic, ammonium and nitrite oxidising bacteria in the microbial granules. The nitrifying activity increased ~10% with the decrease of the COD/N ratio from 20 to 3.3 g/g, while the heterotrophic activity experienced a reduction. At the lowest COD/N ratio tested, heterotrophs became much less dominant, whereas nitrifying populations would be able to compete with them, and became an important component of the aerobic granules.

13.4.3 pH

The pH value is an important environmental factor in microbial growth; however, research of its effect on aerobic granulation has not been widely studied.

The research conducted so far, with aerobic granular biomass for simultaneous organic matter and nitrogen removal, used feeding media with pH values between 7.0 and 8.5. During a cycle of operation, the pH inside the reactor increase in the first minutes of the cycle due to the oxidation of organic matter. Yang et al. (2004) observed an evolution of the pH value from an initial value of 7.5 to 8.5. After full organic matter removal from the liquid media (end of the feast phase), nitrification starts and the pH decreases. However, it had been reported that the optimal pH for nitrification fell into a very narrow range of 7.8–8.0, and a pH value higher than the optimal value would inhibit the activity of nitrifying bacteria due to the pH-enhanced production of free ammonia. Thus, some authors utilised a strategy based on pH control in aerobic granular reactor to maintain the free ammonia concentrations below the inhibitory values for nitrification (Cassidy and Belia 2005; de Kreuk et al. 2005a). However, the simultaneous organic matter and nitrogen removal can be obtained without pH control (Arrojo et al. 2004).

In a biological treatment system, the sludge usually contains fungi, protozoa and bacteria. Most fungi prefer to grow in a low-pH medium, which is unfavourable for the growth of most bacteria. To prevent the drop in pH caused by nitrification and inhibit possible fungal growth, it would be necessary to guarantee a reasonably high alkalinity inside the system by means of controlling the bicarbonate concentration.

13.4.4 Bicarbonate

The presence of enough amount of bicarbonate during nitrification is required to maintain the pH value in an adequate range for nitrifiers and to act as carbon source for bacterial growth. The amount of inorganic carbon needed for ammonium and nitrite oxidation, according to the nitrification process, is of 0.086 mol bicarbonate/mol ammonium oxidised and 0.024 mol bicarbonate/mol nitrite oxidised, respectively (Belser 1984).

The inorganic carbon (IC) is usually provided by the organic matter oxidation by heterotrophic bacteria, which frequently occurs simultaneously to the nitrification process and it ensures that enough IC is available for autotrophic growth. Tokutomi *et al.* (2006) carried out activity assays for ammonia and nitrite oxidizers at different IC concentrations (from 2 to 240 mg IC/L) and found that nitrite oxidizers were not affected while the activity of ammonia oxidisers almost duplicated when the IC concentration increased from 2 to 40 mg IC/L.

13.4.5 Temperature

The effect of temperature changes on the conversion processes and the stability of aerobic granular sludge were studied by de Kreuk *et al.* (2005b). These authors observed that the aerobic granulation was not possible when the reactor was started up at low temperatures (8°C). However, if the reactor was started up at a higher temperature (20°C) and it was lowered afterwards (8°C), no significant effect on the granules stability was observed. As a result of operating the reactor at low temperatures, the biological activity decreased in the outer layers of granules and the oxygen penetration depth could increase, which resulted in a larger active aerobic biomass volume, compensating the decreased activity of individual organisms. Consequently, the denitrifying capacity of the granules decreased at reduced temperatures, resulting in an overall poorer nitrogen removal capacity. Nevertheless, the temperature dependency of the nitrification process was lower for aerobic granules than that usually found for activated sludge (de Kreuk *et al.* 2005b).

13.5 DISTRIBUTION OF MICROBIAL POPULATIONS INSIDE THE GRANULES

Aerobic granules are composed by a large diversity of microorganisms. In order to establish a relationship between the different microbial populations and the macroscopic activities carried out in the granules, it is necessary to identify the main populations present and their location inside the aggregates.

Scanning and transmission electron microscopy (SEM and TEM) are employed to visualise the detailed architecture of aerobic granules. However, the exact structure of aerobic granules is dependent on reactor operation, species selection, and growth morphology. The general observations suggest that aerobic granules have a heterogeneous structure, where different structured layers can be identified: usually an outer shell with high biomass density and an inner core having a relatively low biomass density (Wang et al. 2005).

The external layer is actually made up of large cauliflower-like outgrowths, with voids and channels located on the boundary line between two of them. The channels were found to play a key role in the transport of substrates and metabolites in and out of the granules (Ivanov et al. 2005; Zheng and Yu 2007). These outgrowths consist mainly of bacteria, stalked ciliates or fungal filaments and EPS (a matrix of polymers, including polysaccharides, proteins, glycoproteins, nucleic acids, phospholipids and humic acids), that provide the cohesive material to maintain the bacteria bound to each other (Figure 13.4). The core zone of the granules comprises a dense mixture of bacteria and EPS, and has an inner part that may contain dead cell debris, as it was reported by McSwain et al. (2005). Mass transport limitations will define the size of each of the layers of aerobic, anaerobic, and/or dead biomass within granules (Gapes et al. 2004).

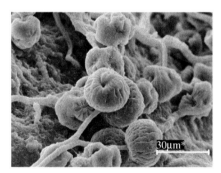

Figure 13.4. SEM image of a bunch of ciliates from the surface of an aerobic nitrifying granule

The nitrifying organisms, that is aerobic ammonia oxidising bacteria (AOB) and nitrite oxidising bacteria (NOB), are important microbial populations in nitrogen removal processes due to their ability to oxidise ammonium and nitrite to nitrite and nitrate, respectively, which can then be reduced to dinitrogen gas by denitrifiers. Members of the genus *Nitrosospira* are generally less abundant than the genus *Nitrosomonas* in activated sludge, biofilms or aerobic granular sludge wastewater treatment plants, from where *Nitrosomonas eutropha*, *Nitrosomonas europaea* and *Nitrosococcus mobilis* have been isolated.

The microbial occurrence of different microorganisms is related to their kinetic characteristics, so the performance of each reactor system can modify the populations distribution. *Nitrosospira* spp. are generally regarded as K-strategists, those that are more abundant in low ammonium conditions, while *Nitrosomonas* spp. are r-strategists, those that proliferate under high substrate concentrations (Schramm *et al.* 1999).

In the case of the NOB, *Nitrospira*-like bacteria were shown to be more abundant than *Nitrobacter* spp. in wastewater treatment plants (Burrell *et al.* 1998) and in a nitrite-oxidising sequencing batch reactor (Juretschko *et al.* 1998). In this case, Schramm *et al.* (1999) suggested that *Nitrospira* are K-strategists, meaning that they exist under low nitrite and oxygen concentrations conditions, while *Nitrobacter*, as an r-strategist, thrives if nitrite and oxygen are present in high concentrations.

The AOB and NOB bacteria populations are distributed in different zones inside the aerobic granules and this distribution can be detected by *in situ* techniques. Confocal laser scanning microscopy (CLSM) combined with the technique of fluorescent *in situ* hybridization (FISH) are very useful tools to study the microbial spatial distribution within the granules due to the wide range of probes designed to identify the most common microbial communities responsible for nitrogen removal. Microelectrodes have also been used to investigate the spatial distributions of various microbial activities in biofilms and in microbial aggregates (de Beer *et al.* 1993; Okabe *et al.* 1999) by means of the measurement of parameters (ammonia, oxygen, etc.) at different depths inside the granules. The combination of these methods provides reliable and direct information about relationships between *in situ* microbial activity and the occurrence of specific micro-organisms in complex microbial consortia as the aerobic granules.

The distribution of nitrifying populations in aerobic granules will differ according to the operation of the system in terms of treated nitrogen loading rate (NLR), solids retention time and wastewater composition (Weber *et al.* 2007); however, a general approximation can be performed. According to the strict aerobic metabolism of AOB and NOB, they should also be located on the outer part of the granule where oxygen is available (Figure 13.5). The oxygen penetration is the limiting factor for both nitrification activity and abundance of the nitrifying population in an aerobic granule or biofilm. The AOB are distributed in densely packed spherical clusters of around 2 μm to 15 μm in diameter, containing probably up to several hundred cells (Egli *et al.* 2003). The depth of this AOB layer is between 100–200 μm (Lemaire *et al.* 2008; Tsuneda *et al.* 2003).

The NOB are also located in the aerobic zone, closely to the AOB clusters as a direct result of the sequential metabolism of ammonia via nitrite to nitrate, but with less dense colonies, and with some single cells and colonies located in the upper anoxic layers (Schramm *et al.* 1996).

Other populations coexist in aerobic nitrifying granules. Although organic matter was not contained in feed solution, heterotrophs could be detected by, growing on cell lysis products in autotrophic nitrifying aerobic granules. Anaerobic bacteria (*Bacteroides* spp.) were observed at a depth of 800–900 μm from the surface of aerobic granules (Tay *et al.* 2002c). Anaerobiosis and cell death in the granules is probably promoted by the formation of polysaccharide plugs in the channels and pores that diminished the mass transfer rate of both nutrients and metabolites.

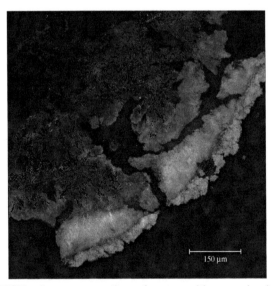

Figure 13.5. FISH of a cross-section of an aerobic granule, brighter zones represent fluorescent signal of probe NEU653 (*Nitrosomonas* spp.)

The knowledge of the type populations and their distribution inside the granules is relevant information when an aerobic granular reactor is operated for nitrogen removal.

13.6 MODELLING OF AEROBIC GRANULAR SYSTEMS

As it has been indicated in previous sections, a wide range of parameters influence the characteristics and performance of aerobic granules. Therefore, mathematical models represent an important and fundamental tool to know in detail the complex processes that take place in aerobic granulation. Furthermore, they can provide a solid foundation for design and operation without the time consumption and materials expense of the experimental approach. Anyway, models, to be valuable, must be calibrated and validated with experimental data.

For the conventional floc-based activated sludge systems, the Activated Sludge Model (ASM) established by the International Water Association (IWA), provides a consistent framework for the description of biological processes. The ASM can be used as a basis to develop the aerobic granule models, but introducing several modifications. The IWA developed four Activated Sludge Models, the ASM1, ASM2, ASM2d and ASM3 (Henze *et al.* 2000). The ASM1 was the structure and platform for further development. The ASM1 allows the simulation of organic matter removal and biological nitrification and denitrification (N-removal) in activated sludge systems. The ASM2 is an extension of the ASM1 and includes biological phosphorous removal processes, and the ASM2d includes denitrifying PAOs. The ASM3 is a new modelling platform based on recent developed knowledge of the activated sludge processes and includes the possibility of following up the concentrations of internal storage compounds. The decay process from ASM1 is replaced by an endogenous respiration process in ASM3, a more realistic approach from a microbiological point of view (van Loosdrecht *et al.* 1999). By working with granules, the accumulation of storage polymers is mandatory due to the feast/famine regime. During the feast period, organic substrates are accumulated in form of storage polymers without ammonia consumption. Therefore, the raw ASM1 approach is not valid and storage processes have to be incorporated in the model (Su and Yu 2006).

The matrix structure and organization of ASM models possibilities an easy integration of these models into several simulation programs: AQUASIM, WEST, Plan-It STOAT... and also the developed of the model using mathematical tools as MATLAB.

In the ASM, biomass is assumed in suspension in the bulk liquid. However, this approach is not suitable when bacteria grow forming colonies that arise in biofilms, aggregates or aerobic granules. The aggregation of bacteria creates a separation between the bulk liquid and the biological media. It can be assumed that the bulk liquid is well mixed and therefore homogeneous; however, in the bacterial phase, biological processes and substrate diffusion velocities will provoke concentration gradients of substrates: DO, COD, ammonia, nitrate, nitrite, phosphate... Those concentration gradients are influenced by many factors, e.g. diffusion coefficients, conversion rates, granule size, biomass spatial distribution, density, etc (Figure 13.6). There are several influences between every factor, thus the effect of separate factors cannot be studied experimentally, and the simulation models can give more insight at a microscopic scale by solving differential equations.

Lübken *et al.* (2005) investigated if model ASM3 could be used as a first simplification to simulate nutrient removal with aerobic granular sludge in a SBR. They modified two model parameters through a calibration procedure: the aerobic endogenous respiration coefficient and the half-saturation coefficient for oxygen, both for the heterotrophic biomass. Simulation tool SIMBA was used due to its possibility of simulate SBR process. The model proved to be capable of describing the performance of a lab-scale SBR reactor concerning COD, ammonium and nitrate in the bulk liquid. However, for describing the processes inside of a granule, biofilm processes should be included in the model.

The mathematical modelling of biofilms has gained increasing interest in the last decade and a wide range of biofilm models of different complexity, applied to different types of benchmark problems, has been developed. Aerobic granules can be viewed as a special form of biofilm, but without carriers for biofilm

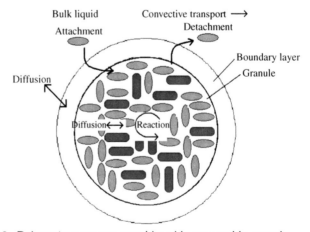

Figure 13.6. Relevant processes considered in an aerobic granule

attachment. For this reason, models of aerobic granular systems have been based on biofilms models. Modelling of biofilm or granular structures is possible using one dimensional models (1-D) or multidimensional models (2-D and 3-D). One dimensional models are simpler and therefore require less computational effort. Nevertheless, in 1-D models some assumptions are made. It is assumed that the substrates gradients are some orders of magnitude higher in the perpendicular direction to the attachment surface than in the plan parallel to it. Important characteristics derived from the dynamics of biofilm structure must be taken into account once assumed 1-D transport. Some examples of properties that must be explicitly defined are: external and internal mass transfer coefficients, changes in pore volume and mobility of bacterial species inside the biofilm matrix (Xavier *et al.* 2005).

Beun *et al.* (2001) described the COD and N removal in a granular sludge batch reactor (GSBR) employing a 1-D model implemented in AQUASIM considering the biomass as spherical particles with a constant diameter of 2.5 mm. In AQUASIM, the spherical granules are simulated as a flat biofilm in Cartesian coordinates (granules are converted in a biofilm without carrier material with a depth equal to the granule radius and a surface equal to the sum of the surface of all the granules). This assumption can be done since the activity in internal layers is low due to substrate limitations. The conversion processes are described using stoichiometry and kinetics derived from previous research, supplemented with parameters from the ASM3. These authors showed that nitrification, denitrification and COD removal can occur simultaneously in a granular sludge SBR. Experimental and model works confirmed the stratification of bacteria species. The developed model revealed that the optimal DO concentration was 40% of air saturation to obtain the best performance by working with a ratio of 5.4 g COD/g N.

Su and Yu (2006) established a generalized model for aerobic granular SBR taking into account the removal of nitrogen and organic compounds, reactor hydrodynamics, oxygen transfer and substrates diffusion. The model was developed in MATLAB using spherical coordinates. Granules in the SBR were classified into various size fractions, and each granule was then sliced up along the radius to obtain an accurate model which takes into account the size distribution of the granules. The maximum specific growth rate of heterotrophs and autotrophs was modified with exponential functions, which parameters were adjusted to describe the experimental nutrient removal observed in the reactor.

Recently, Ni *et al.* (2008) employed a modification of the ASM3 model to describe the simultaneous autotrophic and heterotrophic growth in aerobic granular SBR, implementing the model in AQUASIM. The modifications to the ASM3 are based on the characteristics of an SBR and the difference between granules and flocs, trying to describe the biological reactions occurring in granules. In the ASM3 model, all readily biodegradable substrate is assumed to be initially stored as internal storage products before it is used for growth at the famine phase. However, in systems repeatedly subject to feast and famine conditions (e.g. aerobic granular SBRs), simultaneous aerobic/anoxic storage and growth on organic substrates by the heterotrophs occur under feast conditions (Krishna and van Loosdrecht 1999). Simulation results demonstrated that the influent ammonia concentration determined the composition and distribution of heterotrophic and autotrophic biomass in an aerobic granular SBR.

de Kreuk *et al.* (2007a) introduced the removal of phosphate, in addition to the removal of COD and nitrogen, as a process in the model. They used a 1-D model to study the influence of different parameters (DO concentration, temperature, granule diameter, sludge loading rate and cycle configuration) on the granular SBR operation. In this case, the stoichiometry and kinetics of biological conversions were derived from the Delft bio-P model for activated sludge (Murnleitner *et al.* 1997; van Veldhuizen *et al.* 1999) and presented with the ASM notation (Henze *et al.* 2000). Modifications were performed on the ASM1 with the addition of the metabolic description of the phosphorous accumulating organisms (PAOs) and separate description of ammonium and nitrite oxidizing bacteria activities. The mathematical model was implemented in AQUASIM. Model simulations showed that the ratio between aerobic and anoxic volumes in the granule strongly determines the nitrogen removal efficiency and that oxygen penetration, in combination with the position of autotrophic biomass, played a crucial role in the conversion rate.

The development of multidimensional (2-D or 3-D) models offers more potential to predict local compositions of particulate and dissolved variables. Several approaches to multidimensional modelling of biofilm can be found in the literature (Picioreanu and van Loosdrecht 2003; Xavier *et al.* 2005).

In general, 2-D and 3-D biofilm models could be divided in two classes according to the way of describing the biomass: discrete individual particles or a continuum body. Discrete particle models are the most suitable for extrapolation to granular biomass. Individual-based models (Ibm) are, discrete particle models, widely applied to study effects of spatially multidimensional gradients in biofilms (Kreft *et al.* 2001). Xavier *et al.* (2007) presented a multiscale model of a GSBR describing the complex dynamics of populations and nutrient removal. The multiscale model of a GSBR integrates time dynamics of microbial metabolisms, granule-scale diffusion-reaction with 2-D spatial organization and larger scale sequencing batch reactor operation. The macro scale describes bulk concentrations and effluent composition in six solutes

(DO, acetate, ammonium, nitrite, nitrate, and phosphate). A finer scale, the scale of one granule, describes the two-dimensional spatial arrangement of four bacterial groups, heterotrophs, ammonium oxidisers, nitrite oxidisers, and PAOs, using IbM of the microbial population in the granule, with species-specific kinetic models. The kinetic model for the bioconversions is based on de Kreuk *et al.* (2007a). The results of this research showed that in GSBR the nitrogen removal occurs mostly via alternating nitrification/denitrification rather than simultaneous nitrification/denitrification.

The development of mathematical models able to describe the processes occurring in the granular sludge will allow better understanding the operation of this systems and help in their operation at pilot and industrial scale.

13.7 INDUSTRIAL APPLICATIONS

13.7.1 Laboratory research in aerobic granular sludge

The interest in applying aerobic granular systems is mainly related to the compactness of its design in comparison to conventional activated sludge wastewater treatment systems. The aerobic granulation in SBRs has been extensively researched in laboratory works in terms of settling velocity, size, shape, biomass density, hydrophobicity, physical strength, microbial activity and extracellular polymeric substances (Adav *et al.* 2008). However, most of these previous studies for the cultivation of aerobic granules were based on the use of synthetic feeding, while information associated with urban or industrial wastewater is more limited.

Studies performed with different types of wastewater indicated that this technology is suitable to obtain large efficiencies in terms of COD and nutrients removal. However, a pre- or post-treatment is recommended to fulfil the disposal requirements when important suspended solids concentrations are present (de Bruin *et al.* 2004).

Experiments performed at laboratory scale indicated that in aerobic granular reactors, organic matter removal efficiencies (COD_{rem}) ranged between 80 and 98%, while the reached nitrogen removal efficiencies (N_{rem}) were slightly lower (70–95%) (Table 13.2). The physical properties of the aerobic granular biomass showed values of SVI lower than 60 mL/g VSS, densities higher than 10 g VSS/L and mean feret diameter (D_{feret}) that ranged between 1.0 and 3.5 mm. The feret diameter is defined as the value of the measured distance between theoretical parallel lines that are drawn tangent to the particle profile and perpendicular to the ocular scale.

Table 13.2. Parameters from laboratory scale studies of aerobic granulation with wastewater

Wastewater	OLR_{max} (g COD/ L·d)	NLR (g N/ L·d)	COD_{rem} (%)	N_{rem} (%)	SVI (mL/g VSS)	D_{feret} (mm)	Reference
Laboratory for analysis of dairy products	7.0	0.7	90	70	60	3.5	Arrojo et al. (2004)
Malting	3.2	0.006	80	–	35	–	Schwarzenbeck et al. (2004)
Abattoir	2.6	0.35	98	98	22	1.7	Cassidy and Belia (2005)
Pharmaceutical industry	5.5	0.03	80	–	–	–	Inizan et al. (2005)
Dairy plant	5.9	0.28	90	80	50	–	Schwarzenbeck et al. (2005)
Soybean-processing	6.0	0.3	98.5	–	26	1.2	Su and Yu (2005)
Metal-refinery process	–	1.0	–	95	–	1.1	Tsuneda et al. (2006)
Brewery	3.5	0.24	88.7	88.9	32	2–7	Wang et al. (2007)
Fish canning	1.7	0.18	95	40	30	3.4	Figueroa et al. (2008)

13.7.2 Pilot research in aerobic granular sludge

The first pilot research project using the aerobic granular technology was performed in The Netherlands, in the so called GSBR, in order to demonstrate the applicability of the aerobic granular sludge technology for the treatment of municipal wastewater (de Bruin et al. 2005). The reactor was designed for simultaneous organic matter, nitrogen and phosphorous removal. Several operational philosophies were tested to learn at which conditions granulation occurs with municipal wastewater as a substrate. Fast formation of granules was observed under conditions of extensive COD removal, extensive biological phosphate removal and low nitrate effluent concentrations (de Bruin et al. 2005). The potential of the aerobic granular technology is very promising since several wastewaters from industrial and municipal origin were used to cultivate aerobic granules at pilot scale (de Kreuk et al. 2007b).

Tay et al. (2005) also studied the development of aerobic granules in a pilot-scale sequencing batch reactor seeded with aerobic granules pre-cultured in a small column reactor to provide useful information for scale-up of aerobic granulation technology. It was observed that the seed granules disintegrated in the first few days of operation, but the disintegrated biomass would re-form stable granules latter. On the other hand, the seed granules could be successfully maintained in the laboratory scale reactor without significant disintegration. The different hydrodynamic patterns encountered in the pilot and laboratory scale reactors might be the reason for the observed phenomena.

In spite of the recent intensive research carried out in the aerobic granulation field, there is still a lack of information about some aspects, which could limit the widespread use of aerobic granulation in wastewater treatment:

a) Most works were carried out at laboratory scale and only a few at pilot scale (de Bruin *et al.* 2005; Tay *et al.* 2005). Scale-up of granular systems leads to a modification of the hydrodynamic conditions, which are very important in order to promote the formation and stability of aerobic granules and further information is required in this particular issue.

b) Little information is available regarding the application of aerobic granular technology to the treatment of low-strength domestic wastewater (de Kreuk *et al.* 2006; Liu *et al.* 2007), which is expected to difficult or lengthen the granulation process (Tay *et al.* 2004). Thus, it is necessary to know the effects of the low COD loads on the integrity of the granular structures.

Based on total annual costs, a GSBR systems with only pre- or post-treatment proves to be more attractive than the reference activated sludge alternatives (17% and 6% lower, respectively). However, the GSBR with only pre-treatment cannot meet the present effluent standards for municipal wastewater treatment; mainly because of exceeding the suspended solids effluent standard caused by washout of not well settleable biomass (de Bruin *et al.* 2004). In general, activated sludge plants will have to be extended with a post-treatment step (e.g. sand filtration) or be transformed into a membrane bioreactor. In this case a GSBR variant with pre-treatment as well as post-treatment can be an attractive alternative.

A sensitivity analysis shows that the GSBR technology is less sensitive to land price, because of the compactness, and more sensitive to rain water flow, because of the large impact of the maximum batch volume on the design of the GSBR. The GSBR technology is very compact, which is an important advantage in relation to activated sludge technology, especially in densely populated areas. Because of the high allowable volumetric load the footprint of the GSBR variants is only 25% compared to the references (de Bruin *et al.* 2004). Further study is needed in order to successfully scale up the aerobic granular systems to be applied for the treatment of urban and industrial wastewaters.

REFERENCES

Adav, S.S., Chen, M.Y., Lee, D.J. and Ren, N.Q. (2007a) Degradation of phenol by aerobic granules and isolated yeast *Candida tropicalis. Biotechnol. Bioeng.* **96**, 844–852.

Adav, S.S., Lee, D.J. and Lai, J.Y. (2007b) Biodegradation of pyridine using aerobic granules in the presence of phenol. *Water Res.* **41**, 2903–2910.

Adav, S.S., Lee, D.J. and Lai, J.Y. (2007c) Effects of aeration intensity on formation of phenol-fed aerobic granules and extracellular polymeric substances. *Appl. Microbiol. Biotechnol.* **77**, 175–182.

Adav, S.S, Lee, D.J., Show, K.Y. and Tay, J.H. (2008) Aerobic granular sludge: Recent advances. *Biotechnol. Adv.* **26**, 411–423.

Arrojo, B., Mosquera-Corral, A., Garrido, J.M. and Méndez, R. (2004) Aerobic granulation with industrial wastewater in sequencing batch reactors. *Water Res.* **38**, 3389–3399.

Ballinger, S.J., Head, I.M., Curtis, T.P. and Godley, A.R. (2002) The effect of C/N ratio on ammonia oxidizing bacteria community structure in a laboratory nitrification-denitrification reactor. *Water Sci. Technol.* **46**(1–2), 543–550.

Belser, L.W. (1984) Bicarbonate uptake by nitrifiers: Effects of growth rate, pH, substrate concentration, and mebabolic inhibitors. *Appl. Environ. Microbiol.* **48**, 1100–1104.

Beun, J.J., Hendriks, A., Van Loosdrecht, M.C.M., Morgenroth, E., Wilderer, P.A. and Heijnen, J.J. (1999) Aerobic granulation in a sequencing batch reactor. *Water Res.* **33**, 2283–2290.

Beun, J.J., van Loosdrecht, M.C.M. and Heijnen, J.J. (2001) N-removal in a granular sludge sequencing batch airlift reactor. *Biotechnol. Bioeng.* **75**, 82–92.

Beun, J.J., van Loosdrecht, M.C.M. and Heijnen, J.J. (2002) Aerobic granulation in a sequencing batch airlift reactor. *Water Res.* **36**, 702–712.

Burrell, P.C., Keller, J. and Blackall, L.L. (1998) Microbiology of a nitrite-oxidizing bioreactor. *Appl. Environ. Microbiol.* **64**, 1878–1883.

Campos, J.L., Garrido-Fernandez, J.M., Mendez, R. and Lema, J.M. (1999) Nitrification at high ammonia loading rates in an activated sludge unit. *Biores. Technol.* **68**, 141–148.

Campos J.L., Méndez R. and Lema J.M. (2000) Operation of a nitrifying activated sludge airlift (NASA) reactor without biomass carrier. *Water Sci. Technol.* **41**(4), 113–120.

Cassidy, D.P. and Belia, E. (2005) Nitrogen and phosphorus removal from an abattoir wastewater in a SBR with aerobic granular sludge. *Water Res.* **39**, 4817–4823.

Chen, Y., Jiang, W., Liang, D.T. and Tay, J.H. (2008) Aerobic granulation under the combined hydraulic and loading selection pressures. *Biores. Technol.* **99**, 7444–7449.

Chudoba, J., Grau, P. and Ottava, V. (1973) Control of activated sludge filamentous bulking–II. Selection of microorganism by means of a selector. *Water Res.* **7**, 1389–1406.

Dangcong, P., Bernet, N., Delgenes, J.P. and Moletta, R. (1999) Aerobic granular sludge– a case report. *Water Res.* **33**, 890–893.

de Beer, D., van den Heuvel, J.C. and Ottengraf, S.P.P. (1993) Microelectrode measurements of the activity distribution in nitrifying bacterial aggregates. *Appl. Environ. Microbiol.* **59**, 573–579.

de Bruin, L.M.M., de Kreuk, M.K., van der Roest, H.F.R., Uijterlinde, C. and van Loosdrecht, M.C.M. (2004) Aerobic granular sludge technology: an alternative to activated sludge? *Water Sci. Technol.* **49**(11–12), 1–7.

de Bruin, L.M.M., van der Roest, H.F.R., de Kreuk, M.K. and van Loosdrecht, M.C.M. (2005) Promising results pilot research aerobic granular sludge technology at WWTP Ede. In *Aerobic Granular Sludge*. Water and Environmental Management Series. IWA Publishing, Munich, 135–142.

de Kreuk, M.K. and van Loosdrecht, M.C.M. (2004) Selection of slow growing organisms as a means for improving aerobic granular sludge stability. *Water Sci. Technol.* **49**(11–12), 9–17.

de Kreuk, M.K., Heijnen, J.J. and van Loosdrecht, M.C.M. (2005a) Simultaneous COD, nitrogen, and phosphate removal by aerobic granular sludge. *Biotechnol. Bioeng.* **90**, 761–769.

de Kreuk, M.K., Pronk, M. and van Loosdrecht, M.C.M. (2005b) Formation of aerobic granules and conversion processes in an aerobic granular sludge reactor at moderate and low temperatures. *Water Res.* **39**, 4476–4484.

de Kreuk, M.K., Pronk, M. and van Loosdrecht, M.C.M. (2006) Formation of aerobic granules with domestic sewage. *J. Environ. Eng.* **132**, 694–697.

de Kreuk, M.K., Picioreanu, C., Hosseini, M., Xavier, J.B. and van Loosdrecht, M.C.M. (2007a) Kinetic model of a granular sludge SBR: influences on nutrient removal. *Biotechnol. Bioeng.* **97**, 801–815.

de Kreuk, M.K., Kishida, N., and van Loosdrecht, M.C.M. (2007b) Aerobic granular sludge-state of the art. *Wat. Sci. Technol.* **55**(8–9), 75–81.

Egli, K., Bosshard, F., Werlen, C., Lais, P., Siegrist, H., Zehnder, A.J.B. and van der Meer, J.R. (2003) Microbial composition and structure of a rotating biological contactor biofilm treating ammonium-rich wastewater without organic carbon. *Microb. Ecol.* **45**, 419–432.

Etterer, T. and Wilderer, P.A. (2001) Generation and properties of aerobic granular sludge. *Water Sci. Technol.* **43**(3), 19–26.

Figueroa, M., Mosquera-Corral, A., Campos, J.L and Méndez, R. (2008) Treatment of saline wastewater in SBR aerobic granular reactors. *Water Sci. Technol.* **58**(2), 479–485.

Gapes, D., Wilen, B.M. and Keller, J. (2004) Mass transfer impacts in flocculent and granular biomass from SBR systems. *Water Sci. Technol.* **50**(10), 203–212.

Gaval, G. and Pernell, J.J. (2003) Impact of the repetition of oxygen deficiencies on the filamentous bacteria proliferation in activated sludge. *Water Res.* **37**, 1991–2000.

Henze, M., Gujer, W., Mino, T. and van Loosdrecht, M.C.M. (2000) *Activated sludge models; ASM1, ASM2, ASM2d and A SM3*, IWA Publishing, London.

Hu, L., Wang, J., Wen, X. and Qian, Y. (2005) The formation and characteristics of aerobic granules in sequencing batch reactor (SBR) by seeding anaerobic granules. *Process Biochem.* **40**, 5–11.

Inizan, M., Freval, A., Cigana, J. and Meinhold, J. (2005) Aerobic granulation in a sequencing batch reactor (SBR) for industrial wastewater treatment. *Water Sci. Technol.* **52**(10–11), 335–343.

Ivanov, V., Tay, S.T.L., Liu, Q.S., Wang, X.H., Wang, Z.W. and Tay, J.H. (2005) Formation and structure of granulated microbial aggregates used in aerobic wastewater treatment. *Water Sci. Technol.* **52**(7), 13–19.

Jiang, H.L., Tay, J.H. and Tay, S.T.L. (2002) Aggregation of immobilized activated sludge cells into aerobically grown microbial granules for the aerobic biodegradation of phenol. *Lett. Appl. Microbiol.* **63**, 602–608.

Juretschko, S., Timmermann, G., Schmid, M., Schleifer, K.H., Pommerening-Roser, A., Koops, H.P. and Wagner, M. (1998) Combined molecular and conventional analyses of nitrifying bacterium diversity in activated sludge: Nitrosococcus mobilis and Nitrospira-like bacteria as dominant populations. *Appl. Environ. Microbiol.* **64**, 3042–3051.

Kreft, J.U., Picioreanu, C., Wimpenny, J.W.T. and van Loosdrecht, M.C.M. (2001) Individual-based modelling of biofilms. *Microbiol.* **147**, 2897–2912.

Krishna, C. and van Loosdrecht, M.C.M. (1999) Substrate flux into storage and growth in relation to activated sludge modelling. *Water Res.* **33**, 3149–3161.

Lemaire, R., Webb, R.I. and Yuan, Z. (2008) Micro-scale observations of the structure of aerobic microbial granules used for the treatment of nutrient-rich industrial wastewater. *ISME J.* **2**, 528–541.

Lin, Y.M., Liu, Y. and Tay, J.H. (2003) Development and characteristics of phosphorous-accumulating granules in sequencing batch reactor. *Appl. Microbiol. Biotechnol.* **62**, 430–435.

Liu, Y. and Liu, Q.S. (2006) Causes and control of filamentous growth in aerobic granular sludge sequencing batch reactors. *Biotechnol. Adv.* **24**, 115–127.

Liu, Y. and Tay, J.H. (2002) The essential role of hydrodynamic shear force in the formation of biofilm and granular sludge. *Water Res.* **36**, 1653–1665.

Liu, Q.S., Tay J.H. and Liu Y. (2003) Substrate concentration-independent aerobic granulation in sequential aerobic sludge blanket reactor. *Environ. Technol.* **24**, 1235–1243.

Liu, Y., Yang, S.F., Tai, J-H., Liu, Q.S., Qin, L. and Li Y. (2004) Cell hydrophobicity is a triggering force of biogranulation. *Enzyme Microb. Technol.* **34**, 371–379.

Liu, Y.Q., Moy, B.Y.P. and Tay, J.H. (2007) COD removal and nitrification of low-strength domestic wastewater in aerobic granular sludge sequencing batch reactors. *Enzyme Microb. Technol.* **42**, 23–28.

Lübken, M., Schwarzenbeck, N., Wichern, M. and Wilderer, P.A. (2005) Modelling nutrient removal of an aerobic granular sludge lab-scale SBR using ASM3. In *Aerobic granular sludge*, Water and environmental management series. IWA publishing, Munich.

McSwain, B.S., Irvine, R.L., Hausner, M. and Wilderer, P.A. (2005) Composition and distribution of extracellular polymeric substances in aerobic flocs and granular sludge. *Appl. Environ. Microbiol.* **71**, 1051–1057.

Mishima K. and Nakamura M. (1991) Self-immobilization of aerobic activated sludge–a pilot study of the aerobic upflow sludge blanket process in municipal sewage treatment. *Water Sci. Technol.* **23**(4–6), 981–990.

Moreau, M., Liu, Y., Capdeville, B., Audic, J. M. and Calvez, L. (1994) Kinetic behaviors of heterotrophic and autotrophic biofilm in wastewater treatment processes. *Water Sci. Technol.* **29**(10–11), 385–391.

Mosquera-Corral, A., de Kreuk, M.K., Heijnen, J.J. and van Loosdrecht, M.C.M. (2005a) Effects of oxygen concentration on N-removal in an aerobic granular sludge reactor. *Water Res.* **39**, 2676–2686.

Mosquera-Corral, A., Vázquez-Padín, J.R., Arrojo, B., Campos, J.L. and Méndez, R. (2005b) Nitrifying granular sludge in a sequencing batch reactor. In *Aerobic granular sludge.* Water and Environmental Management Series, IWA Publishing, Munich, 63–70.

Moy, B.Y.P., Tay, J.H., Toh, S.K., Liu, Y. and Tay, S.T.L. (2002) High organic loading influences the physical characteristics of aerobic sludge granules. *Lett. Appl. Microbiol.* **34**, 407–412.

Murnleitner, E., Kuba, T., van Loosdrecht, M.C.M. and Heijnen, J.J. (1997) An integrated metabolic model for the aerobic and denitrifying biological phosphorus removal. *Biotechnol. Bioeng.* **54**, 434–450.

Ni, B.J., Yu, H.Q. and Sun, Y.J. (2008) Modelling simultaneous autotrophic and heterotrophic growth in aerobic granules. *Water Res.* **42**, 1583–1594.

Ohashi, A., Viraj de Silva, D.G., Mobarry, B., Manem, J.A., Stahl, D.A. and Rittmann, B.E. (1995) Influence of substrate C/N ratio on the structure of multi-species biofilms consisting of nitrifiers and heterotrophs. *Water Sci. Technol.* **32**(8), 75–84.

Okabe, S., Satoh, H. and Watanabe, Y. (1999) In situ analysis of nitrifying biofilms as determined by in situ hybridization and the use of microelectrodes. *Appl. Environ. Microbiol.* **65**, 3182–3191.

Picioreanu, C. and van Loosdrecht, M.C.M. (2003) Use of mathematical modelling to study biofilm development and morphology. In *Biofilms in Medicine, Industry and Environmental Biotechnology – Characteristics, Analysis and Control.* IWA Publishing, London, 413–437.

Qin, L., Tay, J.H. and Liu, Y. (2004a) Selection pressure is a driving force of aerobic granulation in sequencing batch reactors. *Process Biochem.* **39**, 579–584.

Qin, L., Liu, Y. and Tay, JH (2004b) Effect of settling time on aerobic granulation in sequencing batch reactor. *Biochem. Eng. J.* **21**, 47–52.

Qin, L., Liu, Y. and Tay, J.H. (2005) Denitrification on poly-β-hydroxybutyrate in microbial granular sludge sequencing batch reactor. *Water Res.* **39**, 1503–510.

Schramm, A., Larsen, L.H., Revsbech, N.P., Ramsing, N.B., Amann, R. and Schleifer, K.H. (1996) Structure and function of a nitrifying biofilm as determined by in situ hybridization and the use of microelectrodes. *Appl. Environ. Microbiol.* **62**, 4641–4647.

Schramm, A., Beer, D., Heuvel, J.C., Ottengraf, S. and Amann, R. (1999) Microscale distribution of populations and activities of *Nitrosospira* and *Nitrospira* spp. along a macroscale gradient in a nitrifying bioreactor: quantification by in situ hybridization and the use of microelectrode. *Appl. Environ. Microbiol.* **65**, 3690–3696.

Schwarzenbeck, N., Erley, R., Mc Swain, B.S., Wilderer, P.A. and Irvine, R.L. (2004) Treatment of malting wastewater in a granular sludge sequencing batch reactor (SBR). *Acta Hydrochim. Hydrobiol.* **32**, 16–24.

Schwarzenbeck, N. and Wilderer, P.A. (2005) Treatment of food industry effluents in a granular sludge SBR. In *Aerobic Granular Sludge.* Water and Environmental Management Series. IWA Publishing, Munich, 95–102.

Schwarzenbeck, N., Borges, J.M. and Wilderer, P.A. (2005) Treatment of dairy effluents in an aerobic granular sludge sequencing batch reactor. *Appl. Microbiol. Biotechnol.* **66**, 711–718.

Su, K.Z. and Yu, H.Q. (2005) Formation and characterization of aerobic granules in a sequencing batch reactor treating soybean-processing wastewater. *Environ. Sci. Technol.* **39**, 2818–2827.

Su, K.Z. and Yu, H.Q. (2006) A generalizad model for aerobic granule-based sequencing batch reactor 1. Model development. *Environ. Sci. Technol.* **40**, 4703–4708.

Sun, F., Yang, C., Li, J. and Yang, Y. (2006) Influence of different substrates on the formation and characteristics of aerobic granules in sequencing batch reactors. *J. Environ. Sci.* **18**, 864–871.

Tay, J.H., Liu, Q.S. and Liu, Y. (2001a) The role of cellular polysaccharides in the formation and stability of aerobic granules. *Lett. Appl. Microbiol.* **33**, 222–226.

Tay, JH, Liu, QS and Liu, Y. (2001b) Microscopic observation of aerobic granulation in sequential aerobic sludge blanket reactor. *J. Appl. Microbiol.* **91**, 168–175.

Tay, J.H., Liu, Q.S. and Liu, Y. (2001c) The effects of shear force on the formation, structure and metabolism of aerobic granules. *Appl. Microbiol. Biotechnol.* **57**, 227–233.

Tay, J.H., Yang, S.F. and Liu, Y. (2002a) Hydraulic selection pressure-induced nitrifying granulation in sequencing batch reactors. *Appl. Microbiol. Biotechnol.* **59**, 332–337.

Tay, J.H., Liu, Q.S. and Liu, Y. (2002b) Characteristics of aerobic granules grown on glucose and acetate in sequential aerobic sludge blanket reactors. *Environ. Technol.* **23**, 931–936.

Tay, J.H., Ivanov, V., Pan, S. and Tay, S.T.L. (2002c) Specific layers in aerobically grown microbial granules. *Lett. Appl. Microbiol.* **34**, 254–257.

Tay, J.H., Pan, S., He, Y.X. and Tay, S.T.L. (2004) Effect of organic loading rate on aerobic granulation. I: reactor performance. *J. Environ. Eng.* **130**, 1094–1101.

Tay, J.H., Liu, Q.S., Liu, Y., Show, K.Y., Ivanov, V. and Tay, S.T.L. (2005) A comparative study of aerobic granulation in pilot- and labratory scale SBRs. In *Aerobic Granular Sludge.* Water and Environmental Management Series. IWA Publishing, Munich, 125–133.

Thanh, B.X., Visvanathan, C.V., Spérandio, M. and Aim, R.B. (2008) Fouling characterization in aerobic granulation coupled baffled membrane separation unit. *J. Memb. Sci.* **318**, 334–339.

Tijhuis, L., van Loosdrecht, M.C.M. and Heijenen, J.J. (1994) Formation and growth of heterotrophic aerobic biofilms on small suspended particles in airlift reactors. *Biotechnol. Bioeng.* **44**, 595–608.

Toh, S.K., Tay, J.H., Moy, B.Y.P., Ivanov, V. and Tay, S.T.L. (2003) Size-effect on the physical characteristics of the aerobic granule in a SBR. *Appl. Microbiol. Biotechnol.* **60**, 687–695.

Tokutomi, T., Kiyokawa, T., Shibayama, C., Harada, H. and Ohashi, A. (2006) Effect of inorganic carbon on nitrite accumulation in an aerobic granule reactor. *Water Sci. Technol.* **53**(12), 285–294.

Trinet, F., Hein, R., Amar, D., Chang, H.T. and Rittmann, B.E. (1991) Study of biofilm and fluidization of bioparticles in a three-phase fluidized bed reactor. *Water Sci. Technol.* **23**, 1347–1354.

Tsuneda, S., Nagano, T., Hoshino, T., Ejiri, Y., Noda, N. and Hirata, A. (2003) Characterization of nitrifying granules produced in an aerobic upflow fluidized bed reactor. *Water Res.* **37**, 4965–4973.

Tsuneda, S., Ogiwara, M., Ejiri, Y. and Hirata, A. (2006) High-rate nitrification using aerobic granular sludge. *Wat. Sci. Technol.* **53**(3), 147–154.

van Benthum, W.A.J., Garrido-Fernandez, J.M., Tijhuis, L., van Loosdrecht, M.C.M. and Heijnen, J.J. (1996) Formation and detachment of biofilms and granules in a nitrifying biofilm airlift suspension reactor. *Biotech. Progress,* **12**, 764–772.

van Loosdrecht, M.C.M. and Henze, M. (1999) Maintenance, endogeneous respiration, lysis, decay and predation. *Wat. Sci. Technol.* **39**(1), 107–117.

van Veldhuizen, H.M., van Loosdrecht, M.C.M. and Heijnen, J.J. (1999) Modelling biological phosphorus and nitrogen removal in a full scale activated sludge process. *Water Res.* **33**, 3459–3468.

Wang, Q., Du, G. and Chen, J. (2004) Aerobic granular sludge cultivated under the selective pressure as a driving force. *Process Biochem.* **39**, 557–563.

Wang, Z.W., Liu, Y. and Tay, J.H. (2005) Distribution of EPS and cell surface hydrophobicity in aerobic granules. *Appl. Microbiol. Biotechnol.* **69**, 469–473.

Wang, S.G., Liu, X.W., Gong, W.X., Gao, B.Y., Zhang, D.H. and Yu, H.Q. (2007) Aerobic granulation with brewery wastewater in a sequencing batch reactor. *Bioresour. Technol.* **98**, 2142–2147.

Weber, S.D., Ludwig, W., Schleifer, K.H. and Fried, J. (2007) Microbial composition and structure of aerobic granular sewage biofilms. *Appl. Environ. Microbiol.* **73**, 6233–6240.

Wilderer, P.A., Irvine, R.L. and Goronszy M.C. (2001) Sequencing batch reactor technology. *Scientific and Technical Report*, IWA Publishing, London.

Xavier, J.B., Picioreanu, C. and van Loosdrecht, M.C.M. (2005) A framework for multidimensional modelling of activity and structure of multispecies biofilms. *Environ. Microbiol.* **7**, 1085–1103.

Xavier, J.B., de Kreuk, M., Picioreanu, C. and van Loosdrecht, M.C.M. (2007) Multi-scale individual-based model of microbial and bioconversion dynamics in aerobic granular sludge. *Environ. Sci. Technol.* **41**, 6410–6417.

Yang, S.F., Liu, Y. and Tay, J.H. (2003) A novel granular sludge sequencing batch reactor for removal of organic and nitrogen from wastewater. *J. Biotechnol.* **106**, 77–86.

Yang, S.F., Tay, J.H. and Liu, Y. (2004) Inhibition of free ammonia to the formation of aerobic granules. *Biochem. Eng. J.* **17**, 41–48.

Yang, S.F., Tay, J.H. and Liu, Y. (2005) Effect of Substrate Nitrogen/Chemical Oxygen Demand Ratio on the Formation of Aerobic Granules. *J. Environ. Eng.* **131**, 86–92.

Zheng, Y.M., Yu, H.Q. and Sheng, G.P. (2005) Physical and chemical characteristics of granular activated sludge from a sequencing batch airlift reactor. *Process Biochem.* **40**, 645–650.

Zheng, Y.M. and Yu, H.Q. (2007) Determination of the pore size distribution and porosity of aerobic granules using size-exclusion chromatography. *Water Res.* **41**, 39–46.

Index

Printed in the United Kingdom by
Lightning Source UK Ltd., Milton Keynes
140800UK00001BA/69/P